Symmetric Galerkin Boundary Element Method

Alok Sutradhar · Glaucio H. Paulino ·
Leonard J. Gray

Symmetric Galerkin Boundary Element Method

Alok Sutradhar
Ohio State University
Department of Surgery
410 W 10th Ave, Columbus OH
USA
alok.sutradhar@osumc.edu

Prof. Glaucio H. Paulino
University of Illinois at
Urbana-Champaign
Newmark Laboratory
205 North Mathews Avenue
Urbana IL 61801-2352
USA
paulino@uiuc.edu

Leonard J. Gray
Oak Ridge National Laboratory
POB 2008, Bldg 6012 MS 6367
Oak Ridge
TN 37831-6367
graylj1@ornl.gov

ISBN 978-3-540-68770-2   ISBN 978-3-540-68772-6 (eBook)

Library of Congress Control Number: 2008928276

© 2008 Springer-Verlag Berlin Heidelberg

This work is subject to copyright. All rights are reserved, whether the whole or part of the material is concerned, specifically the rights of translation, reprinting, reuse of illustrations, recitation, broadcasting, reproduction on microfilm or in any other way, and storage in data banks. Duplication of this publication or parts thereof is permitted only under the provisions of the German Copyright Law of September 9, 1965, in its current version, and permission for use must always be obtained from Springer. Violations are liable to prosecution under the German Copyright Law.

The use of general descriptive names, registered names, trademarks, etc. in this publication does not imply, even in the absence of a specific statement, that such names are exempt from the relevant protective laws and regulations and therefore free for general use.

Printed on acid-free paper

9 8 7 6 5 4 3 2 1

springer.com

To our families

CONTENTS

Foreword   xiii

Preface   xv

1 Introduction   1

  1.1 Boundary Element Method   1

      1.1.1 Approximations and Solution   2

      1.1.2 The Green's function $G(P,Q)$ 4

      1.1.3 Singular and Hypersingular Integrals   5

      1.1.4 Numerical Solution: Collocation and Galerkin   6

      1.1.5 Symmetric Galerkin BEM   7

  1.2 An Application Example: Automotive Electrocoating   8

      1.2.1 Engineering Optimization   9

      1.2.2 Electrocoating Simulation   10

  1.3 Visualization   11

      1.3.1 Virtual Reality   12

      1.3.2 CAVE: Cave Automatic Virtual Environment   14

      1.3.3 The MechVR   14

  1.4 Other Boundary Techniques   15

      1.4.1 Singular Integration   16

      1.4.2 Meshless and Mesh-Reduction Methods   16

  1.5 A Brief History of Galerkin BEM   19

viii CONTENTS

2 Boundary Integral Equations | 23

2.1 Boundary Potential Equation	23
2.2 Boundary Flux Equation	28
2.3 Elasticity	30
2.4 Numerical Approximation	33
2.4.1 Approximations	33
2.4.2 Collocation	35
2.4.3 Galerkin Approximation	37
2.4.4 Symmetric-Galerkin	39
2.5 Hypersingular Integration: an example	39
2.5.1 Collocation: $\mathcal{C}^1$ Condition	40
2.5.2 Galerkin: $\mathcal{C}^0$	42

3 Two Dimensional Analysis | 45

3.1 Introduction	45
3.2 Singular Integrals: Linear Element	48
3.2.1 Coincident Integration	48
3.2.2 Coincident: Symbolic Computation	51
3.2.3 Adjacent Integration	53
3.2.4 Cancellation of $\log(\varepsilon^2)$	56
3.2.5 Adjacent: shape function expansion	57
3.2.6 Numerical Tests	58
3.3 Higher Order Interpolation	60
3.3.1 Integral of G	61
3.3.2 Integral of $\partial G/\partial \mathbf{n}$ and $\partial G/\partial \mathbf{N}$	62
3.3.3 Integral of $\partial^2 G/\partial \mathbf{N} \partial \mathbf{n}$	63
3.4 Other Green's functions	63
3.5 Corners	64
3.6 Nonlinear Boundary Conditions	65
3.7 Concluding Remarks	67

4 Three Dimensional Analysis | 69

4.1 Preliminaries	69
4.2 Linear Element Analysis	73
4.2.1 Nonsingular Integration	74
4.2.2 Coincident Integration	75
4.2.3 Coincident CPV integral	82
4.2.4 Edge Adjacent Integration	83
4.2.5 Vertex Adjacent Integration	88
4.2.6 Proof of Cancellation	90
4.3 Higher Order Interpolation	92

		CONTENTS	ix

	4.4	Hypersingular Boundary Integral: Quadratic Element	93
	4.4.1	Coincident Integration	94
	4.4.2	r Expansion	94
	4.4.3	First Integration	95
	4.4.4	Edge Integration	96
4.5		Corners	97
4.6		Anisotropic Elasticity	97
	4.6.1	Anisotropic Elasticity Boundary Integral Formulation	99
	4.6.2	T Kernel: Coincident Integration	100
	4.6.3	Spherical Coordinates	105
	4.6.4	Second integration	106
	4.6.5	Edge Adjacent Integration	106

5 Surface Gradient — 109

	5.1	Introduction	109
	5.2	Gradient Equations	112
	5.2.1	Limit Evaluation in two dimensions	114
	5.2.2	Example: Surface Stress	118
	5.2.3	Limit Evaluation in three dimensions	121
	5.3	Hermite Interpolation in Two Dimensions	123
	5.3.1	Introduction	123
	5.3.2	Hermite Interpolation	124
	5.3.3	Iterative Solution	125

6 Axisymmetry — 129

	6.1	Introduction	129
	6.2	Axisymmetric Formulation	131
	6.3	Singular Integration	134
	6.3.1	Adjacent Integration	135
	6.3.2	Coincident Integration	135
	6.3.3	Axis singularity	136
	6.3.4	Log Integral Transformation	136
	6.3.5	Analytic integration formulas	138
	6.4	Gradient Evaluation	139
	6.4.1	Gradient Equations	139
	6.4.2	Coincident Integration	140
	6.5	Numerical Results	142

7 Interface and Multizone — 145

	7.1	Introduction	145
	7.2	Symmetric Galerkin Formulation	147

x CONTENTS

7.3	Interface and Symmetry	149
	7.3.1 Multiple Interfaces	151
	7.3.2 Corners	152
	7.3.3 Free interface	152
	7.3.4 Computational Aspects	152
7.4	Numerical Examples	153
7.5	Remarks	155

8 Error Estimation and Adaptivity — **157**

8.1	Introduction	157
8.2	Boundary Integral Equations	158
8.3	Galerkin Residuals and Error Estimates	160
8.4	Self Adaptive Strategy	161
	8.4.1 Local Error Estimation	162
	8.4.2 Element Refinement Criterion	162
	8.4.3 Global Error Estimation	163
	8.4.4 Solution Algorithm for Adaptive Meshing	164
8.5	Numerical Example	164
8.6	BEAN Code	167

9 Fracture Mechanics — **171**

9.1	Introduction	171
9.2	Fracture parameters: Stress intensity factors (SIFs) and T-stress	172
9.3	SGBEM Formulation	173
	9.3.1 Basic SGBEM formulation for 2D elasticity	173
	9.3.2 Fracture analysis with the SGBEM	175
9.4	On Computational Methods for Evaluating Fracture Parameters	178
9.5	The Two-state Interaction Integral: M-integral	179
	9.5.1 Basic Formulation	179
	9.5.2 Auxiliary Fields for T-stress	180
	9.5.3 Determination of T-stress	182
	9.5.4 Auxiliary Fields for SIFs	183
	9.5.5 Determination of SIFs	184
	9.5.6 Crack-tip elements	185
	9.5.7 Numerical implementation of the M-integral	186
9.6	Numerical Examples	187
	9.6.1 Infinite plate with an interior inclined crack	187
	9.6.2 Slanted edge crack in a finite plate	191
	9.6.3 Multiple interacting cracks	192
	9.6.4 Various fracture specimen configurations	193

10 Nonhomogenous media — 197

10.1 Introduction — 197
10.2 Steady State Heat Conduction — 198
    10.2.1 On the FGM Green's function — 199
    10.2.2 Symmetric Galerkin Formulation — 199
    10.2.3 Treatment of Singular and Hypersingular Integrals — 202
10.3 Evaluation of singular double integrals — 203
    10.3.1 Coincident Integration — 204
    10.3.2 Edge Adjacent Integration — 210
    10.3.3 Vertex Adjacent Integration — 212
    10.3.4 Numerical Example — 214
10.4 Transient heat conduction in FGMs — 216
    10.4.1 Basic Equations — 217
    10.4.2 Green's Function — 218
    10.4.3 Laplace Transform BEM (LTBEM) Formulation — 219
    10.4.4 Numerical Implementation of the 3D Galerkin BEM — 221
    10.4.5 Numerical Inversion of the Laplace Transform — 222
    10.4.6 Numerical Examples — 223
10.5 Concluding Remarks — 224

11 BEAN: Boundary Element ANalysis Program — 227

11.1 Introduction — 227
11.2 Main Control Window: BEAN — 228
    11.2.1 Menu — 228
    11.2.2 File — 228
    11.2.3 Geometry — 228
    11.2.4 Boundary Conditions (BCs) — 229
    11.2.5 Analysis — 230
    11.2.6 Results — 230
11.3 BEANPlot — 231
    11.3.1 Menus — 231
    11.3.2 Curves — 232
11.4 BEANContour — 232
    11.4.1 Menus — 232
11.5 General Instructions — 234
11.6 Troubleshooting — 235
    11.6.1 Error Message Meanings — 236
11.7 Sample Problems — 237

Appendix A: Mathematical Preliminaries and Notations — 241

A.1 Dirac Delta function — 241

xii CONTENTS

A.2	Kronecker Delta function	242
A.3	Derivative, Gradient, Divergence and Laplacian	242
A.4	Divergence theorem	243
A.5	Stokes theorem	243
A.6	Green's Identities	243
A.7	Fourier and Laplace transform	244
A.8	Free Space Green's function	244

Appendix B: Gaussian Integration   **247**

B.1	Gaussian rule for logarithmic singularities	247
B.2	Gaussian rule for One-dimensional non-singular integration	247

Appendix C: Maple Codes for treatment of hypersingular integral   **251**

C.1	Maple Script: Coincident	251
C.2	Maple Script: Edge Adjacent	253
C.3	Maple Script: Vertex Adjacent	254

References   257

Topic Index   275

FOREWORD

The boundary element methodology (BEM) may be regarded as one of the main developments in computational mechanics, fostered by the diffusion of computers, in the second half of the 20th century. If a sort of competition is attributed to research streams oriented to engineering applications, it can be said that triumph was achieved by the finite element methodology, primarily because of its versatility (which, among other effects, led to a dissemination of commercial general-purpose software in the community of practitioners, an effect not generated by BEM).

However, as this book evidences, there are several meaningful problems of practical interest, for the numerical solutions of which BE approaches are significantly advantageous. As an example, additional to the variety of such situations dealt with by the Authors, I would like to mention here the customary assessment of "added masses" for traditional dynamic analysis of a dam, by solving preliminarily a (linear, tridimensional) potential problem over the reservoir with results of practical interest confined on part of the boundary only.

Within the BE area, the "symmetric Galerkin (SG) method", systematically treated in this book, does deserve more attention and further efforts, both in research institutions and in industrial environments. Although it implies some peculiar conceptual and mathematical difficulties, the SGBEM turns out to be a timely, attractive and promising subject, as the Authors, who are internationally acknowledged leading researchers in the field, show in this book.

It is well known that the BE methodology has remote deep roots in the history of mechanics. The central integral equation was established in 1886 by the young descendent of Alessandro Volta, Carlo Somigliana (then 26 year old, at the beginning of his scientific career, which ended when he was 95, after his last twenty years in Milan). Somigliana investigated consequences of the reciprocity theorem in elasticity due to his master Enrico Betti and, in particular, the involvement in it of the fundamental solution found in 1848 by William Thomson, when he was 24 (not yet Lord Kelvin, Baron of Largs), at the beginning of his professorship in Glasgow (which lasted 53 years!). A few years later (1891) Michele Gebbia, in Palermo, established his two-point fundamental solution, more intriguing than Kelvin's kernel due to its hyper-singularity, and later he related it to the works of the mathematicians Green and Fredholm (with some mild polemical hints to the latter).

As for the SGBEM it is worth noticing first that Boris Galerkin (1871-1945) personified great, fruitful synergy between mathematics and engineering by means of his crucial roles both in Saint Petersburg Mathematical Society and in design of dams and hydro-and thermo- electrical power-plants (and his one-and-half years in prison as a student under the Czars witness his passion for social justice).

The concept of symmetry has been recurrent in science, from Pythagoras' canons of music and Vitruvius' theory of architecture up to present particle physics, through emphasis on it in works of great mathematicians like Abel and Galois.

The synergistic combination of the symmetric integral equation couple in elasticity with the solution approximations by Galerkin's weighted residual approach gave rise to the SGBEM, now extensively expounded in this book.

The Authors discuss variants and alternative approaches within, and also outside, the BE methodology and elucidate potentialities and limitations of the SGBEM. Among the pros, it may be mentioned here the fact that, when symmetry reflects essential features of the physical problem, its preservation in passing from continuum to discrete problem formulations sometimes permits to achieve theoretical results in a simpler algebraic context.

Readers are likely to particularly appreciate in the book various parts devoted to recent or current developments, desirable but unexpected in a textbook, namely e.g.: a fairly detailed survey of visualization techniques and relevant software; a whole chapter on error estimation and adaptivity; another chapter on non-homogeneous media like functionally graded materials, nowadays fashionable subject of research in various technologies and especially in micro-technologies.

As a conclusion of this brief foreword, it can be stated that this is a timely book: it probably fills a niche in computationally mechanics and might promote a revival of research on BE methods and on their fruitful engineering applications.

GIULIO MAIER

Professor Emeritus of Structural Engineering
Technical University (Politecnico) of Milan, Italy
January 2007

PREFACE

As a number of good books covering the boundary element method have appeared relatively recently, we are perhaps obligated to justify the appearance of this volume. In this preface we would therefore like to delineate how this text differs from its predecessors and what we hope that it can add to the field.

The first distinguishing feature is that, as indicated by the title, this book is concerned solely with a Galerkin approximation of boundary integral equations, and more specifically, symmetric Galerkin. Most books on boundary elements deal primarily with more traditional collocation methods, so we hope that this volume will complement existing material. The symmetric Galerkin approximation is an accurate and versatile numerical analysis method, possessing the attractive feature of producing a symmetric coefficient matrix. Moreover, it is based upon the ability - that Galerkin provides - to handle the hypersingular (as well as the standard singular) boundary integral equation by means of standard continuous elements (*i.e.* no special elements are needed). Thus, a second noteworthy aspect of this book is that singular and hypersingular equations are introduced together and are treated numerically by means of a unified approach, as elaborated below.

A primary reason that this can be accomplished, and as well an important theme of this book, is that all singular integrals - weakly singular, singular (Cauchy principal value), and hypersingular - can be handled using the same basic concept and algorithms. These algorithms are based upon a mathematically rigorous definition of the integrals as *limits to the boundary* , certainly a unique feature of this book. We hope that our readers, especially students, will find this direct approach more intuitive than (in our prejudiced view) the somewhat sleight-of-hand removal of divergences in the (more or less) standard principal value and Hadamard finite part treatments of the singular and hypersingular integrals. Moreover, as the limit based singular integration methods are largely independent of the particular Green's function, it suffices to separately discuss the analyses for two and three-dimensional problems. While most texts have separate chapters devoted to specific applications (potential theory, elasticity, etc.), here we take a different approach to emphasize that all equations can be treated by fundamentally the same methods leading to a variety of applications.

The consolidation of the basic analysis procedures into two chapters also leaves us room to include some important topics not usually covered in other texts. These include surface gradient evaluation (a key advantage of limit based singular integration methods), error estimation (once again based upon the ability to easily treat hypersingular equations), and non-homogeneous (*e.g.*, graded) materials. Finally, a symmetric Galerkin Matlab© educational program called BEAN, which stands for Boundary Element ANalysis, is available for download, and our hope is that this will assist in using the book as classroom material and also in learning how to program the symmetric Galerkin boundary element method.

The book is arranged into 11 Chapters. A brief summary of the organization is as follows.

Chapter 1 provides an introduction to the boundary element method with special emphasis on the symmetric Galerkin formulation. The basic aspects of the integral equation formulation are introduced, and the advantages of this technique are discussed in the context of a specific industrial application, the electrodeposition of paint in automotive manufacturing. This application is integrated with advanced visualization techniques, including virtual reality. Other integral equation methods, such as meshless and mesh-reduction techniques, are briefly discussed, and the chapter concludes with a succinct history of the Galerkin approximation.

Chapter 2 introduces boundary integral equations and their numerical approximations. For the Laplace equation, the integral equations for surface potential and for surface flux are derived, which involve the Green's function and its first and second derivatives. These functions are divergent when the source and field points coincide, the singularity becoming progressively stronger with higher derivatives, and thus the evaluation of (highly) singular integrals is of paramount importance. As noted above, the fundamental approach adopted in this book is to define and evaluate all singular integrals as '*limits to the boundary*'. This approach unifies the numerical analysis of the integral equations and affects almost every aspect of the book, most especially the evaluation of gradients examined in Chapter 5.

Chapters 3 & 4 are the core of the book, presenting the numerical implementation of a Galerkin boundary integral analysis in two and three dimensions, respectively. The analysis techniques presented in these Chapters represent well the book philosophy outlined above. The primary task is the evaluation of singular integrals, and for the hypersingular integral it is necessary to isolate the divergent terms and to prove that they cancel. The methods are first described in the simplest possible setting, a piecewise linear solution of the Laplace equation. Subsequently, higher order curved interpolation and more complicated Green's functions are considered. As noted above, the limit to the boundary approach provides a consistent scheme for defining all singular integrals, and moreover results in direct semi-analytical evaluation algorithms. Symbolic computation is utilized to simplify the work involved in carrying out the limit process and related analytical integration, and example Maple© codes are provided.

Chapter 5 addresses the evaluation of surface gradients. A significant advantage of the boundary limit approach is that it leads to a highly accurate and efficient scheme for computing surface derivatives: only local singular integrals need to be evaluated, *i.e.* a complete boundary integration is not required. The key is to exploit the limit definition, writing the gradient equation as a difference of interior and exterior limits. In many applications, most notably moving boundary problems, the knowledge of boundary derivatives is necessary, *e.g.*, the potential gradient for Laplace problems or the complete stress tensor in elasticity. A specific, and in many ways typical, example of a free boundary problem is discussed, a coupled level set-boundary element analysis for modeling two-dimensional breaking waves over sloping beaches. This example demonstrates the advantage of the gradient techniques for the general class of moving front problems.

Chapter 6 considers three-dimensional axisymmetric problems. As an example, the boundary integral equation for the axisymmetric Laplace equation is solved by employing modified Galerkin weight functions. The alternative weights smooth out the singularity of the Green's function at the symmetry axis, and restore symmetry to the formulation. The modified weight functions, together with a boundary limit definition, also result in a relatively simple algorithm for the post-processing of the surface gradient.

Chapter 7 presents a symmetric Galerkin boundary integral method for interface and multizone problems. This type of problem arises, for example, in applications such as composite materials (bi-material interfaces) and geophysical simulations (internal boundaries). In the present formulation, the physical quantities are known to satisfy continuity conditions across the interface, but no boundary conditions are specified. The algorithm described herein achieves a symmetric matrix of reduced size.

Chapter 8 addresses error estimation and adaptivity from a practical viewpoint. The so-called *a-posteriori* error estimators are used to guide the associated adaptive mesh refinement procedure. The estimators make use of the "hypersingular residuals", originally developed for error estimation in a standard collocation approximation, and later extended to the symmetric Galerkin setting. This leads to the formulation of Galerkin residuals, which are natural to the symmetric Galerkin boundary integral approach, and forms the basis of the present error estimation scheme. The error estimation and adaptive procedure are implemented in the educational, user-friendly, symmetric Galerkin Matlab© code **BEAN** mentioned above, which solves problems governed by the Laplace equation.

Chapter 9 is dedicated to one of the most important and successful application areas for boundary integral methods: fracture analysis. Problems involving fracture and failure arise in many critical engineering areas, and boundary integral methods have inherent advantages for these calculations. It is therefore essential that an efficient and effective symmetric Galerkin approximation be developed for this class of problems, and this chapter demonstrates that this is indeed the case.

Chapter 10 focuses on the development of symmetric Galerkin formulations for nonhomogeneous problems of potential theory. Specifically, a formulation and corresponding implementation for heat conduction in three dimensional functionally graded materials are presented. The Green's function of the actual problem is used to develop a boundary-only formulation without any domain discretization, in which the thermal conductivity varies exponentially in one coordinate. A transient

xviii PREFACE

implementation using the Laplace transform Galerkin boundary element method is also provided.

Chapter 11 is dedicated to the educational computer code Boundary Element ANalysis, including its graphical user interface. BEAN is a user-friendly adaptive symmetric Galerkin BEM code to solve the two-dimensional Laplace Equation. The Chapter outlines the specific procedures to set up the problems, and the steps to utilize BEAN's post-processing capabilities. The book website contains additional related material, as discussed below. In an effort to make the book as self-contained as possible, three appendices are provided, covering mathematical preliminaries, some Gaussian integration tables, and the symbolic Maple© codes. As discussed above, the Maple© codes are an integral part of the book and should be used in conjunction with Chapters 2-4.

A web site at `http://www.ghpaulino.com/SGBEM_book` associated with this book will be maintained, primarily as a means of allowing access to the codes employed in the text. It contains the BEAN code, including the graphical user interface. The complete source-code is provided, together with a library of practical examples, and a video tutorial which demonstrates step-by-step how to use the software. The source code can be used for instructional purposes, as well as the building block for new applications. In addition, the symbolic Maple© codes, for singular and hypersingular integrations, are also provided. These codes supplement the explanations of Chapters 2 through 4. We hope that readers will also use the web site to report misprints, errors, and other comments/suggestions about the text.

Finally, this book would not have been possible without the patience of our wives and children. We also owe a tremendous debt to the many colleagues and students, most of whom are now good friends, with whom we have collaborated over the past twenty years. They are too numerous to mention, but they know who they are, and we thank them for many enjoyable hours of struggling to get on the right path, hunting down insidious bugs, and on occasion, finding gold.

We hope that you enjoy the book.

ALOK SUTRADHAR, G. H. PAULINO, L. J. GRAY

February, 2008

CHAPTER 1

INTRODUCTION

Synopsis: Following the well-known country preacher's advice about sermons, beginning with 'first tell'em what you're gonna tell'em', this chapter is an expository overview of the contents of this monograph. The basic aspects of the boundary element method will be introduced with special emphasis on the symmetric Galerkin boundary element method, and the advantages of this technique are discussed in the context of one specific industrial application, the electrodeposition of paint on automobiles. The numerical solution of boundary integral equations can be carried out in many ways, and this chapter concludes with a discussion of the specific choices employed throughout this book.

1.1 BOUNDARY ELEMENT METHOD

The boundary element method (BEM) can be categorized as a part of modern scientific computing, a collection of numerical techniques for solving some basic (but certainly not all) partial differential equations. It has emerged as a powerful numerical technique for solving a wide variety of computational engineering and science problems. Examples of application of the BEM can be drawn for example, from fields of elasticity [28], geomechanics [13, 18], structural mechanics [273], electromagnetics [227], acoustics [153], hydraulics, low-Reynolds number hydrodynamics, biomechanics, off-shore structures [28], and cathodic protection [28]. A distinguishing feature of this approach, as opposed to commonly employed 'volume methods' such as finite elements [304] or finite differences [262], is that only the boundary of

The Symmetric Galerkin Boundary Element Method. By Sutradhar, Paulino and Gray **1**
ISBN 978-3-540-68770-2 ©2008 Springer-Verlag Berlin Heidelberg

2 INTRODUCTION

the domain is discretized, thereby reducing the dimension of the problem by one. In three dimensions, only the bounding surface is discretized and in two dimensions, only the boundary contour is discretized as shown in Figure 1.1. The BEM is based upon transforming the differential equation into an integral equation. This integral equation is valid everywhere – inside the domain, on the boundary, and even exterior to the domain – but it is usually the on-boundary form that is of interest. A wide variety of algorithms have been employed to numerically approximate the solution of this boundary integral equation, and this book focuses on one particular relevant method, symmetric-Galerkin, or more generally, just Galerkin.

A major emphasis in this book will be that boundary integral numerical techniques are largely independent of the particular differential equation being solved. As indicated by the titles for Chapters 3 and 4, there are differences in treating problems in two and three dimensions, but even here the underlying approaches are essentially the same. Thus, to introduce this subject it suffices to consider what is possibly the simplest situation, the three-dimensional Laplace equation $\nabla^2 \phi = 0$ posed in the domain $\mathcal{V} \subset \mathcal{R}^3$ with boundary Σ. The corresponding integral equation (to be derived in Chapter 3) for the potential function $\phi(x, y, z)$ can be written as

$$\mathcal{P}(P) \equiv \chi(P)\phi(P) + \int_{\Sigma} \left(\frac{\partial G}{\partial \mathbf{n}}(P, Q)\phi(Q) - G(P, Q)\frac{\partial \phi}{\partial \mathbf{n}}(Q) \right) \, dQ = 0 \ . \quad (1.1)$$

where $G(P, Q)$ is a known function (the Green's function, to be derived later), and P and Q are called the source point and the field point, respectively. The function $\chi(P)$ is the characteristic function of the open domain $\mathcal{V}$,

$$\chi(P) = \{ \begin{array}{cc} 1 & P \in \mathcal{V} \\ 0 & P \notin \mathcal{V} \cup \Sigma \end{array} , \quad (1.2)$$

so the *interior* $P \in \mathcal{V}$ and *exterior* $P \notin \mathcal{D} \cup \Sigma$ are readily defined. However, the desired boundary equation, $P \in \Sigma$, is a more delicate issue, due to the important fact that $G(P, Q)$ is singular when $P = Q$. The case $P \in \Sigma$ is therefore left unresolved until Section 1.1.3 where some of the other important features of Eq.(1.1) are discussed.

1.1.1 Approximations and Solution

We first define what we seek to obtain from a numerical solution of Eq.(1.1). As with all partial differential equations, it is necessary to specify some additional information. For instance, in Laplace equation, this supplemental information is boundary values, and the boundary conditions are most commonly given as, for every boundary point Q, the value of $\phi(Q)$ *or* its normal derivative

$$\phi_n(Q) \equiv \frac{\partial}{\partial \mathbf{n}}\phi(Q) = \nabla \phi \bullet \mathbf{n} = \left(\frac{\partial \phi}{\partial x}, \frac{\partial \phi}{\partial y}, \frac{\partial \phi}{\partial z} \right) \bullet \mathbf{n} \ . \quad (1.3)$$

In this expression, $\mathbf{n} = \mathbf{n}(Q)$ is the unit *outward* normal at this point. However, in some applications neither function value is given, instead a relationship between the values of the potential and its derivative is known, an example being the problem discussed in Section 1.2.1. The goal of a boundary element solution is, first and

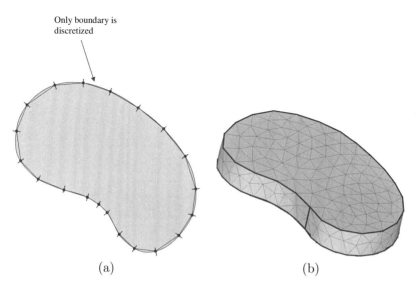

Figure 1.1 Discretization: (a) for 2D problem line elements are used, (b) for 3D problem surface elements are used.

foremost, to complete the knowledge of the two basic boundary functions, potential and flux (in the case of elasticity, displacement and traction). Thus, generally speaking, BEM is most suitable for applications that only require a boundary solution. Subsequent to solving for the boundary functions, interior values can be computed from Eq.(1.1), but this may be an expensive computation.

Note that Eq.(1.1) provides an infinite number of equations, one for every choice of the point P, to balance the infinite number of unknown boundary values. Numerical approximations are used to reduce this to a finite system of equations that can be solved. A major point of departure between the boundary and volume methods is therefore that the numerical approximations in BEM are limited to the boundary: the approximations are the interpolation of Σ and the boundary functions. An important consequence is that, as illustrated in Figure 1.1, only the boundary must be discretized, and depending upon the application (again to be illustrated by the problem discussed in Section 1.2.1) this can provide significant advantages. What is less obvious at this point is that these boundary interpolations are the only approximations, *no* approximation of the governing differential equation has been invoked; this point will be discussed further in the next section. There are some additional numerical errors, numerical quadrature and the linear algebra computation to be specific, but ideally these errors should be negligibly small.

Another novel aspect of BEM is that the boundary potential and its normal derivative are treated on an equal footing. Thus, unknown values of the derivative are solved for directly; by contrast, the standard finite element method obtains an approximate potential everywhere in the domain, and then must differentiate this approximation to obtain any needed derivatives. This differentiation necessarily involves a loss of accuracy, and thus BEM is appropriate for applications that require accurate derivatives. The ability to obtain accurate surface derivatives is

4 INTRODUCTION

an important feature of BEM, and this will be discussed further in Section 1.2.1 and Chapter 5.

1.1.2 The Green's function $G(P, Q)$

The transformation from differential to integral equation is made possible by the **Green's function** $G(P, Q)$. For the Laplace equation in three dimensions, the 'free space' Green's function or fundamental solution is simply given by

$$G(P, Q) = \frac{1}{4\pi r} , \tag{1.4}$$

and the corresponding normal derivative is

$$\frac{\partial}{\partial \mathbf{n}} G(P, Q) = -\frac{1}{4\pi} \frac{\mathbf{n} \cdot \mathbf{R}}{r^3} . \tag{1.5}$$

Here $r = \|\mathbf{R}\| = \|Q - P\|$ is the distance between the integration variable (and hence boundary point) Q and the point P. As the term *fundamental solution* implies, the most important attribute of G is that for fixed P, $G(P, Q)$ satisfies the Laplace equation everywhere in Q except, as can be noted from Eq.(1.4), at $Q = P$. However, as G is symmetric in its two argument, $G(P, Q) = G(Q, P)$, the roles of P and Q can be reversed. The exception at $Q = P$ is critically important: the divergence of the Green's function and its normal derivative at $Q = P$ means that the numerical implementation of Eq.(1.1) for $P \in \Sigma$ must define and evaluate the singular integrals, and this task will in fact dominate the discussion of the numerics. Note that the singularity in the normal derivative is worse than for G, and thus these integrals will therefore be more difficult to handle than for G.

This normal derivative of G is nevertheless also a solution of the governing differential equation, and thus Eq.(1.1) can be interpreted as stating that any particular solution $\phi(P)$ of the Laplace equation can be written as a linear combination of these basic solutions. The coefficient functions in this linear combination are in fact the physical functions, the potential and its normal derivative. Once these functions are known on the boundary (albeit approximately), the representation of $\phi(P)$ given by Eq.(1.1) is in fact an *exact* solution of the Laplace equation in $\mathcal{V}$; this is again in marked contrast to the piecewise polynomial finite element solution, which provides a good approximation of ϕ but in no way satisfies the differential equation. The BEM solution is, however, not an exact solution of the problem at hand, as the boundary conditions can only be enforced approximately.

1.1.2.1 Fast Methods It should be noted that Eq.(1.1) states that the value of the potential at P is related to the values of potential and flux *everywhere* on the boundary. A negative aspect of the integral equation is therefore that the discrete matrix equation that will result is fully populated (dense). For a standard boundary integral implementation having M unknowns, the computational effort to construct the matrix, and the memory required to store it, scale as $\mathcal{O}(M^2)$. By contrast, the matrix systems generated by finite elements or finite differences will necessarily be larger, but they are sparse and relatively inexpensive to assemble.

The $\mathcal{O}(M^2)$ scaling of storage and operations is a drawback for integral equation methods, necessarily limiting the size of the problem that can be tackled. However, an important characteristic of the Green's function is that, as can be seen from

Eq.(1.4), $G(P, Q) \to 0$ for $r \to \infty$. That is, the influence of the point Q on P decays with the distance between them. This has led to the development of *Fast Methods*, e.g., the work by Greengard and Roklin [124] on the Fast Multipole Method. There are at this point a variety of fast techniques, including methods based upon wavelet expansions [245], Fast Fourier Transforms [224, 225], and low rank approximations (Adaptive Cross Approximation [237]).

Although these methods differ in how the key approximations are carried out, they share some basic features. The coefficient matrix A is never fully assembled: the solution of the equation $Ax = b$ is accomplished with an iterative solver, which only requires the ability to compute an arbitrary matrix vector product Ax. In this computation the 'near field' contributions are computed directly with standard methods, but the 'far field' (r large) integrals are computed approximately and rapidly. This reduces the scaling with size to (very roughly) $\mathcal{O}(M)$, and this allows integral equations to be applied to large scale problems. A discussion of Fast Methods is beyond the scope of this book, but fortunately we can refer the reader to the recent monograph by Rjasanow and Steinbach [237].

1.1.3 Singular and Hypersingular Integrals

As noted above, Eq.(1.1) provides an equation for when P is either interior or exterior to the domain, but the essential case $P \in \Sigma$ requires further discussion. The extension that will allow P to be on the boundary cannot be trivial, as this must somehow make sense of the singular integrals. The approach adopted in this book is to define the boundary equation as a limit as P approaches the boundary, of either the interior or exterior equation; it will turn out that the two limits of Eq.(1.1) result in the same boundary equation. The singular integrals are therefore defined in terms of this limit, and while this definition is physically reasonable and intuitively simple, it is necessary to establish that the limits exist, and moreover that they can be readily computed. Chapters 3 and 4 will largely be concerned with the development of effective hybrid analytic/numeric methods for evaluating these singular integral limits for two- and three-dimensional problems.

There are other techniques available in the literature for defining and computing singular integrals [275]. One motivation for adopting the boundary limit is that, in addition to Eq.(1.1), it is also necessary to be able to handle the corresponding equation for the normal derivative of ϕ,

$$\mathcal{F}(P) \equiv \chi(P)\frac{\partial \phi}{\partial \mathbf{N}}(P) + \int_{\Sigma} \left(\frac{\partial^2 G}{\partial \mathbf{n} \partial \mathbf{N}}(P, Q)\phi(Q) - \frac{\partial G}{\partial \mathbf{N}}(P, Q)\frac{\partial \phi}{\partial \mathbf{n}}(Q) \right) \, dQ = 0 \, ,$$

$$(1.6)$$

where $\mathbf{N} = \mathbf{N}(P)$ is the unit normal at P, and derivatives are taken with respect to the coordinates of P. Note that this equation has been formally obtained by differentiating Eq.(1.1), and then interchanging the derivative with the integral. Due to the singularity of the Green's function, this interchange is legal if P is off the boundary, and thus interior and exterior equations are mathematically correct. As might be expected, the boundary limit analysis for this equation is challenging.

Eq.(1.6) is essential for achieving the symmetry in symmetric-Galerkin. The singularity in the second derivative of the Green's function, termed *hypersingular*, will require a careful analysis. A significant benefit of the boundary limit approach

6   INTRODUCTION

is that one set of singular integration techniques suffices i.e., the same methods apply equally to G and its first and second derivatives.

In addition to the equation for surface flux, Eq.(1.6), it is important for many applications to be able to calculate any directional derivative of ϕ, *i.e.* the gradient of ϕ. A notable class of problems requiring gradient evaluation is *moving boundary problems*, as the critical surface velocity is often a function of the gradient. A second key advantage of a boundary limit definition is that combining interior and exterior equations will lead to a highly effective algorithm for computing the surface gradient, either $\nabla\phi$ or surface stress in elasticity (Chapter 5).

1.1.4   Numerical Solution: Collocation and Galerkin

There are two basic procedures that are generally used to reduce the continuous integral equations Eq.(1.1) and Eq.(1.6) to a finite system. The simpler procedure is *collocation*, wherein the boundary integral equations are explicitly enforced at a finite set of points. In its simplest form, these collocation points are chosen to be the nodes used to discretize the boundary. Thus, a collocation approximation of Eq.(1.1) can be simply stated as

$$\mathcal{P}(P_k) = 0 \tag{1.7}$$

where P_k, $1 \leq k \leq M$ are the selected boundary points. If the boundary potential and flux are interpolated from their values at these M points, $\phi(P_k)$ and $\phi_n(P_k)$, then the boundary conditions usually provide M of these $2M$ numbers. The point-wise enforcement of Eq.(1.1) in Eq.(1.7) then provides the M equations needed to solve for the unknown values. Collocation necessarily leads to non-symmetric matrices.

In contrast to collocation, the *Galerkin* approach does not require that the integral equations be satisfied at any point. Instead the equations are enforced in a weighted average sense,

$$\int_{\Sigma} \psi_k(P)\mathcal{P}(P)\,\mathrm{d}P = 0$$
$$\int_{\Sigma} \psi_k(P)\mathcal{F}(P)\,\mathrm{d}P = 0\,. \tag{1.8}$$

where $\psi_k(P)$ are the chosen weight functions. The needed M equations can be generated by an appropriate choice of M weights. The strict definition of Galerkin is that the weight function $\psi_k(P)$ is composed of the shape functions that are non-zero at the node P_k, the shape functions being the local basis functions used to interpolate the boundary functions. For a linear interpolation, the Galerkin weight functions are illustrated in Fig. 1.2.

In mathematical terminology, collocation is a 'strong' solution, the equations being satisfied at the specified points, whereas Galerkin is a 'weak' solution. The requirement that the equations hold in integrated sense has a geometric interpreta-tion: the approximate Galerkin solution is obtained by projecting the exact solution onto the subspace consisting of all functions which are a linear combination of the shape functions. The Galerkin solution is therefore the linear combination that is the 'closest' to the exact solution.

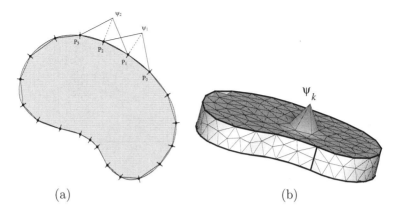

Figure 1.2 Illustration of the Galerkin weight functions for 2D and 3D BEM.

In general, Galerkin is more accurate than collocation, and also provides a more elegant treatment of boundary corners. However, the primary advantage of Galerkin is that the treatment of hypersingular integrals is actually much simpler than with collocation. As will be discussed in detail in Chapter 3, the Galerkin form of the hypersingular integral exists if the interpolation of the boundary potential is continuous ($\mathcal{C}^0$). For collocation, a smoother interpolation (differentiable $\mathcal{C}^1$) is required for the existence of the integral. Standard elements, e.g., linear or quadratic, are continuous across element boundaries, and it is quite a bit more complicated to implement a differentiable interpolation [33, 284].

1.1.5 Symmetric Galerkin BEM

With Galerkin, P and Q are treated equally, and thus it is possible to produce a symmetric coefficient matrix, the symmetric Galerkin boundary element method (SGBEM). This algorithm is based upon the symmetry properties of the Green's function,

$$\begin{aligned} G(P,Q) &= G(Q,P) \\ \nabla_Q G(P,Q) &= -\nabla_P G(P,Q) = \nabla_P G(Q,P) \\ \frac{\partial^2 G}{\partial \mathbf{N} \partial \mathbf{n}}(P,Q) &= \frac{\partial^2 G}{\partial \mathbf{N} \partial \mathbf{n}}(Q,P) \,. \end{aligned} \quad (1.9)$$

It follows that if the potential is specified everywhere on the boundary, then the Galerkin form of the potential equation will yield a symmetric matrix; a similar statement holds for flux boundary data and the flux equation. For a general mixed boundary value problem, symmetry results if each equation is applied on its appropriate surface. Thus the potential equation is applied on the Dirichlet portion of the boundary and the flux equation is applied on the Neumann portion of the boundary.

8 INTRODUCTION

After discretization, the set of equations Eq.(1.8) can be written in block-matrix form as $[H]\{\phi\} = [G]\left\{\frac{\partial\phi}{\partial\mathbf{n}}\right\}$, and in block-matrix these equations become

$$
\begin{bmatrix} H_{11} & H_{12} \\ H_{21} & H_{22} \end{bmatrix}
\begin{Bmatrix} \phi_{bv} \\ \phi_u \end{Bmatrix}
=
\begin{bmatrix} G_{11} & G_{12} \\ G_{21} & G_{22} \end{bmatrix}
\begin{Bmatrix} \left(\frac{\partial\phi}{\partial\mathbf{n}}\right)_u \\ \left(\frac{\partial\phi}{\partial\mathbf{n}}\right)_{bv} \end{Bmatrix}. \tag{1.10}
$$

Symmetry of the coefficient matrix for a general mixed boundary value problem is achieved by the following simple arrangement. The BIE is employed on the Dirichlet surface, and the HBIE equation is used on the Neumann surface. The first row represents the BIE written on the Dirichlet surface, and the second row represents the HBIE written on the Neumann surface. Similarly, the first and the second columns arise from integrating over Dirichlet and Neumann surfaces, respectively. The subscripts in the matrix therefore denote known boundary values (bv) and unknown (u) quantities. Rearranging Eq.(10.21) into the form $[\mathbf{A}]\{x\} = \{b\}$, one obtains

$$
\begin{bmatrix} -G_{11} & H_{12} \\ G_{21} & -H_{22} \end{bmatrix}
\begin{Bmatrix} \left(\frac{\partial\phi}{\partial\mathbf{n}}\right)_u \\ \phi_u \end{Bmatrix}
=
\begin{Bmatrix} -H_{11}\phi_{bv} + G_{12}\left(\frac{\partial\phi}{\partial\mathbf{n}}\right)_{bv} \\ H_{21}\phi_{bv} - G_{22}\left(\frac{\partial\phi}{\partial\mathbf{n}}\right)_{bv} \end{Bmatrix}. \tag{1.11}
$$

The symmetry of the coefficient matrix, $G_{11} = G_{11}^T$, $H_{22} = H_{22}^T$, and $H_{12} = G_{21}^T$, now follows from the properties of the kernel functions (see Eq.(1.9)).

The advantages of Galerkin, of course, come at a price: although, as shown in Fig. 1.2, the weight functions have local support, the additional boundary integration with respect to P is nevertheless computationally expensive. These costs can be reduced somewhat by exploiting symmetry, but Galerkin will nevertheless require more computation time than collocation.

1.2   AN APPLICATION EXAMPLE: AUTOMOTIVE ELECTROCOATING

As noted earlier, boundary integral equations have been employed to solve scientific and engineering problems in a variety of fields. A principal advantage is that the boundary-only formulation, compared to domain-based methods, leads to reduced problem size and problem setup time. There can also be important advantages for specific types of problems. For example, for problems posed in infinite domains, domain-based methods like FEM usually truncate the domain with some sort of artificial boundary; in BEM, the boundary condition at infinity is embedded in the Green's function, and the problem can be solved by discretizing solely the finite boundary. Similarly, BEM is advantageous for moving boundary problems (*e.g.*, free surface flow and crack propagation), as remeshing the evolving boundary is far easier than remeshing the volume.

Although BEM may provide advantages for certain applications, the problem might nevertheless be solved using a volume method. There are, however, applications for which BEM is more or less essential, and the purpose of this section is to present one such application. Moreover, this problem exploits many of the unique aspects of the integral formulation, and this discussion will be a concrete example of the issues discussed above.

The engineering application to be described is the electrocoating of automobiles. The discussion begins with an explanation of the problem and why simulations are essential.

1.2.1 Engineering Optimization

A critical step in the automotive manufacturing process is the deposition of a thin layer of rust-preventing paint onto the entire metal frame. Failure to coat vehicles properly has a very high economic cost -warrantee expenses to replace rusted parts - and thus ensuring a satisfactory coating is an essential part of engineering car design.

The coating is applied by an electrochemical process: the car frame is submerged in a large paint tank, and a voltage is applied between the frame and the tank anodes. The liquid paint is designed to be electrochemically active, and the amount of paint deposited is at any given time proportional to the local current density. The ability to accurately simulate the electrodeposition process would enable designers to answer a number of important manufacturing questions.

Car Frame Design. To prevent rust, the paint coating must be of a specified minimum thickness everywhere on the frame. However, it is difficult for sufficient current to reach recessed areas of the highly complicated frame geometry, and too little paint may be deposited. The design engineers must therefore place holes strategically in the frame to increase the current in difficult to reach areas, but the size and location of holes cannot be chosen arbitrarily. The counterbalancing constraint is the need for structural integrity, the car must pass crashworthiness tests.

This complex design optimization problem has generally been solved by trial-and-error, involving the construction and testing of prototypes. This however is time consuming, expensive, and due to the small number of possible trials, uncertain of success. Simulations would be faster, cheaper, and many more prospective hole patterns could be examined.

Tank Design and Operation. A second role for simulations is to increase operational efficiency and reduce costs by improving tank design: the number of anodes, anode positioning, and the voltage history for each anode. For most of these design problems, physical experiments are simply not an option: time in commercial tanks is not available, and the costs of set-up (moving anodes), materials, and measuring of the prototypes is simply too costly.

A related operational problem is that anodes frequently go out of service. As it is too expensive to stop the production line to replace one anode, or even a handful of anodes, manufacturers will continue to run with non-functioning electrodes. The question that needs to be answered, and can be answered via computational simulation, is at what point does the paint coverage suffer and it becomes necessary to close the line and fix the tank.

A third goal of tank design, also with potentially significant economic impact, is to minimize overpainting. While ensuring the minimum coverage is the primary goal, overpainting is also expensive, due to the the high cost of the paint. This is another design constraint that can be effectively studied through simulation.

10 INTRODUCTION

1.2.2 Electrocoating Simulation

As motivated above, electrodeposition simulation is crucial for automotive manufacturing. The modeling, and why integral equations are vital, will now be discussed. The governing equation is the Laplace equation, $\nabla^2\phi = 0$, for the electric potential ϕ in the paint tank (and exterior to the car and anode surfaces). The primary quantity of interest is the current density on the car frame, as this determines the rate of paint deposition, and the current density is proportional to the surface flux (normal derivative of the potential).

The boundary conditions on the anode are known potential (voltage), and on the tank walls the current density is zero. The boundary conditions on the car frame are more complicated, and are discussed below. The aspects of this problem that demand a boundary integral formulation are discussed below.

Boundary Condition and Boundary Solution. BEM is effective for problems that only require a boundary solution, and in this case it is only the surface flux (on the car frame) that is of interest. Equally important in this case is that while the tank (zero flux) and anodes (applied voltage) have standard boundary conditions, on the car frame the known condition is of Robin type,

$$\frac{\partial\phi}{\partial\mathbf{n}}(P) = F\left(\phi(P)\right) , \tag{1.12}$$

where F can depend upon many things. In particular, the physical properties of the paint. What is important is that the boundary integral formulation, due to the explicit treatment of the flux, can enforce this boundary condition relatively easily.

Meshing. The car frame geometry is highly complex, as can be seen in Figure 1.3 and Figure 1.4, and moreover there are in general numerous anodes in a typical tank. A surface mesh is therefore very much simpler to produce than a volume discretization, and a volume mesh would involve a huge number of nodes and elements. Moreover, the surface mesh for the tank and anodes is independent of the vehicle, and this provides two important simplifications. First, as the tank configuration generally does not change for different products on the assembly line, the tank mesh can be re-used. Second, during deposition, the car frame moves continuously through the tank: this would be quite impossible to simulate with a standard volume formulation, as a completely new mesh would be required at every time step. A boundary formulation can accommodate vehicle movement without remeshing, the coordinates of the car frame are simply shifted in time.

It should also be noted that the car frame metal is relatively thin, and for the electric field solution the thickness can safely be approximated as zero. Thus, the car is a complex "crack-like" geometry, and the availability of effective Galerkin methods for working with the hypersingular flux equation is quite important.

Design Iterations. As discussed above, the primary role of the simulations is design optimization, and here too it is highly advantageous to work only with a surface mesh. Altering the hole geometry on the frame requires only 'local' changes in the boundary discretization, whereas modifying a volume mesh is necessarily a much more complicated task. A similar statement applies equally as well to the meshing requirements needed for the multiple runs required for tank design optimization.

VISUALIZATION 11

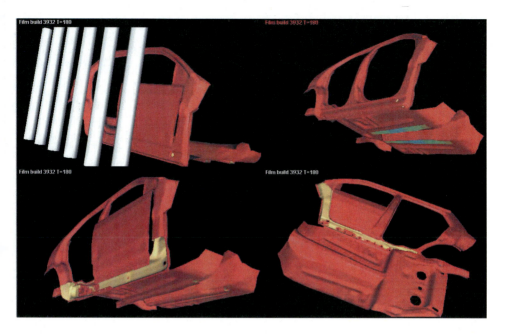

Figure 1.3 Four different views of paint deposited on a car frame after 180 seconds. The white cylinders in the first picture (upper left corner) are anodes.

Large Scale Computation. Due to the complex geometry, these simulations will be a large scale calculations for either a volume or a surface formulation. However, accurate representation of the car surface would demand a highly refined volume mesh, and thus the problem size would be huge. A boundary integral formulation can take advantage of fast methods to reduce the computational resources required.

1.3 VISUALIZATION

Scientific visualization [179] for representing data is an essential component of scientific computing. For problems with complicated geometries or that are time dependent (or, as with electrocoating simulation, both) the results cannot be understood by simply looking at the enormous quantity of numbers. Similarly, errors in the calculation, that might for example arise due to either a poorly constructed or incorrect mesh, are most easily seen through a visual representation of the data. Visualization of the mesh itself can be beneficial for constructing and verifying that the complicated surface discretization is correct. Note that for the electrocoating problem, many different simulations, and thus many different discretizations, will be required.

The above snapshots of the deposition process give just the barest indication of the visualization tools that are required to analyze and understand the results from these simulations. For time dependent problems, in general, animations created from the time history data can impart a physical understanding of the process being modeled. For the ultimate purposes of the electrocoat calculations, solving the difficult design optimization problems discussed above, these insights can

12   INTRODUCTION

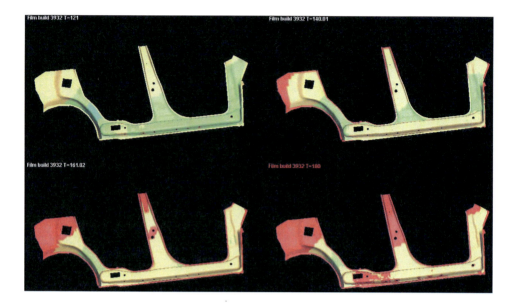

Figure 1.4 Paint deposited at four different time steps on the inside of a rocker panel, often the most difficult area to paint.

potentially be highly useful for choosing new trial geometries. Thus, visualization is more than just a way of presenting results, it is an integral part of scientific computation that is very useful in BEM.

Scalar data consists of single data associated with each location in the dataset of the BEM analysis. Various techniques that can be used for scalar are mainly, contours, glyphs and clipping. Figure 1.5 illustrates the techniques showing their use. Contours are surfaces/lines of constant scalar values. We can either set up contours for specific values or specify a scalar range and the number of contours to be generated in the range. Glyphing is a visualization technique that represents data by using symbols, or glyphs. The symbols may be a cone, a sphere or multivariated glyphs which can be scaled, colored and oriented along a direction. The glyphs are copied at each point of the input dataset. Clipping can be used to visualize the details of interior of a body. For example, the contour plot of the flux on the surface of the cavity can be visualized clearly if it is clipped using a plane. Also multiple clippers can be used too. Figure 1.6 (a) and (b) show the two clippers that were used to generate Figure 1.6(c). By using the clippers, the interior results of temperature and also the temperature on the crack surfaces can be conveniently visualized.

1.3.1 Virtual Reality

Virtual Reality (VR) has been successfully explored in many fields of science and engineering [254]. Thus the use of VR in conjunction with boundary element methods offers new avenues to advance and to make such methods more popular and accessible. In general, VR makes possible multisensory, three-dimensional modeling of scientific data [30, 254]. While the emphasis is on visualization, other senses are

VISUALIZATION 13

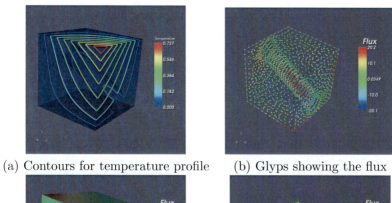

(a) Contours for temperature profile (b) Glyps showing the flux

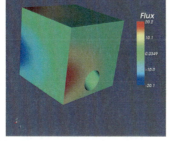 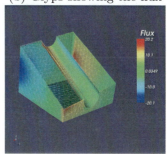

(c) Flux results without clippers (d) Flux results with clippers

Figure 1.5 Visualization techniques for scalars

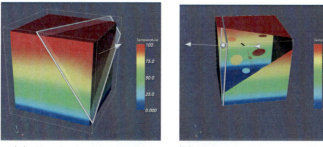

(a) Cracks inside a cube (b) Interior using one clipper

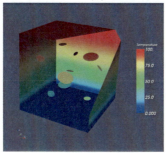

(c) Using two clippers

Figure 1.6 Visualizing the results on the crack surfaces using the two clippers.

14   INTRODUCTION

also added to enhance what the user can visualize. Also, sharing data and visualization with other scientists are integral parts of the growing collaboration and connection among researchers. Providing results using a new type of visualization based on VRML gives the opportunity for researchers to navigate the visualization in an interactive 3D virtual environment over the internet [147]. It is a new interface paradigm that uses computers and human-computer interfaces to create the effect of a three-dimensional world in which the user interacts directly with virtual objects [30]. The three-dimensional display and interaction capabilities of VR allow for significantly enhanced three-dimensional perception and interaction over three-dimensional computer graphics. In order to understand complex phenomena from a time-dependent data set in a complicated three-dimensional structure, VR allows rapid and intuitive exploration of the volume containing the data, enabling the phenomena at various places in the volume to be explored, as well as provide simple control to manipulate and interact through interfaces integrated into the environment.

1.3.2 CAVE: Cave Automatic Virtual Environment

The Cave Automatic Virtual Environment (CAVE) is an immersive virtual reality facility designed for the exploration of and interaction with spatially engaging environments. A virtual environment constructed in the CAVE places the user inside the simulation and thereby provides the level of understanding necessary to accurately assess the design of the system being modeled [26]. A typical CAVE is a projection-based VR system that surrounds the viewer with 4 screens. The screens are arranged in a cube made up of three rear-projection screens for walls and a down-projection screen for the floor; that is, a projector overhead points to a mirror, which reflects the images onto the floor (see Figure 1.7). A viewer wears stereo shutter glasses and a six-degrees-of-freedom head-tracking device. As the viewer moves inside the CAVE, the correct stereoscopic perspective projections are calculated for each wall. A second sensor and buttons in a wand held by the viewer provide interaction with the virtual environment. For instance, the use of such immersive environmental for BEM electrocoating simulation would allow the user to go inside the "virtual tank" and to actively participate and interact with engineering process of electrodeposition in space and time.

1.3.3 The MechVR

MechVR (**M**echanics **V**irtual **R**eality) is a preliminary VR application software that was developed in Paulino's research group. It can visualize BEM (and FEM) data in the CAVE. It is preferable to visualize such data in a 3D interactive environment and allow researchers to share information in an effective and easy way. In this application, data sets obtained from mechanics computations (e.g. BEM simulations), are taken as data source, and by using VTK [244] the data is processed and rendered in the CAVE. **MechVR** gives users the ability to create and modify a set of visualization tools for exploring the response field from a given dataset. They also have the ability to change their viewpoint in the virtual environment, manipulate the dataset. For instance, isosurfaces, slices, and hyperstreamlines are generated to represent certain stress/flux component and associated tensors. A user interface is provided in the CAVE environment that allows new images to be generated according to user-defined parameters. Figure 1.7 shows the author (Dr. Sutradhar) using the **MechVR** inside the CAVE at NCSA (National Center for Supercomputing Applications www.ncsa.uiuc.edu).

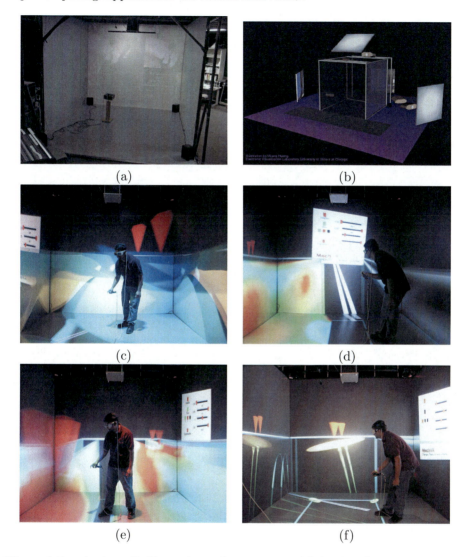

Figure 1.7 A schematic illustration and a cartoon model showing the projectors and the overall assembly. Demonstration of the results of a stress field in MechVR inside CAVE

1.4 OTHER BOUNDARY TECHNIQUES

As the title suggests, this book makes no attempt to cover all boundary integral methods. In fact, precisely the opposite, a very specific way of reducing the integral equations to a finite system, based on Galerkin method, is the sole focus. Moreover, major topics are omitted - fast methods, time dependent problems, mathematical theory, to name a few. Many viable numerical techniques are likewise not discussed.

16 INTRODUCTION

To take just one example, while only direct boundary integral equations, Eq.(1.1) and Eq.(1.6), are employed herein, there is a large literature on an indirect method called the Method of Fundamental Solutions [88, 104].

As a way of highlighting the choice of topics and algorithms, and moreover emphasizing that there are always alternative methods available, this section will very briefly discuss some of the techniques that will *not* appear in subsequent chapters.

1.4.1 Singular Integration

As noted above, a dominant aspect of the numerical implementation of a direct boundary integral equation is singular integration. The boundary limit evaluation that will be utilized in this book has attractive features: the techniques are the same irrespective of the particular Green's function and which derivative of G is being integrated, and second, an effective algorithm for surface gradient evaluation depends upon adopting this limit definition. The generality of the singular integration treatment has an important consequence for the organization of this book: instead of the commonly employed chapter division by partial differential equation (Laplace, elasticity, etc.), it suffices to divide the discussion into two- and three-dimensional analyses.

There are, however, many other singular integration techniques. In particular, for the hypersingular integral, the first algorithms were methods based upon the Hadamard Finite Part definition [129]. Numerical techniques based upon transformations that reduce the order of the singularity (Telles [276] and Duffy [181] transformations) have been widely used. Another principal algorithm for hypersingular integrals exploits Stokes' Theorem (integration by parts) to reformulate the integral [90, 165, 170]. This approach is quite effective, even for the difficult case of anisotropic elasticity [41], and moreover Stokes' can be employed to reformulate all boundary integrations, the basis for the Boundary Contour Method.

1.4.2 Meshless and Mesh-Reduction Methods

As with singular integration, it would be wrong to leave the impression that Galerkin and collocation are the only available options for the reduction of the integral equation to a finite system. Boundary integral analysis is an active field of research, and several new numerical approaches have recently been put forward. The goal in the following is not to be exhaustive or detailed about alternative *mesh-reduction* and *meshless* methods, but to give a brief indication of their existence, and to provide the interested reader with pointers to the recent literature.

1.4.2.1 Boundary Contour Method. Typically in BEM for a 3D problem, only the surface is discretized, thus reducing the dimensionality of the problem by one. Based upon work by Lutz [170], the boundary contour method (BCM) achieves a further reduction in dimension. As proposed by Lutz [170] for the 2-D Laplace equation, Nagarajan *et al.* [194] for 2-D linear elasticity, for 3-D linear elasticity [195], and Phan *et al.* [220] for Stokes' flow, the BCM, involves no numerical quadrature in 2-D and only numerical evaluation of line integrals (over the closed bounding contours of the usual surface elements) in 3-D.

The central idea of the BCM is to exploit the divergence-free property of the boundary integral kernels to analytically convert, via Stokes' Theorem, a surface integral to a line integral over the contour that bounds the surface element. This is essentially the same procedure employed for the hypersingular integrations mentioned above, and hypersingular boundary contour formulations for two-dimensional [221] and three-dimensional [192] linear elasticity have been proposed. A symmetric Galerkin BCM for 2-D linear elasticity appears in [201].

1.4.2.2 Boundary Node Method. Motivated by a flurry of work on the development of meshless finite element methods [9, 19, 80, 197], Mukherjee and co-workers proposed a boundary meshfree method called the Boundary Node Method [43–47, 152, 193]. A meshfree method, as the name implies, does not require an explicit mesh, in the traditional sense, to discretize the problem boundary and define the approximate boundary functions.

In the Boundary Node Method, the spatial discretization (meshing) and the boundary function interpolations are not coupled. Instead, a set of scattered boundary nodes is employed as the basis for a 'diffuse' moving least-squares (MLS) interpolation [158] to represent the boundary functions. The surface integration is carried out based upon a surface cell partition of the boundary, see Fig. 1.8. As shown in this figure, a key difference between a cell decomposition of Σ and a traditional mesh is that the cells are 'unstructured', i.e., the boundaries of the cells can intersect arbitrarily. The integration over cells is performed by using regular Gauss quadrature, except for the cell where the integration is singular

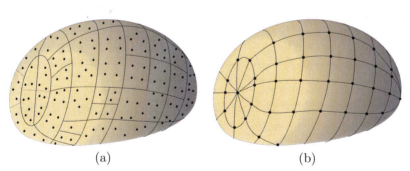

(a)   (b)

Figure 1.8 (a) Boundary Node Method with the nodes and the cells (b) BEM with nodes and elements

Adaptive procedures based on the "hypersingular residuals", developed for error estimation in the mesh-based collocation boundary element method (BEM) and symmetric Galerkin BEM by Paulino et al. [[208], [210], [209]], can be extended to the meshless BNM setting. Dirichlet problem on a cube, which is analysed where the Laplace's equation is solved using the BNM, and the (hypersingular) residuals are obtained using the HBNM. The iterative cell design cycle is shown in Figure 1.9(a) and Figures 1.9(b). It is noted that the cell refinement should begin at the corners of the cube where the error in $\partial u/\partial n$ is the largest.

1.4.2.3 Boundary Cloud Method. A key difficulty in the Boundary Node Method is the construction of the moving least-squares interpolations. For 2-D problems,

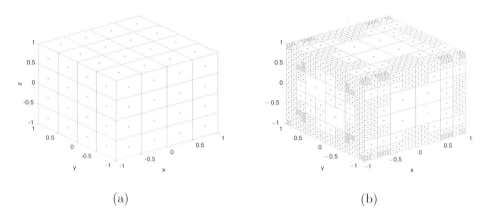

Figure 1.9 Cell configurations on the surface of a cube (a) Initial configuration # 2 (96 surface cells), (b) third step (1164 cells)

where the boundary is 1-D, (x, y) coordinates cannot be used to define the approximations, a cyclic coordinate is used in the MLS approach. For 3-D problems, corresponding 2-D curvilinear coordinates (s_1, s_2) are used to construct interpolation functions, but the definition of these coordinates is highly nontrivial for complex geometries.

The Boundary cloud method proposed by Li and Aluru [164] also employs a scattered point approach for constructing interpolation functions, and a cell structure for carrying out the boundary integration. However, regular Cartesian coordinates (x, y) for 2-D or (x, y, z) for 3-D are employed to construct Hermite-type interpolation functions. In addition, the boundary cell integrations are classified as singular, nearly singular and nonsingular, and different techniques are employed to evaluate the integrals: direct analytical integration for singular, a Nystrom scheme combined with a singular value decomposition technique for near-singular, and regular Gauss quadrature for nonsingular.

1.4.2.4 Local BIE.
In the Local Boundary Integral Equation [259, 303] methodology, a cloud of nodal points is once again employed, but now points are also distributed throughout the volume, as illustrated in Fig. 1.10. All nodal points belong in regular sub-domains (*e.g.*, circles for two dimensional problems) centered at the corresponding collocation points When non-linear elastic problems or elastic problems with body forces are considered, the displacement field at these sub-domains is described through the same surface integral equation used in the conventional static elastic BEM, accompanied by volume integrals arising from the non-linear terms and/or the body forces. The displacements on the boundary and in the volume are usually approximated by a moving least square scheme. The LBIE, however, is not a true meshless method, as it requires evaluation of integrals over surfaces that are surfaces of standard boundary elements.

There are several other variants of the methods described above, *e.g.*, the Hp cloud method [199, 200] and the Boundary Point Interpolation Method [169]. More

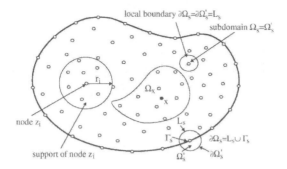

Figure 1.10 Local boundaries, the supports of nodes and the domain of definition of the MLS approximation in LBIE.

information concerning meshless methods can be found in the book by Mukherjee [191].

1.5 A BRIEF HISTORY OF GALERKIN BEM

The mathematical foundations of the boundary element method include the theorems of Gauss, Green and Stokes – they allow the basic reduction from volume differential equation to boundary integral equation [55]. The rigorous analysis of the method (a subject not touched upon here) [243] exploits the modern tools of functional analysis, itself an outgrowth of the early work on the study of integral equations. The brief history of Galerkin and symmetric-Galerkin (SG) methods presented below is far from complete, the review article by Bonnet et al. [25] provides a better summary up to 1998.

Initial works by Jaswon and Symm [140] on potential theory inspired Rizzo to develop a boundary integral formulation for elastostatics and a corresponding numerical (collocation) solution [235]. Rizzo's work paved the way for the modern development of the boundary integral equation method with subsequent works by Cruse [63] in elastodynamics, three-dimensional elastostatics and fracture [64]. Encouraged by Cruse's work, Lachat and Watson [155] developed the method further, introducing concepts from the finite element method, including shape functions, Gauss quadrature, and linear algebra techniques. Brebbia introduced a weighted residual formulation, and in 1977, the boundary integral equation method was established as the "Boundary element method" in three simultaneous publications [55]. In the following year, a book on the method was published (Brebbia, 1978 [28]). Details on the history of BEM can be found in the article by Cheng and Cheng [55].

Spurred by the rapid development of computer technology in the 1970s, the BEM was beginning to have an impact, although almost all of the early BEM computations employed a collocation approximation. In 1977 Bui [31] produced a Galerkin formulation for fracture analysis, and in 1979, Sirtori [256] proposed the symmetric Galerkin formulation for linear elasticity, though these papers did not receive much attention until the late eighties. However, it was recognized that collocation BEM failed to retain many of the nice properties of finite elements: sign-definite symmetric matrix operators, a variational formulation, and proofs of convergence and stability. In addition, a collocation implementation for fracture analysis had to resort to higher order $\mathcal{C}^1$ elements (*e.g.*, Overhauser or Hermite elements) to adequately deal with the hypersingular kernel, and these elements proved difficult to handle, especially in three dimensions. These negative aspects of collocation fueled the interest in the symmetric Galerkin boundary element formulation, and subsequent development was primarily carried out in Europe.

Hartmann *et al.* (1985) [132] extended the SG formulation to beams and Kirchhoff plates, while Maier and Polizzotto (1987) [176] developed SGBEM for elastoplastic solids. In the late eighties and early nineties the SGBEM literature grew at a faster pace, led by Maier and co-workers. They extensively studied elastoplasticity, gradient elasticity, cohesive crack simulation and plasticity problems using the SGBEM [174, 175, 257]. Time-dependent problems were also addressed, including elastodynamics, by Maier *et al.* [173]. Different computational implementations appeared, such as those by Holzer [136], Bonnet and Bui [24], and Kane and Balakrishna [142]. In the books by Kane (1994) [143] and Bonnet (1995) [22], one chapter was devoted to the fundamentals of the SGBEM.

One of the main hurdles in three dimensional computations is the treatment of hypersingular integrals, thus numerous works have focused on this topic. As this is a primary subject of this book, here it is just mentioned that the two main approaches have been based upon Hadamard finite parts, including Andra [6], Carini and Salvadori [37], Carini *et al.* [36], Haas and Kuhn [172], Frangi and Novati [94], Frangi and Guiggiani [93], and Salvadori [241, 242]), among others; and direct limit methods by Gray *et al.* [109, 112] for homogeneous materials, and Sutradhar *et al.* [271] for graded materials.

The early work by Bui [31] on the important topic of fracture has, of course, been followed by an extensive literature. Afterwards, an implementation for fracture analysis in plane orthotropic elasticity was reported by Gray and Paulino [118]. A weak form integral equation for three dimensional fracture analysis was proposed by Li, Mear and Xiao [165]. Frangi *et al.* developed a similar crack analysis based up integration by parts [95], and applied it to crack propagation [91] and fatigue crack growth. Salvadori [240] presented a crack formulation for cohesive interface problems, and Sutradhar and Paulino [270] presented a SG formulation to evaluate T-stress and stress intensity factors (SIFs) using the interaction integral method. Xu, Lie and Cen [296] presented a 2D crack propagation analysis using quasi-higher order elements. Phan and his coworkers presented SIF analysis with frictional contact sliding at discontinuity and junctions, SIF calculations for crack-inclusion interaction problems, and a SGBEM based quasi-static crack-growth prediction tool to investigate crack-particle interactions [151, 218, 222, 223, 290].

Galerkin methods have found a broad range of engineering applications. For instance, a symmetric Galerkin boundary element formulation for two-dimensional, steady and incompressible flow was developed by Capuana et al. [34]. A Stokes problem with general boundary condition with slip condition was reported by Reidinger and Steinbach [231]. Other works (quite incomplete list) using SGBEM include, for example, steady state harmonic solution of the Navier equation by Perez-Gavilan and Aliabadi [215], a fully symmetric formulation for interface and multi-zone problems by Maier [173] and by Gray and Paulino [118], lower bound shakedown analysis by Zhang *et al.* [299], dynamic soil-structure interaction by Lehman and Antes [161], problems of finite elasticity with hyperelastic material and incompressible materials by Polizotto [226], shear deformable plates by Perez [217], dynamic frequency domain viscoelastic problems subjected to steady-state time-harmonic loads by Perez-Gavilan and Aliabadi [216], analysis of Kirchhoff elastic plates by Frangi and Bonnet [92], and a Galerkin residual technique for error estimation and effective adaptive mesh refinement procedure by Paulino and Gray [117].

There are now many books covering the subject of BEM. For further reading, and different views on the numerics of integral equations, books on the fundamentals of BEM include Banerjee [13], Becker [17], Bonnet [22], Brebbia *et al.* [28, 29], Chandra and Mukherjee [40], Gaul *et al.* [99], Hall [131], Kane [143], and Paris and Cañas [205]. In particular, the recent book by Rjasanow and Steinbach [237] covers a vital topic that is not dealt with herein, the development of Fast integral equation methods.

CHAPTER 2

BOUNDARY INTEGRAL EQUATIONS

Synopsis: This chapter introduces boundary integral equations and their numerical approximations. For potential theory, *i.e.*, the Laplace equation, the boundary integral equations for surface potential and for surface flux are derived, these equations involve the Green's function $G(P, Q)$ (fundamental solution) and its first and second derivatives. These functions are divergent when $P = Q$, the singularity becoming progressively worse with higher derivatives, and thus the definition and evaluation of (highly) singular integrals is of paramount importance. The fundamental approach adopted in this book is to define all singular integrals as a 'limit to boundary'. To illustrate this concept with a concrete example, two explicit calculations are carried out. The integrals of the second derivative of $G(P, Q)$, termed *hypersingular*, arising from the two basic approximations – **collocation** and **Galerkin** – are considered. The important observation from these calculations is that the collocation hypersingular integral is *not* finite unless a differentiability condition is met, whereas the corresponding Galerkin calculation is finite under a weaker continuity requirement.

2.1 BOUNDARY POTENTIAL EQUATION

Although the boundary integral formulations for different partial differential equations will necessarily change due to the presence of different Green's functions, these equations all *look* the same. Moreover, the Green's function behavior at the singular point $P = Q$ is essentially the same for all problems (even, roughly

The Symmetric Galerkin Boundary Element Method. By Sutradhar, Paulino and Gray **23**
ISBN 978-3-540-68770-2 ©2008 Springer-Verlag Berlin Heidelberg

speaking, for the more complex Green's function arising in anisotropic elasticity, (Section 4.6). Thus, for the most part, it suffices to discuss the simplest boundary integral equation, that for the Laplace equation.

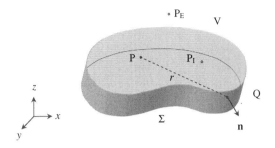

Figure 2.1 A volume V with boundary Σ, the source point P and the field point Q and the interior and exterior points P_I and P_E.

The three-dimensional (isotropic) Laplace equation for the function $\phi(x,y,z)$ is

$$\nabla^2 \phi = \nabla \bullet \nabla \phi = \left(\frac{\partial^2}{\partial x^2} + \frac{\partial^2}{\partial y^2} + \frac{\partial^2}{\partial z^2}\right)\phi = 0 \,, \qquad (2.1)$$

where this equation is assumed to hold for $\mathbf{x} = (x,y,z)$ in an open volume V having boundary Σ (see Fig. 2.1). The function $\phi(\mathbf{x})$ is called a *potential function*. Problems governed by the Laplace equation are found in a wide range of applications e.g. steady state heat transfer, electrostatics (for example, EEG modeling), electrochemical problems [292], magnetostatics, ideal fluid flow, flow in porous media [167] and many others.

The surface flux corresponding to the potential ϕ is

$$-\kappa \frac{\partial \phi}{\partial \mathbf{n}}(Q) = -\kappa \nabla \phi \bullet \mathbf{n}(Q) \qquad (2.2)$$

where $\mathbf{n}(Q)$ is the unit outward normal at the boundary point Q and the terminology for the constant κ depends upon the specific application, e.g., thermal conductivity for heat transfer problems. In many instances it suffices to ignore the constant κ; however including this parameter will be important for the discussion of multi-region problems, Chapter 7.

The derivation of the boundary integral equation that is equivalent to Eq.(2.1) begins by noting that

$$\int_V f(\mathbf{x}) \nabla^2 \phi(\mathbf{x}) \, dV = 0 \qquad (2.3)$$

for any sufficiently well behaved function $f(x)$ defined in the volume. Integrating by parts, *i.e.*, applying the divergence theorem (see the Appendix A), we obtain

$$0 = \int_\Sigma \left[\phi(Q)\frac{\partial f}{\partial \mathbf{n}}(Q) - f(Q)\frac{\partial \phi}{\partial \mathbf{n}}(Q)\right] dQ - \int_V \phi(\mathbf{x}) \nabla^2 f(\mathbf{x}) \, dV \,. \qquad (2.4)$$

In order to get rid of the volume integral, we wish to choose $f(\mathbf{x})$ so that it also satisfies the Laplace equation. These *fundamental solutions* can be taken as

$$G(P,Q) = \frac{1}{4\pi r} \qquad (2.5)$$

where

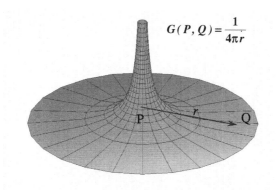

Figure 2.2 Green's function $G(P,Q)$

$$r = \|\mathbf{R}\| = \|Q - P\| = \sqrt{(x_Q - x_P)^2 + (y_Q - y_P)^2 + (z_Q - z_P)^2} \ . \tag{2.6}$$

is the distance between between $P = (x_P, y_P, z_P)$ and $Q = (x_Q, y_Q, z_Q)$. These functions are called the 'free space' Green's function (see Fig. 2.2) and physically correspond to the electrostatic potential at the *field point* Q given a point charge at P. Thus, $G(P,Q)$ is also called the 'point source potential', and for each *fixed* P, it satisfies the Laplace equation as a function for Q, for $Q \neq P$. A derivation of Eq.(2.5) is given in the end of the chapter.

The Green's function is clearly singular when $Q = P$ and, in fact, satisfies the equation

$$\nabla^2 G(P, Q) = -\delta(P, Q) \ , \tag{2.7}$$

where the derivative is with respect to Q and $\delta(P,Q)$ is the Dirac Delta function. As a consequence, for $P \in V \cup \Sigma$ and $f(\mathbf{x}) = G(P, \mathbf{x})$, the above use of the Divergence Theorem is not permissible. However, for $P = P_E$ exterior to V, i.e., $P_E \notin V \cup \Sigma$, substituting $G(P, \mathbf{x}) = f(\mathbf{x})$ in Eq.(2.4) yields the *exterior* integral equation

$$\int_\Sigma \left[\phi(Q) \frac{\partial G}{\partial \mathbf{n}}(P_E, Q) - G(P_E, Q) \frac{\partial \phi}{\partial \mathbf{n}}(Q) \right] \mathrm{d}Q = 0 \ , \tag{2.8}$$

where

$$\frac{\partial G}{\partial \mathbf{n}}(P, Q) = -\frac{1}{4\pi} \frac{\mathbf{n} \cdot \mathbf{R}}{r^3} \ . \tag{2.9}$$

Note that this normal derivative is with respect to the coordinates of Q, and that the singularity at $Q = P$ now behaves as r^{-2}, worse than the r^{-1} for $G(P,Q)$.

The *interior equation*, namely for $P_I \in V$,

$$\phi(P_I) + \int_\Sigma \left[\phi(Q) \frac{\partial G}{\partial \mathbf{n}}(P_I, Q) - G(P_I, Q) \frac{\partial \phi}{\partial \mathbf{n}}(Q) \right] \mathrm{d}Q = 0 \tag{2.10}$$

can, informally, be seen to arise from substituting the delta function Eq.(2.7) into the volume integral Eq.(2.4) and by the definition of δ,

$$\int_V \phi(\mathbf{x}) \nabla^2 f(\mathbf{x}) \, \mathrm{d}V - \int_V \phi(\mathbf{x}) \, \delta(P_I, Q) \, \mathrm{d}V = -\phi(P_I) \ . \tag{2.11}$$

This result is more properly derived by deleting from V a sphere S_ε of radius ε centered at P_I, as shown in Fig. 2.3. The Divergence Theorem can then be legitimately applied to the punctured volume having boundary $\Sigma \backslash S_\varepsilon$, resulting in

$$\int_\Sigma \left[\phi(Q) \frac{\partial G}{\partial \mathbf{n}}(P_I, Q) - G(P_I, Q) \frac{\partial \phi}{\partial \mathbf{n}}(Q) \right] \mathrm{d}Q = \tag{2.12}$$
$$\int_{S_\varepsilon} \left[\phi(Q) \frac{\partial G}{\partial \mathbf{n}}(P_I, Q) - G(P_I, Q) \frac{\partial \phi}{\partial \mathbf{n}}(Q) \right] \mathrm{d}Q \,.$$

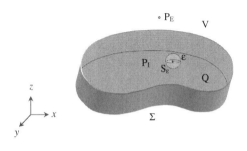

Figure 2.3 A volume V with boundary Σ, and the punctured volume S_ε.

Consider now the integral over S_ε as $\varepsilon \to 0$. Transferring to spherical coordinates, it is first seen that $G(P_I, Q)$ is integrable (weakly singular) at P_I, and thus as the surface area goes to zero, this integral must vanish in the limit $\varepsilon \to 0$. Second, for the integral with $\partial G/\partial \mathbf{n}$, employing a Taylor expansion of ϕ at the point P_I, it is easily seen that only the constant $\phi(P_I)$ term contributes in the limit. Having simplified the integral, direct evaluation of the now yields

$$\lim_{\varepsilon \to 0} \phi(P_I) \int_{S_\varepsilon} \frac{\partial G}{\partial \mathbf{n}}(P_I, Q) \,\mathrm{d}Q = -\phi(P_I) \,, \tag{2.13}$$

and once again we obtain Eq.(2.10).

The goal is now to obtain a boundary-only statement, allowing $P \in \Sigma$. As just noted, $G(P, Q)$ is only weakly singular at $P = Q$, and thus this integral exists for $P \in \Sigma$. However, the singularity for the normal derivative of the Green's function is worse, and this integral is not immediately finite. The standard practice in the literature is to invoke a 'Cauchy Principal Value', but we prefer, for reasons mentioned in the previous chapter, and discussed in detail later, to define this singular integral as a *limit to the boundary*. The *boundary* integral equations, when $P \in \Sigma$ are therefore simply defined as

$$\lim_{P_E \to P} \int_\Sigma \left[\phi(Q) \frac{\partial G}{\partial \mathbf{n}}(P_E, Q) - G(P_E, Q) \frac{\partial \phi}{\partial \mathbf{n}}(Q) \right] \mathrm{d}Q = 0 \tag{2.14}$$

$$\phi(P) + \lim_{P_I \to P} \int_\Sigma \left[\phi(Q) \frac{\partial G}{\partial \mathbf{n}}(P_I, Q) - G(P_I, Q) \frac{\partial \phi}{\partial \mathbf{n}}(Q) \right] \mathrm{d}Q = 0 \tag{2.15}$$

Of course, this simplicity is deceptive: for this approach to be useful it will be necessary to show that the limits exist and, moreover, that they can be readily computed. The development of appropriate boundary limit algorithms for evaluating singular integrals in two and three dimensions are the primary subjects of Chapters 3 and 4.

While not immediately obvious, the interior and exterior boundary equations are indeed identical. The weakly singular integral of $G(P,Q)$ is finite for P on the boundary, and thus the interior and exterior limits are clearly the same. However, the integral involving the derivative of the Green's function is *discontinuous* crossing the boundary, and the jump in this integral balances the presence of the additional $\phi(P)$ term in the interior limit equation. We make note of this here primarily to contrast this with the behavior of the surface gradient equations, examined in detail in Chapter 5.

The interior limit form of Eq.(2.10) is clearly an expression for potential, stating that the potential at $P \in V$ can be calculated by integrating the surface potential and surface flux around the boundary. Thus, once all boundary values are known, ϕ can be computed in the interior (if needed), and the problem is effectively solved. The role of the boundary integral statements is therefore to provide equations needed to solve for the unknown boundary values of potential and flux.

Cauchy Principal Value. The Cauchy Principal Value (CPV) definition for the integral of the normal derivative of the Green's function is a standard technique in boundary integral analysis. This method will *not* be used herein, but contrasting this approach with a boundary limit evaluation can be quite instructive. For a more detailed discussion of the CPV we refer the reader to presentations found in most boundary element texts, *e.g.*, [191]; in addition, the article [127] contains an extensive analysis.

The CPV integral that arises in a two-dimensional analysis is in essence the one-dimensional integral

$$\int_{-a}^{b} \frac{1}{t}\, dt \qquad a, b > 0 \ . \tag{2.16}$$

In the CPV sense, this improper integral is defined by deleting a *symmetric neighborhood* $\{-\varepsilon, \varepsilon\}$ around the singular point $t = 0$ (see Fig. 2.4), and taking the limit $(\varepsilon > 0)$

$$\lim_{\varepsilon \to 0} \left\{ \int_{-a}^{-\varepsilon} \frac{1}{t}\, dt + \int_{\varepsilon}^{b} \frac{1}{t}\, dt \right\} = \lim_{\varepsilon \to 0} \{\log(\varepsilon/a) + \log(b/\varepsilon)\}$$

$$= \log(b/a) \tag{2.17}$$

Thus, CPV *starts* with a singular integral, and to obtain a finite value, it exploits the fact that the function diverges symmetrically to $\pm\infty$ on either 'side' of the singularity.

An important distinction of the boundary limit method is that the integrals are *never* improper: with P off the boundary, one is always dealing with integrals that exist. However, it must be shown that limiting value exists (and is readily computable). In the equivalent boundary limit analysis, the integral in Eq.(2.16) takes the form

$$\lim_{\varepsilon \to 0} \int_{-a}^{b} \frac{t}{t^2 + \varepsilon^2}\, dt = \lim_{\varepsilon \to 0} \frac{1}{2} \log((b^2 + \varepsilon^2)/(a^2 + \varepsilon^2))$$

$$= \log(b/a) \tag{2.18}$$

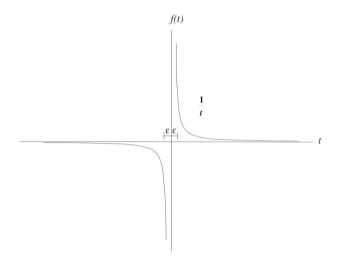

Figure 2.4 CPV: Symmetric Neighborhood around the singular point

A second observation worth noting is that the limit is taken *after* an integration. This is in fact key to the methods developed in Chapters 3 and 4, performing sufficient analytic integration that the limit process can be carried out.

Note that while CPV relies on a somewhat artificial device, the deletion of a symmetric neighborhood around the singular point, there is no such requirement in the boundary limit. The limit is based upon the physically reasonable assumption of continuity of potential at the boundary. Most importantly, the boundary limit works equally well for the boundary integral equation for surface flux: as discussed in the next section, a second derivative of the Green's function appears in this equation. The singularity in this function, termed *hypersingular*, is worse than for the first derivative, and the analogous one-dimensional integral is

$$\int_{-a}^{b} \frac{1}{t^2}\, dt \qquad a, b > 0, \tag{2.19}$$

(see Section 2.5.1). In this case the integrand is strictly positive, and no CPV-type cancellation is possible, which explains why other procedures (most commonly Hadamard Finite Part) have been employed. A primary attraction of the singular integration procedures based upon the boundary limit is that they apply to *all* singular integrals. Chapters 3 and 4 will in fact primarily detail the treatment of the hypersingular integral, the implementation for the less singular integrals will then be quite straightforward. A final argument in favor of the limit approach must await the discussion of surface gradient evaluation, Chapter 5.

2.2 BOUNDARY FLUX EQUATION

The majority of problems that have a boundary integral formulation can be solved using solely the equation that corresponds to the surface potential equation for Laplace, *e.g.*, the displacement equation in elasticity. This is, however, not the case for one important class of applications, those in which the domain volume contains a *crack* boundary. The reason for this, together with methods for solving fracture problems, will be taken up in detail in Chapter 9; for now we simply wish to point out that an effective solution of crack problems requires an additional equation, one for surface flux (in elasticity, surface traction). This equation is also quite useful for other reasons, most importantly it is an essential ingredient for obtaining a symmetric coefficient matrix in the symmetric-Galerkin approximation.

One rationale for defining the surface potential equation as a boundary limit, Eq.(2.14), is that it is a simple matter to obtain a mathematically correct statement of the corresponding boundary integral equations for surface flux. For $P \notin \Sigma$, the Green's function quantities in the integrals are well behaved, and thus Eq.(2.14) can be differentiated with respect to P by first moving the derivative underneath the integral, and then taking the limit. We thus obtain exterior and interior surface *flux* equations

$$\lim_{P_E \to P} \int_\Sigma \left[\phi(Q) \frac{\partial^2 G}{\partial \mathbf{N} \partial \mathbf{n}}(P_E, Q) - \frac{\partial G}{\partial \mathbf{N}}(P_E, Q) \frac{\partial \phi}{\partial \mathbf{n}}(Q) \right] \mathrm{d}Q = 0 \qquad (2.20)$$

$$\frac{\partial \phi}{\partial \mathbf{N}}(P) + \lim_{P_I \to P} \int_\Sigma \left[\phi(Q) \frac{\partial^2 G}{\partial \mathbf{N} \partial \mathbf{n}}(P_I, Q) - \frac{\partial G}{\partial \mathbf{N}}(P_I, Q) \frac{\partial \phi}{\partial \mathbf{n}}(Q) \right] \mathrm{d}Q = 0 \ ,$$

where $\mathbf{N} = \mathbf{N}(P)$ is the unit outward normal at P and $\partial \mathbf{N}$ indicates a derivative with respect to the coordinates of P. The kernel functions are

$$\frac{\partial G}{\partial \mathbf{N}}(P, Q) = \frac{1}{4\pi} \frac{\mathbf{N} \cdot \mathbf{R}}{r^3} \qquad (2.21)$$

$$\frac{\partial^2 G}{\partial \mathbf{N} \partial \mathbf{n}}(P, Q) = \frac{1}{4\pi} \left[\frac{\mathbf{n} \cdot \mathbf{N}}{r^3} - 3 \frac{(\mathbf{n} \cdot \mathbf{R})(\mathbf{N} \cdot \mathbf{R})}{r^5} \right] \ , \qquad (2.22)$$

and it can be seen that the *hypersingular* kernel, Eq.(2.22), behaves as r^{-3} at the singular point.

It should be noted that if the potential equation is defined as a Cauchy Principal Value, one has, in effect, already evaluated the singular part of the $\partial G/\partial n$ integral via a limit process. Thus, moving the derivative under the integral involves interchanging limit processes, and due to the singularity of the kernel function, this interchange is definitely not allowed. This partly explains why early in its development, hypersingular evaluation was a highly contentious issue.

As with the potential equations, the two limit equations for surface flux are identical on the boundary: the *hypersingular* integral involving two derivatives of the Green's function will turn out to be continuous crossing the boundary while, as before, the first derivative integral is not. Again, it will be necessary to demonstrate that the limits can be calculated, this task being obviously somewhat more difficult.

As the interior and exterior limits result in the same equations, and the limit process is the same in each case, either form can be employed. However, as the exterior limit lacks the 'free term' outside the integral, it is slightly more convenient to compute with this equation. Thus, until we come to consider the equation for surface gradient, we will work exclusively with the two exterior limit equations

30 BOUNDARY INTEGRAL EQUATIONS

$\mathcal{P}(P)$ for potential and $\mathcal{F}(P)$ for flux, defined as

$$\mathcal{P}(P) \equiv \lim_{P_E \to P} \int_\Sigma \left[\phi(Q)\frac{\partial G}{\partial \mathbf{n}}(P_E, Q) - G(P_E, Q)\frac{\partial \phi}{\partial \mathbf{n}}(Q) \right] dQ = 0 \qquad (2.23)$$

$$\mathcal{F}(P) \equiv \lim_{P_E \to P} \int_\Sigma \left[\phi(Q)\frac{\partial^2 G}{\partial \mathbf{N}\partial \mathbf{n}}(P_E, Q) - \frac{\partial G}{\partial \mathbf{N}}(P_E, Q)\frac{\partial \phi}{\partial \mathbf{n}}(Q) \right] dQ = 0$$

In the following, unless needed for clarity or emphasis, we will not explicitly write the boundary limit, it being understood that the singular integrals are interpreted in this sense.

Two dimensions. In two dimensions the Laplace boundary integral equations for potential and flux retain precisely the same form as above,

$$\int_\Gamma \left[\phi(Q)\frac{\partial G}{\partial \mathbf{n}}(P_E, Q) - G(P_E, Q)\frac{\partial \phi}{\partial \mathbf{n}}(Q) \right] dQ = 0 \qquad (2.24)$$

$$\int_\Gamma \left[\phi(Q)\frac{\partial^2 G}{\partial \mathbf{N}\partial \mathbf{n}}(P_E, Q) - \frac{\partial G}{\partial \mathbf{N}}(P_E, Q)\frac{\partial \phi}{\partial \mathbf{n}}(Q) \right] dQ = 0 \ , \qquad (2.25)$$

only now the integrals are line integrals over the boundary curve Γ. The Green's function kernels in this case are

$$\begin{aligned}
G(P, Q) &= -\frac{1}{2\pi}\log(r) \\
\frac{\partial G}{\partial \mathbf{n}}(P, Q) &= -\frac{1}{2\pi}\frac{\mathbf{n} \cdot \mathbf{R}}{r^2} \\
\frac{\partial^2 G}{\partial \mathbf{N}\partial \mathbf{n}}(P, Q) &= -\frac{1}{2\pi}\left[-\frac{\mathbf{N} \cdot \mathbf{n}}{r^2} + 2\frac{(\mathbf{n} \cdot \mathbf{R})(\mathbf{N} \cdot \mathbf{R})}{r^4} \right] \ ,
\end{aligned} \qquad (2.26)$$

and the behavior at $r = 0$ is analogous to three dimensions, *e.g.*, $\log(r)$ is integrable.

2.3 ELASTICITY

One of the themes of this book is that the numerical implementation of a boundary integral equation is, for the most part, independent of the particular partial differential equation being solved. Said another way, the algorithms for singular integration – the chief task – are essentially the same irrespective of the particular fundamental solution. Some complicated Green's functions, *e.g.*, anisotropic elasticity (Chapter 4), may require some additional techniques to evaluate the singular integrals, but for most formulations, there are no significant differences from the simplest situation, the Laplace equation.

Nevertheless, elasticity problems are a primary application of boundary integral methods, and elasticity will be specifically discussed in Chapter 9 on fracture. Thus, it is necessary to introduce here the boundary integral elasticity formulation. In addition, elasticity, unlike scalar Laplace, is a vector problem and it is useful to present at least one vector formulation.

The three-dimensional isotropic elasticity equations, derived from Newton's law, state that the divergence of the stress tensor is zero [190]. Thus,

$$\nabla \cdot \sigma = \sigma_{ij,j} = 0 \ , \qquad (2.27)$$

where the summation convention is employed and a subscript after the comma denotes partial differentiation with respect to that coordinate. An isotropic solid can be defined by two material parameters, the Poisson ratio ν and the shear modulus μ, and setting $\gamma = \nu/(1 - 2\nu)$, the stress and strain tensors are defined as

$$
\begin{aligned}
\sigma_{ij} &= 2\mu \left(\epsilon_{ij} + \gamma\epsilon_{kk}\delta_{ij}\right) \\
\epsilon_{ij} &= \frac{1}{2}\left(u_{i,j} + u_{j,i}\right) ,
\end{aligned}
\tag{2.28}
$$

$\mathbf{u}$ being the displacement vector. Following the procedure employed to derive the Laplace integral equation, we multiply Eq.(2.27) by functions g_i, integrate over the volume,

$$
0 = \int_V g_i(\mathbf{x})\sigma_{ij,j}(\mathbf{x})\mathrm{d}V ,
\tag{2.29}
$$

and then integrate by parts. Summing these three equations results in

$$
0 = \int_\Sigma \left[g_i(Q)\tau_i(Q) - t_i(Q)u_i(Q)\right]\mathrm{d}x + \int_V S_i(\mathbf{x})u_i(\mathbf{x})\mathrm{d}V ,
\tag{2.30}
$$

where

$$
\begin{aligned}
S_1 &= \mu \left(\nabla^2 g_1 + (1 + 2\gamma)(g_{1,11} + g_{2,12} + g_{3,13})\right) \\
S_2 &= \mu \left(\nabla^2 g_2 + (1 + 2\gamma)(g_{1,21} + g_{2,22} + g_{3,23})\right) \\
S_3 &= \mu \left(\nabla^2 g_3 + (1 + 2\gamma)(g_{1,13} + g_{2,23} + g_{3,33})\right)
\end{aligned}
\tag{2.31}
$$

and the traction vector $\tau(\xi) = \{t_i(\xi)\}$ is obtained by applying the traction operator to g_i:

$$
\begin{aligned}
t_1 &= \mu \left(n_1(2g_{1,1} + 2\gamma\nabla\bullet g) + n_2(g_{1,2} + g_{2,1}) + n_3(g_{1,3} + g_{3,1})\right) \\
t_2 &= \mu \left(n_1(g_{1,2} + g_{2,1}) + n_2(2g_{2,2} + 2\gamma\nabla\bullet g) + n_3(g_{2,3} + g_{3,2})\right) \\
t_3 &= \mu \left(n_1(g_{1,3} + g_{3,1}) + n_2(g_{2,3} + g_{3,2}) + n_3(2g_{3,3} + 2\gamma\nabla\bullet g)\right) .
\end{aligned}
\tag{2.32}
$$

As before, the goal is to remove the volume integral from Eq.(2.30), and in analogy with the Laplace equation we set, for $1 \leq k \leq 3$,

$$
\begin{pmatrix} S_1 \\ S_2 \\ S_3 \end{pmatrix} = -\delta(P,Q) \begin{pmatrix} \delta_{1k} \\ \delta_{2k} \\ \delta_{3k} \end{pmatrix} ,
\tag{2.33}
$$

where $\delta(P,Q)$ and δ_{jk} are the Dirac and Kronecker Delta functions, respectively. The solution of these equations gives the displacement at Q given a point load at P in the direction k, and is known as the Kelvin solution.

This fundamental displacement tensor $U_{kj}(P,Q)$ is

$$
U_{kj}(P,Q) = \frac{1}{16\pi\mu(1 - \nu)r} \left[(3 - 4\nu)\delta_{kj} + r_{,k}r_{,j}\right] ,
\tag{2.34}
$$

where $r_{,i} = \partial r/\partial x_i$ (see Fig. 2.5). Differentiating with respect to ξ we obtain the traction kernel

$$
T_{kj}(P,Q) = -\frac{1}{8\pi(1 - \nu)r^2} \left[\{(1 - 2\nu)\delta_{kj} + 3r_{,k}r_{,j}\}\frac{\partial r}{\partial \mathbf{n}} - (1 - 2\nu)\{n_j r_{,k} - n_k r_{,j}\}\right] .
$$

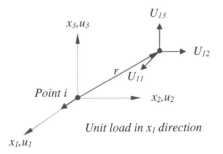

Figure 2.5 Geometric Interpretation of the components of the fundamental solution.

From Eq.(2.30), the exterior boundary integral equation for surface displacement is given by [64, 235]

$$\mathcal{U}(P) \equiv \int_\Sigma [T_{kj}(P,Q)u_j(Q) - U_{kj}(P,Q)\tau_j(Q)]\,dQ = 0 \qquad (2.35)$$

Note that it is understood that for $P_k \in \Sigma$, these equations are defined as a limit to the boundary. The stress equation, from which one gets an equation for surface traction by multiplying by the appropriate normal vector, is obtained by differentiating the displacement equation with respect to P. This results in the interior

$$\mathcal{S}_I(P) \equiv$$
$$\sigma_{lk}(P) + \int_\Sigma [S_{lkm}(P,Q)u_m(Q) - D_{lkm}(P,Q)\tau_m(Q)]\,dQ = 0 \qquad (2.36)$$

and exterior

$$\mathcal{S}_E(P) \equiv \int_\Sigma [S_{lkm}(P,Q)u_m(Q) - D_{lkm}(P,Q)\tau_m(Q)]\,dQ = 0 \qquad (2.37)$$

stress equations. The kernels $D_{lkm}(P,Q)$ (singular) and $S_{lkm}(P,Q)$ (hypersingular) for the stress equation result from differentiating U_{kj} and T_{kj} and are given by

$$D_{lkm} = \frac{1}{8\pi(1-\nu)r^2}\left[(1-2\nu)\{\delta_{lm}r_{,k} + \delta_{km}r_{,l} - \delta_{lk}r_{,m}\} + 3r_{,l}r_{,k}r_{,m}\right]$$

$$S_{lkm} = \frac{\mu}{4\pi(1-\nu)r^3}\Bigg[\qquad\qquad\qquad\qquad\qquad (2.38)$$
$$3\frac{\partial r}{\partial \mathbf{n}}\left(\{1-2\nu\}\delta_{lk}r_{,m} + \nu(\delta_{km}r_{,l} + \delta_{lm}r_{,k}) - 5r_{,l}r_{,k}r_{,m}\right)$$
$$+(1-2\nu)(3n_m r_{,l}r_{,k} + n_k\delta_{lm} + n_l\delta_{km})$$
$$+3\nu(n_l r_{,k}r_{,m} + n_k r_{,l}r_{,m}) - (1-4\nu)n_m\delta_{lk}\Bigg].$$

In addition to being 3 × 3 matrices, the Kelvin solution and its derivatives are lengthier expressions than the corresponding formulas for the Laplace equation. Nevertheless, they are still rational algebraic expressions, and the singular integration methods developed for the Laplace equation will carry over directly to elasticity. The important point is that the order of the singularities at $r = 0$ is precisely the same for both differential equations.

Two Dimensions. The fundamental displacement tensor $U_{kj}(\xi, \mathbf{y})$ in two dimensions is

$$U_{kj}(\xi, \mathbf{y}) = \frac{1}{8\pi\mu(1-\nu)} \left[-(3-4\nu)\delta_{kj}\log(r) + r_{,k}r_{,j}\right], \qquad (2.39)$$

and the corresponding derivatives of this function are

$$T_{kj}(\xi, \mathbf{y}) = -\frac{1}{4\pi(1-\nu)r} \left[\{(1-2\nu)\delta_{kj} + 2r_{,k}r_{,j}\}\frac{\partial r}{\partial \mathbf{n}} - (1-2\nu)\{n_j r_{,k} - n_k r_{,j}\}\right].$$

$$D_{lkm} = \frac{1}{4\pi(1-\nu)r}\left[(1-2\nu)\{\delta_{lm}r_{,k} + \delta_{km}r_{,l} - \delta_{lk}r_{,m}\} + 2r_{,l}r_{,k}r_{,m}\right]$$

$$S_{lkm} = \frac{\mu}{2\pi(1-\nu)r^2}\Bigg[\qquad\qquad (2.40)$$

$$2\frac{\partial r}{\partial \mathbf{n}}\left(\{1-2\nu\}\delta_{lk}r_{,m} + \nu\left(\delta_{km}r_{,l} + \delta_{lm}r_{,k}\right) - 4r_{,l}r_{,k}r_{,m}\right)$$

$$+(1-2\nu)\left(2n_m r_{,l}r_{,k} + n_k\delta_{lm} + n_l\delta_{km}\right)$$

$$+2\nu\left(n_l r_{,k}r_{,m} + n_k r_{,l}r_{,m}\right) - (1-4\nu)n_m\delta_{lk}\Bigg].$$

2.4 NUMERICAL APPROXIMATION

Having defined the boundary integral equations, we now begin the consideration of their numerical solution. Analytic solutions of the integral equations are no easier to obtain than for the original differential equation, and thus it is necessary to reduce the continuous equations to a discrete system of linear equations that can be solved.

The main purpose of this section is to introduce the Galerkin method that will be used throughout the book. To do this we must first discuss the scheme for the two main approximations, and hence the two main sources of error in the calculation: interpolation of the boundary and the interpolation of the boundary functions. Moreover, before describing Galerkin, we first give a brief description of the simpler *collocation* method.

2.4.1 Approximations

One of the most convenient ways of accomplishing the necessary approximations is an *isoparametric* method, in which the boundary and boundary functions are represented through the same set of simple *shape functions* defined on a parameter space. The best way to describe this is through an example, and we choose the 3D linear triangular element that will be employed extensively in Chapter 4.

The boundary surface is approximated as a sum of small surface patches called elements, $\Sigma = \bigcup E_i$ (see Fig. 1.1), and each element is defined as a mapping from a fixed parameter domain in $\mathcal{R}^2$. This parameter space the can be selected more or less arbitrarily; however, for the Galerkin integration over the linear triangular element it will turn out to be convenient to choose an equilateral triangle as shown in

Fig. 2.6. The parametric variables will be called $\{\eta, \xi\}$, and the equilateral triangle with vertices $v_1 = (-1, 0), v_2 = (1, 0), v_3 = (0, \sqrt{3})$ is defined via $-1 \leq \eta \leq 1$, $0 \leq \xi \leq \sqrt{3}(1 - |\eta|)$.

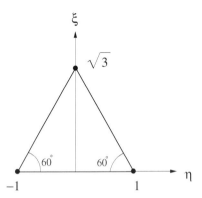

Figure 2.6 Isoparametric equilateral triangular linear element in $\{\eta, \xi\}$ space, where $-1 \leq \eta \leq 1$, $0 \leq \xi \leq \sqrt{3}(1 - |\eta|)$.

The surface and function interpolations will be constructed based upon the three linear shape functions

$$\begin{aligned}
\psi_1(\eta, \xi) &= \frac{\sqrt{3}(1 - \eta) - \xi}{2\sqrt{3}} \\
\psi_2(\eta, \xi) &= \frac{\sqrt{3}(1 + \eta) - \xi}{2\sqrt{3}} \\
\psi_3(\eta, \xi) &= \frac{\xi}{\sqrt{3}},
\end{aligned} \qquad (2.41)$$

defined by the property that $\psi_l(v_j) = \delta_{lj}$. A linear triangular element E_i is defined by three nodal points $\{Q_j = (x_j, y_j, z_j)\}$, oriented so as to give an exterior normal, and the linear interpolation of these nodes defines the approximate boundary patch. The mapping from parameter space to E_i is then easily written as

$$\Sigma_i(\eta, \xi) = \sum_{j=1}^{3} (x_j, y_j, z_j) \psi_j(\eta, \xi) , \qquad (2.42)$$

and the corresponding approximate surface potential and flux on this element are

$$\phi(\eta, \xi) = \sum_{j=1}^{3} \phi(Q_j) \psi_j(\eta, \xi) \qquad (2.43)$$

$$\frac{\partial \phi}{\partial \mathbf{n}}(\eta, \xi) = \sum_{j=1}^{3} \frac{\partial \phi}{\partial \mathbf{n}}(Q_j) \psi_j(\eta, \xi) .$$

Note that $\phi(\eta, \xi)$ is a convenient shorthand notation: the function ϕ is in reality defined on the surface Σ, and not on the parameter space. Thus by $\phi(\eta, \xi)$ we really mean $\phi(\Sigma_i(\eta, \xi))$.

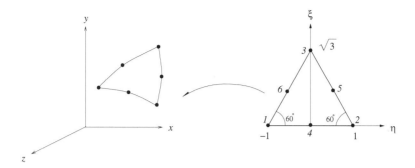

Figure 2.7 The equilateral triangle parameter space $\{\eta, \xi\}$ is mapped onto the quadratic triangular element in 3D.

A linear approximation suffices for many, if not most, applications. However, for the development of an effective approximation at a crack front, a quadratic interpolation is highly useful. For completeness, we therefore list here the quadratic shape functions, again for the equilateral triangle parameter space (see Fig. 2.7). These functions are once again constructed so that $\psi_l(v_j) = \delta_{lj}$, $1 < l, j < 6$.

$$
\begin{align}
\psi_1(\eta, \xi) &= \frac{(\xi + \sqrt{3}\eta - \sqrt{3})(\xi + \sqrt{3}\eta)}{6} \\
\psi_2(\eta, \xi) &= \frac{(\xi - \sqrt{3}\eta - \sqrt{3})(\xi - \sqrt{3}\eta)}{6} \\
\psi_3(\eta, \xi) &= \frac{(2\xi - \sqrt{3})\xi}{3} \\
\psi_4(\eta, \xi) &= \frac{(\xi + \sqrt{3}\eta - \sqrt{3})(\xi - \sqrt{3}\eta - \sqrt{3})}{3} \\
\psi_5(\eta, \xi) &= -2\frac{(\xi - \sqrt{3}\eta - \sqrt{3})\xi}{3} \\
\psi_6(\eta, \xi) &= -2\frac{(\xi + \sqrt{3}\eta - \sqrt{3})\xi}{3}
\end{align}
\tag{2.44}
$$

2.4.2 Collocation

In collocation the boundary integral equations are enforced at specified points. In its simplest form, these collocation points are chosen to be the nodes used to discretize the boundary, see Fig. 2.7. Thus, for example, if there are M nodes on the boundary, the exterior limit potential equation (we could work equally well with the flux equation) can be used to generate the M collocation equations

$$
\begin{pmatrix} \mathcal{P}(P_1) \\ \mathcal{P}(P_2) \\ \vdots \\ \mathcal{P}(P_M) \end{pmatrix} = 0 , \tag{2.45}
$$

36   BOUNDARY INTEGRAL EQUATIONS

where $\mathcal{P}$ is defined in Eq.(2.23). If the corresponding $M \times 1$ vectors of boundary values of potential and flux are denoted by

$$\left[\phi(P_j)\right] \qquad \left[\frac{\partial \phi}{\partial \mathbf{n}}(P_j)\right] \ , \tag{2.46}$$

the collocation equations (after the numerical approximations discussed below) eventually become the matrix equation

$$\mathcal{H}\left[\phi(P_j)\right] = \mathcal{G}\left[\frac{\partial \phi}{\partial \mathbf{n}}(P_j)\right] \ , \tag{2.47}$$

where $\mathcal{H}$ and $\mathcal{G}$ are $M \times M$ matrices. Of the $2M$ surface quantities in this equation, M are given by the boundary conditions; rearranging columns, collecting unknowns as the vector x, a linear system $Ax = b$ is obtained and x can be determined.

The elements of the $\mathcal{H}$ matrix, and a similar discussion for $\mathcal{G}$, arise from approximating the integrals

$$\int_\Sigma \phi(Q) \frac{\partial G}{\partial \mathbf{n}}(P_k, Q) \, \mathrm{d}Q \tag{2.48}$$

for $1 \le k \le M$. Specifically, using the approximations for Σ and ϕ discussed above we can write

$$
\begin{aligned}
\int_\Sigma \phi(Q) \frac{\partial G}{\partial \mathbf{n}}(P_k, Q) \, \mathrm{d}Q \ &\approx \ \sum_{E_l} \int_{E_l} \phi(Q) \frac{\partial G}{\partial \mathbf{n}}(P_k, Q) \, \mathrm{d}Q \\
&= \ \sum_{E_l} \sum_j \phi(Q_j) \int \psi_j(\eta, \xi) J(\eta, \xi) \frac{\partial G}{\partial \mathbf{n}}(P_k, Q_l(\eta, \xi)) \, \mathrm{d}\xi \, \mathrm{d}\eta \\
&= \ \sum_{E_l} \sum_j \phi(Q_j) \mathcal{I}_{kj}^l \tag{2.49} \\
&= \ \sum_j \phi(Q_j) \sum_{E_l} \mathcal{I}_{kj}^l = \sum_j \mathcal{H}_{kj} \phi(Q_j)
\end{aligned}
$$

where $J(\eta, \xi)$ is the jacobian of the surface mapping (constant for a linear interpolation). Thus, the matrix elements $\mathcal{H}_{kj}$ (and similarly for $\mathcal{G}_{kj}$) are given by

$$\mathcal{H}_{kj} = \sum_{E_l} \mathcal{I}_{kj}^l \ , \tag{2.50}$$

the integrals being of the form

$$\mathcal{I}_{kj}^l = \int \psi_j(\eta, \xi) J(\eta, \xi) \frac{\partial G}{\partial \mathbf{n}}(P_k, Q_l(\eta, \xi)) \, \mathrm{d}\xi \, \mathrm{d}\eta \ , \tag{2.51}$$

and the notation $Q_l(\eta, \xi)$ indicates that this surface mapping is defined by the element E_l. Note that everything in the above integral is a known quantity and can, assuming the singular integrals have finite limiting values, be computed. Moreover, for collocation, these integrations have to be carried out for every node/element pair $\{P_k, E_l\}$.

Singular integrals therefore arise when $P_k \in E_l$. Also note that the $\mathcal{H}$ and $\mathcal{G}$ matrices are dense and non-symmetric, unlike the sparse systems obtained using finite element methods. In fact, these matrices, generally speaking, have no exploitable

NUMERICAL APPROXIMATION   **37**

structure. The best that can be said is that as the Green's function kernels generally die off for large r, the singular integrals tend to be the dominant terms. These terms are located near the diagonal, and this observation is potentially useful for an iterative method for solving the linear system. Finally, it is useful to observe that $\mathcal{H}$ and $\mathcal{G}$ depend only on geometry, not on the boundary conditions.

2.4.3 Galerkin Approximation

In contrast to collocation, the Galerkin approach does not require that the boundary integral equations be satisfied at any point. Instead, the equations are enforced in a weighted average:

$$\int_\Sigma \hat{\psi}_k(P)\mathcal{P}(P)\,\mathrm{d}P \;=\; 0 \qquad\qquad (2.52)$$

$$\int_\Sigma \hat{\psi}_k(P)\mathcal{F}(P)\,\mathrm{d}P \;=\; 0$$

In mathematical terminology, we give up the 'strong' requirement that the integral equations are actually satisfied at any given point, in exchange for a 'weak solution' in which the equations hold in an integrated sense. This requirement has a nice geometric interpretation: the approximate Galerkin solution is the exact solution projected onto the subspace consisting of all functions which are linear combinations of the shape functions. The Galerkin solution is therefore the linear combination which is the 'closest' to the exact solution.

In the standard Galerkin procedure, the weight functions $\hat{\psi}_k(P)$ are composed of all shape functions that are non-zero at a node P_k.

For the double integral over the surface, two sets of parameters will be needed. We will use η, ξ for the outer P integration and η^*, ξ^* for the inner Q integral. Repeating for Galerkin the previous calculation that led to the expression for the collocation $\mathcal{H}$ matrix elements, we obtain

$$\int_\Sigma \hat{\psi}_k(P) \int_\Sigma \phi(Q)\frac{\partial G}{\partial \mathbf{n}}(P,Q)\,\mathrm{d}Q$$

$$= \sum_{E_m}\sum_{E_l}\int_{E_m} \hat{\psi}_k(P)\int_{E_l}\phi(Q)\frac{\partial G}{\partial \mathbf{n}}(P,Q)\,\mathrm{d}Q$$

$$= \sum_{E_m,E_l}\sum_j \phi(Q_j)\int \hat{\psi}_k(\eta,\xi)J_p(\eta,\xi)\int \psi_j(\eta^*,\xi^*)J_q(\eta^*,\xi^*)G^{m,l}_{,\mathbf{n}}\,\mathrm{d}\xi^*\,\mathrm{d}\eta^*\,\mathrm{d}\xi\,\mathrm{d}\eta$$

$$= \sum_{E_m,E_l}\sum_j \phi(Q_j)\mathcal{I}^{m,l}_{kj} \qquad\qquad (2.53)$$

$$= \sum_j \phi(Q_j)\sum_{E_m,E_l}\mathcal{I}^{m,l}_{kj} = \sum_j \mathcal{H}_{kj}\phi(Q_j)$$

where the shorthand

$$G^{m,l}_{,\mathbf{n}} = \frac{\partial G}{\partial \mathbf{n}}(P_m(\eta,\xi),Q_l(\eta^*,\xi^*)) \qquad\qquad (2.54)$$

has been introduced. The matrix elements for Galerkin are therefore given by

$$\mathcal{H}_{kj} = \sum_{E_m,E_l}\mathcal{I}^{m,l}_{kj}\,, \qquad\qquad (2.55)$$

where the component integrals are of the form

$$\mathcal{I}_{kj}^{m,l} = \int \hat{\psi}_k(\eta,\xi) J_p(\eta,\xi) \int \psi_j(\eta^*,\xi^*) J_q(\eta^*,\xi^*) G_{,\mathbf{n}}^{m,l} \, d\xi^* \, d\eta^* \, d\xi \, d\eta \ . \quad (2.56)$$

Although more complicated than for collocation, everything in this expression is once again known, dependent solely on the problem geometry. As an integration is now required for every pair of elements $\{E_m, E_l\}$, singular integrals will occur when the elements are *coincident* $E_m = E_l$, or *adjacent*, sharing either an edge or a vertex. Thus, for three dimensional problems, there are four possible configurations for the P and Q elements that must be considered, as illustrated in Fig. 2.8:

- *Non-singular case*, when the source point P and the field point Q lie on distinct elements, that do not share a common vertex or edge.
- *Coincident case*, when the source point P and the field point Q lie in the same element;
- *Edge adjacent case*, when two elements share a common edge; and
- *Vertex adjacent case*, when a vertex is the only common node between the two elements.

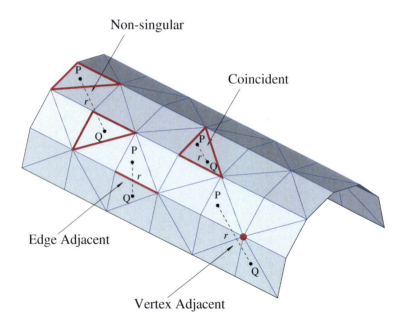

Figure 2.8 Four different cases considered for integration: (a) non-singular; (b) coincident; (c) edge adjacent; and (d) vertex adjacent.

In two dimensions there are only two singular cases, coincident and common vertex.

Although Galerkin clearly requires more computational effort, there are significant benefits. In addition to the very important concern that hypersingular integrals can be handled far more simply than collocation (see Section 2.5.1), Galerkin is generally more accurate, can lead to a symmetric coefficient matrix, and provides a more natural and efective algorithm for dealing with boundary corners and edges.

2.4.4 Symmetric-Galerkin

Unlike collocation, there is some structure to the Galerkin matrices. Of the four matrices arising from the four possible kernel functions (two for the potential equation, two for the flux equation), two matrices are symmetric. The kernel functions

$$G(P,Q) \;=\; G(Q,P) \tag{2.57}$$
$$\frac{\partial^2 G}{\partial \mathbf{N}\partial \mathbf{n}}(P,Q) \;=\; \frac{\partial^2 G}{\partial \mathbf{N}\partial \mathbf{n}}(Q,P) \;,$$

are symmetric, and as P and Q are treated on an equal footing in the double boundary integration, this property is passed onto the resulting matrices. As a consequence, if the problem is purely Dirichlet (ϕ specified everywhere on Σ), use of the potential equation leads to a symmetric coefficient matrix for the unknown flux. For a Neumann problem, the flux equation similarly produces a symmetric matrix.

These facts lead to a general prescription: if the potential equation is invoked on the part of Σ having Dirichlet boundary conditions, and the negative of the flux equation is employed on the Neumann surface, the coefficient matrix for this mixed problem is symmetric. This follows from observing that the first order derivatives are related,

$$\nabla_q G(P,Q) = -\nabla_p G(P,Q) = \nabla_p G(Q,P) \;. \tag{2.58}$$

Multiplying one of the equations by -1 is needed to account for the negative sign when columns are rearranged to bring all unknowns to one side.

2.5 HYPERSINGULAR INTEGRATION: AN EXAMPLE

Limit evaluation of the singular integrals, collocation or Galerkin, is relatively straightforward and unexciting, except for the hypersingular integral. To get a feel for the behavior of this integral, and to contrast collocation and Galerkin, it is instructive to examine a simple setting that allows for complete evaluation of the integrals.

This section will first demonstrate that collocation of the hypersingular integral

$$\lim_{P_E \to P_0} \int_\Sigma \phi(Q) \frac{\partial^2 G}{\partial \mathbf{N}\partial \mathbf{n}}(P_E,Q) \, \mathrm{d}Q \tag{2.59}$$

is not a simple matter, and in fact *cannot* be accomplished for the linear interpolation discussed above. Specifically, we show that existence of the boundary limit requires that $\phi(P)$ be differentiable ($\mathcal{C}^1$) at the collocation point P_0 (the limit can be either interior or exterior). Second, the corresponding Galerkin integral will be shown to be finite.

The need for the differentiability condition for collocation can be understood by the following hand-waving argument. One way to attack the evaluation of the hypersingular integral would be to use integration by parts, moving the second derivative off of the Green's function and onto the function $\phi(Q)$. This would return the singularity in the Green's function quantity to the more manageable r^{-2}. However, in order to do this, $\phi(Q)$ must be differentiable. Regarding Galerkin, note that the extra outer integral with respect to P can be viewed as a counterbalance to the second derivative (with respect to $\mathbf{N}(P)$) that created the hypersingular kernel, and thus the Galerkin form should exist under the weaker continuity requirement. Standard boundary integral interpolations are continuous crossing from one element to another, but in general not differentiable. This C^1 condition is therefore a serious impediment for a collocation computation, and consequently a principal argument in favor of Galerkin.

There is another good motivation for carrying out the calculations below: they illustrate, in a very concrete fashion, the general *limit to the boundary* and analytic integration procedures that will be used in Chapter 3 and Chapter 4.

2.5.1 Collocation: C^1 Condition

To establish the C^1 condition, evaluation of the hypersingular integral will be attempted for a boundary node P on a flat boundary segment. It will be seen that, without differentiability, the limit, and hence the integral, fails to exist. A similar calculation can be carried out in three dimensions [107].

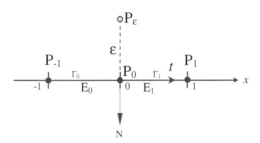

Figure 2.9 Limit to the boundary for collocating the hypersingular integral.

Consider, as shown in Fig.2.9, a flat boundary segment lying along the x−axis, consisting of the interval $[-1, 1]$. This segment is divided into the two elements $E_0 = [-1, 0]$ and $E_1 = [0, 1]$, outward normals $\mathbf{n} = \mathbf{N} = (0, -1)$, with nodes $P_{-1} = (-1,\ 0)$, $P_0 = (0,\ 0)$, and $P_1 = (1,\ 0)$.

We seek to collocate the hypersingular integral at the point P_0. The interior limit to the boundary is effected by replacing P_0 with the point

$$P_\epsilon = P_0 - \epsilon \mathbf{N} , \qquad (2.60)$$

where $\epsilon > 0$. With P off the boundary the integral exists, and then the limit

$$\lim_{\varepsilon \to 0} -\frac{1}{2\pi} \int_{E_0 + E_1} \phi(Q) \left\{ -\frac{\mathbf{N} \cdot \mathbf{n}}{r^2} + 2\frac{(\mathbf{n} \cdot R)(\mathbf{N} \cdot R)}{r^4} \right\} dQ \qquad (2.61)$$

can be considered. On E_0 the linear interpolations for the boundary and for ϕ are

$$
\begin{aligned}
\Gamma_0(t) &= (-1,0)\,\psi_1(t) + (0,0)\,\psi_2(t) = (-1+t,0) \qquad\qquad (2.62)\\
\phi(\Gamma_0(t)) &= \phi(P_{-1})\,\psi_1(t) + \phi(P_0)\,\psi_2(t) = \phi(P_{-1}) + t\,(\phi(P_0) - \phi(P_{-1}))\\
r^2 &= (-1+t)^2 + \varepsilon^2
\end{aligned}
$$

where the linear shape functions $0 < t < 1$ are

$$
\begin{aligned}
\psi_1(t) &= 1 - t & (2.63)\\
\psi_2(t) &= t \; . & (2.64)
\end{aligned}
$$

Similarly for the element E_1,

$$
\begin{aligned}
\Gamma_1(t) &= (0,0)\,\psi_1(t) + (1,0)\,\psi_2(t) = (t,0)\\
\phi(\Gamma_1(t)) &= \phi(P_0)\,\psi_1(t) + \phi(P_1)\,\psi_2(t) = \phi(P_0) + t\,(\phi(P_1) - \phi(P_0)) \quad (2.65)\\
r^2 &= t^2 + \varepsilon^2
\end{aligned}
$$

Integrating first the constant term from ϕ over E_0 (cf. Eq.(2.62) and Eq.(2.61)) we find, with a coefficient of $\phi(P_{-1})/2\pi$,

$$
\begin{aligned}
\int_0^1 &\frac{1}{(-1+t)^2 + \epsilon^2}\, dt - \int_0^1 \frac{2\epsilon^2}{((-1+t)^2 + \epsilon^2)^2}\, dt\\[2mm]
&= \frac{\tan^{-1}(1/\varepsilon)}{\varepsilon} - \left(\frac{1}{1+\varepsilon^2} + \frac{\tan^{-1}(1/\varepsilon)}{\varepsilon} \right) \qquad (2.66)\\[2mm]
&= -\frac{1}{1+\varepsilon^2} = -1
\end{aligned}
$$

Note that as $\tan^{-1}(1/\varepsilon) \to \pi/2$, a $1/\varepsilon$ term appears separately in each term, but they fortunately cancel by themselves. This is in fact a general feature of the hypersingular integration that will show up in the general analysis in subsequent chapters.

For the more interesting linear term, coefficient of $(\phi(P_0) - \phi(P_{-1}))/2\pi$, integrating over E_0 yields

$$
\begin{aligned}
\int_0^1 t &\left[\frac{1}{(-1+t)^2 + \epsilon^2} - \frac{2\epsilon^2}{((-1+t)^2 + \epsilon^2)^2} \right] dt\\[2mm]
&= \log(\varepsilon^2)/2 - \log(1+\varepsilon^2)/2 \qquad\qquad (2.67)\\[2mm]
&= \log(\varepsilon^2)/2
\end{aligned}
$$

For E_1 the constant term is again finite, while the linear term has a coefficient of $(\phi(P_1) - \phi(P_0))/2\pi$ and resulting integral

$$
\begin{aligned}
\int_0^1 t &\left[\frac{1}{t^2 + \epsilon^2} - \frac{2\epsilon^2}{(t^2 + \epsilon^2)^2} \right] dt\\[2mm]
&= -\log(\varepsilon^2)/2 + \log(1+\varepsilon^2)/2 - \frac{1}{1+\varepsilon^2} \qquad (2.68)\\[2mm]
&= -\log(\varepsilon^2)/2 - 1
\end{aligned}
$$

42 BOUNDARY INTEGRAL EQUATIONS

For the hypersingular integral to have a finite limiting value, the two $\log(\epsilon^2)$ terms must cancel, and thus it follows that

$$\phi_0 - \phi_{-1} = \phi_1 - \phi_0 \qquad (2.69)$$

The expressions on the left and right are in fact the tangential derivatives of $\phi(P_0)$ on E_0 and E_1, respectively, as defined by the linear interpolations. Thus for a non-$\mathcal{C}^1$ interpolation at P_0 the hypersingular integral does not exist.

More precisely, without a differentiable interpolation of ϕ at P_0, there is no way to define the integral in Eq.(2.61) so that the flux is *continuous from the interior*. Although it is possible to implement $\mathcal{C}^1$ interpolations (*e.g.*, Overhauser elements), they are in general difficult and computationally expensive, especially in 3D. What has generally been done in the literature is to invoke the Hadamard Finite Part (Ioakimidis [138], Hadamard [129], Ioakimidis [139], Kaya & Erdogan [144]). This definition will assign a value to this integral, essentially by ignoring the $\log(\epsilon^2)$ singularity, but lacking continuity to the boundary, the physical relevance of this definition is dubious.

Finally, if collocating the hypersingular integral is difficult at a boundary point where the surface is perfectly flat, it can be expected that at a boundary corner/edge, the situation will be even more complicated. We will return to this issue in Chapter 3 when we consider the Galerkin corner treatment.

2.5.2 Galerkin: $\mathcal{C}^0$

Another observation stemming from the computation in the previous section is to note that obtaining a finite value for the hypersingular integral demands integrating over *a complete neighborhood* of the singular point P. With or without differentiability, the individual integrals over E_1 and E_0 are always divergent, the linear term from the shape functions will always produce a $\log(\varepsilon^2)$ contribution. The best that can be hoped for is a cancellation of the divergences when the two integrals are summed, *i.e.* the total singular integral at P_0 is computed.

As Galerkin can be viewed as collocating at points P interior to an element – namely at the Gauss points for the P integration – it might appear that this requirement of a complete neighborhood is always met. However, for a node on the edge of an element, half of its neighborhood is contained in the coincident integral, and the remaining half neighborhood executed in the corresponding adjacent edge integration. It is therefore possible that the hypersingular kernel is sufficiently singular that the Galerkin hypersingular coincident and adjacent edge integrals will be separately divergent. This is indeed the situation, with once again $\log(\varepsilon^2)$ terms appearing in both integrals. Nevertheless, under the weaker condition of continuity, $\mathcal{C}^0$, the Galerkin hypersingular integral is finite, the divergent terms cancel. Standard boundary integral approximations (linear, quadratic, *etc.*) are continuous crossing element boundaries, and can therefore be safely employed for hypersingular equations.

It is important to note that this presumed cancellation of divergent terms cannot be trusted to the numerics. A direct evaluation of the hypersingular Galerkin integrals (coincident and adjacent edge) must explicitly identify the divergent terms, so that they can be exactly cancelled. This can in fact be achieved in general, but will first be illustrated below by re-doing the above simple collocation calculation in Galerkin form.

HYPERSINGULAR INTEGRATION: AN EXAMPLE **43**

Considering just the two elements E_0 and E_1 in Fig. 2.9, the Galerkin formulation gives rise to four integrals: two coincident and two adjacent (one with E_0 as the outer integral and one with E_1). We treat only the coincident (E_0, E_0) and adjacent with (E_0, E_1) (E_0 as the outer P integral), the remaining two integrals are similar. As with the collocation calculation, attention is focused on the equation written at the node P_0. In terms of Galerkin, this means that, for E_0, we only examine the integral for the weight function $\psi_2(P)$, the shape function that is equal to 1 at P_0 ($t = 1$).

In reference to the above discussion on complete neighborhoods, note that adding the two integrals (E_0, E_0) and (E_0, E_1) results in a complete inner Q integral that covers P_0. Thus, we will eventually find this sum to be finite.

2.5.2.1 Coincident E_0, E_0 integration We seek to evaluate

$$\lim_{\varepsilon \to 0} \int_{E_0} \psi_2(P) \int_{E_0} \phi(Q) \frac{\partial^2 G}{\partial \mathbf{N} \partial \mathbf{n}}(P_\varepsilon, Q) \, \mathrm{d}Q \, \mathrm{d}P \ , \tag{2.70}$$

and unlike the collocation calculation in Section 2.5.1, the Galerkin double integral requires two parametric variables. Using $s < 1$, for the parameter for the inner Q integral, t for P, we have

$$\begin{aligned} Q(s) &= (-1 + s, 0) \\ \phi(Q(s)) &= \phi(P_{-1})\psi_1(s) + \phi(P_0)\psi_2(s) \\ P_\varepsilon(t) &= (-1 + t, \varepsilon) \\ r^2 &= (s - t)^2 + \varepsilon^2 \ . \end{aligned} \tag{2.71}$$

As just noted, we choose to examine the equation at P_0, and thus $\psi_2(t) = t$. Every point P has now been shifted off the boundary, and the singular point for the Q integral is $s = t$. The integrals with the two shape functions $\psi_j(s)$, $j = 1, 2$ can be evaluated analytically, resulting in

$$\begin{matrix} -\frac{1}{4\pi} & j = 1 \\ \frac{1}{4\pi}\left(1 + \log(\varepsilon^2)\right) & j = 2 \end{matrix} \tag{2.72}$$

Thus, the coincident integral multiplying $\phi(P_0)$ is divergent, whereas the zero in the weight function at $t = 0$ is sufficient to kill the divergence at P_{-1}. Of course, had we considered the weight function $1 - t$, the opposite situation would result (and subsequent cancellation of the divergent term relying on the element to the left of E_0). It is useful to note the different origins of the collocation and Galerkin $\log(\varepsilon^2)$ terms: for collocation the divergence comes from the linear term in the expansion of the potential at P_0, and therefore has a coefficient that involves both $\phi(P_{-1})$ and $\phi(P_0)$. For Galerkin however, the coefficient simply involves the value of ϕ at the singular point, reflecting the fact that the divergence now arises from the constant term in the shape function.

We now show that adding in the adjacent integral results in a finite quantity.

2.5.2.2 Adjacent integration For E_1 as the inner Q integral, namely

$$\lim_{\varepsilon \to 0} \int_{E_0} \psi_2(P) \int_{E_1} \phi(Q) \frac{\partial^2 G}{\partial \mathbf{N} \partial \mathbf{n}}(P_\varepsilon, Q) \, \mathrm{d}Q \, \mathrm{d}P \ . \tag{2.73}$$

we now have

$$\begin{aligned} Q(s) &= (s, 0) \\ \phi(Q(s)) &= \phi(P_0)\psi_1(s) + \phi(P_1)\psi_2(s) \\ r^2 &= (s - t + 1)^2 + \varepsilon^2 \ . \end{aligned} \qquad (2.74)$$

The kernel function is therefore singular only when $s = 0$ and $t = 1$, which corresponds to $Q = P = P_0$. Once again doing the complete integral analytically, we find

$$\begin{array}{ll} -\frac{1}{4\pi}\left(1 + \log(\varepsilon^2) - 4\log(2)\right) & j = 1 \\ \frac{1}{4\pi} & j = 2 \end{array} \qquad (2.75)$$

Thus, as $\psi_2(s = 0) = 0$, the $j = 2$ integral is finite, whereas for $j = 1$ (again having coefficient $\phi(P_0)$) the adjacent integral is also divergent. However, the sum of coincident and adjacent is perfectly finite. The Galerkin form of the hypersingular integral therefore exists for the C^0 linear interpolation.

In general, it will not be necessary or possible – especially in three dimensions – to do the complete integral analytically. It is only necessary to do sufficient analytic integration to produce the divergent terms, and allow setting $\varepsilon = 0$ in the well-behaved remainder.

Finally, note that in three dimensions, the individual adjacent vertex integrals are also carried out over a partial neighborhood of the singular point. However in this case the singularity is limited to one point in a four dimensional parameter space, and the singularity is therefore sufficiently weak that the vertex integrals will turn out to be separately finite.

CHAPTER 3

TWO DIMENSIONAL ANALYSIS

Synopsis: The numerical implementation of a Galerkin boundary integral analysis in two dimensions is presented. The primary task, in three dimensions as well as two, is the evaluation of singular integrals, and this will be the primary focus of chapter. The methods are first described in the simplest possible setting, a piecewise linear solution of the Laplace equation. Subsequently, higher order curved interpolation and more complicated Green's functions are considered. The three key features of the singular integration analysis are (a) the definition of the integrals as limits from the exterior of the domain, (b) the combination of analytical and numerical evaluation procedures, and (c) the development of a consistent scheme for evaluating all weak, strong (CPV), and hypersingular integrals. Symbolic computation is exploited to simplify the work involved in carrying out the limit process and the analytic integration, and example Maple codes are provided.

3.1 INTRODUCTION

We begin with a recapitulation of the basic components of a boundary integral equation analysis developed in the previous chapter. In Galerkin form, the exterior potential and flux boundary integral equations for the Laplace equation are given

The Symmetric Galerkin Boundary Element Method. By Sutradhar, Paulino and Gray
ISBN 978-3-540-68770-2 ©2008 Springer-Verlag Berlin Heidelberg

by

$$\int_\Gamma \psi_k(P) \int_\Gamma \left[\phi(Q)\frac{\partial G}{\partial \mathbf{n}}(P,Q) - G(P,Q)\frac{\partial \phi}{\partial \mathbf{n}}(Q) \right] \, \mathrm{d}Q \, \mathrm{d}P \;=\; 0 \quad (3.1)$$

$$\int_\Gamma \psi_k(P) \int_\Gamma \left[\phi(Q)\frac{\partial^2 G}{\partial \mathbf{N}\partial \mathbf{n}}(P,Q) - \frac{\partial G}{\partial \mathbf{N}}(P,Q)\frac{\partial \phi}{\partial \mathbf{n}}(Q) \right] \, \mathrm{d}Q \, \mathrm{d}P \;=\; 0 \,.$$

The two dimensional Green's function and its derivatives are given by Eq.(2.26), repeated again here for convenience,

$$
\begin{aligned}
G(P,Q) &= -\frac{1}{2\pi}\log(r) \\
\frac{\partial G}{\partial \mathbf{n}}(P,Q) &= -\frac{1}{2\pi}\frac{\mathbf{n}\cdot\mathbf{R}}{r^2} \\
\frac{\partial^2 G}{\partial \mathbf{N}\partial \mathbf{n}}(P,Q) &= -\frac{1}{2\pi}\left[-\frac{\mathbf{N}\cdot\mathbf{n}}{r^2} + 2\frac{(\mathbf{n}\cdot R)(\mathbf{N}\cdot R)}{r^4} \right] .
\end{aligned}
$$

Here Γ is the boundary of the two dimensional domain $\mathcal{D} \subset \mathcal{R}^2$, $r = \|Q - P\|$, and $\mathbf{n} = \mathbf{n}(Q)$, $\mathbf{N} = \mathbf{N}(P)$ are the unit outward normals at Q and P.

To simplify notation we have dropped the boundary limit in Eq.(3.1), it being understood that the singular integrals are defined in terms of the limit $P_E \to P$, P_E exterior to the domain $\mathcal{D}$. A particular advantage of this definition is that, as will be established below, the potential and flux equations can be treated by the same methods. Discretization, *i.e.*, reducing these equations to a finite system that can be solved, involves decomposing the surface as a union of simple elements, and then approximating surface potential and flux on these elements. This process was described in Chapter 2, and it was seen that the basic task was to compute integrals of the form

$$\int_{E_P} \psi_k(P) \int_{E_Q} \psi_j(Q)\mathcal{K}(Q,P) \, dQ \, dP \qquad (3.2)$$

for every pair of elements $\{E_P,\ E_Q\}$. Here $\{\psi_m\}$ denotes the chosen shape functions and $\mathcal{K}(Q,P)$ is a kernel function, either the Green's function, its first derivative or second derivative.

The inner Q and outer P line integrals become one dimensional integrals over the parameter space employed to define the shape functions (and hence the interpolations). This parameter space will be chosen as the unit interval $[0,1]$, and the parameters for the P and Q integrations will be denoted by t and s, respectively (see Figure 3.1). In terms of the shape functions (*e.g.*, linear, quadratic), the basic approximations are

$$
\begin{aligned}
Q(s) &= \sum_{j=1}^{M} \psi_j(s)Q_j \\
\phi(Q(s)) &= \sum_{j=1}^{M} \psi_j(s)\phi(Q_j) \qquad (3.3) \\
\frac{\partial \phi}{\partial \mathbf{n}}(Q(s)) &= \sum_{j=1}^{M} \psi_j(s)\frac{\partial \phi}{\partial \mathbf{n}}(Q_j) \,,
\end{aligned}
$$

where $\{Q_j\}$ are the M nodes defining E_Q. Similarly we have

$$P(t) = \sum_{k=1}^{M} \psi_k(t) P_k \;, \qquad (3.4)$$

where $\{P_k\}$ are the nodes defining the E_P. The integrals to be computed, Eq.(3.2), can therefore be expressed as

$$\int_0^1 \psi_k(t) J_p(t) \int_0^1 \psi_j(s) J_q(s) \mathcal{K}(Q(s), P(t)) \, ds \, dt \;, \qquad (3.5)$$

where $1 \leq k, j \leq M$ and J_p and J_q are the Jacobians

$$J_q(s) = \left\| \frac{dQ}{ds} \right\| \qquad J_p(t) = \left\| \frac{dP}{dt} \right\| \;. \qquad (3.6)$$

Note that one major simplification of the linear element is that the jacobians will be constants.

For the nonsingular integrals, i.e., E_P and E_Q non-intersecting, Gauss quadrature (see the Appendix B) can be used to compute the integrals. Thus,

$$\int_0^1 \psi_k(t) J_p(t) \int_0^1 \psi_j(s) J_q(s) \mathcal{K}(s,t) \, ds \, dt = \qquad (3.7)$$

$$\sum_{m=1}^{n_g} w_m \psi_k(\xi_m) J_p(\xi_m) \sum_{l=1}^{n_g} w_l \psi_j(\xi_l) J_q(\xi_l) \mathcal{K}(Q(\xi_l), P(\xi_m)) \;,$$

where n_g is the number of points for the chosen Gauss rule, $\{\xi_l, w_l\}$ the associated Gauss points and weights.

Gauss integration is inappropriate for the singular integrals, and in two dimensions there are two types of singular integrals, illustrated schematically in Fig. 3.1. The *coincident* case is $E_P = E_Q$ and *adjacent* refers to E_P and E_Q sharing a common node.

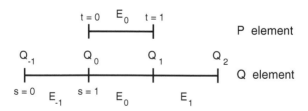

Figure 3.1 For $E_P = E_0$, the coincident singular integration is $E_Q = E_0$, and the adjacent singular integrations are $E_Q = E_1$ and $E_Q = E_{-1}$.

We hope to convince the reader that Galerkin singular integral evaluation, often perceived as a far more complicated undertaking than for collocation, is a reasonably straightforward process. The analytic integration required for carrying out the boundary limit can be somewhat painful, but this task can be eased by employing symbolic computation programs such as Maple. We will therefore display some of the Maple scripts used to perform these operations.

48 TWO DIMENSIONAL ANALYSIS

Symbolic computation is not absolutely necessary for handling the Laplace singular integrals. With integration tables, and a moderate amount of work, it is certainly possible to develop the formulas by hand. However, for more complicated kernels, symbolic computation becomes more essential. Even for the simple Laplace case, compared to checking hand calculations, it is probably faster and easier to assure that the Maple coding is correct. Moreover, once the Maple procedures have been established for one Green's function, it is then relatively easy to implement another equation by changing these kernel formulas.

3.2 SINGULAR INTEGRALS: LINEAR ELEMENT

The primary reason that a linear element is simple to work with is that the distance function r^2 is a quadratic polynomial in the parameter variables. With this quadratic, the Green's function expressions can be integrated (partially) analytically, this exact integration being essential for evaluating the boundary limit.

The linear element moreover provides the key for treating higher order interpolations. Although the Green's function expressions are now too complicated to be integrated directly, the integrands can be split into two parts: the 'singular part' of the kernel will have the same form as in a linear interpolation, and can be integrated analytically, while the remainder will be nonsingular and can be safely evaluated numerically. Thus, the analysis for a higher order approximation will require this one additional task of splitting the integrals.

A linear element is defined by two nodes, $E_0 = \{Q_0, Q_1\}$, the order or orientation defining an outward normal: the interior of the volume is on the left as the boundary curve is traversed. The two linear shape functions are then simply

$$
\begin{aligned}
\psi_1(t) &= 1 - t \\
\psi_2(t) &= t \,,
\end{aligned}
\tag{3.8}
$$

$0 < t < 1$, and the approximations are

$$
\begin{aligned}
Q(s) &= \psi_1(s)Q_0 + \psi_2(s)Q_1 \\
\phi(Q(s)) &= \phi(Q_0)\,\psi_1(s) + \phi(Q_1)\,\psi_2(s) \\
\frac{\partial\phi}{\partial\mathbf{n}}(Q(s)) &= \frac{\partial\phi}{\partial\mathbf{n}}(Q_0)\,\psi_1(s) + \frac{\partial\phi}{\partial\mathbf{n}}(Q_1)\,\psi_2(s)
\end{aligned}
\tag{3.9}
$$

We treat the two types of singular integrals separately.

3.2.1 Coincident Integration

Although the coincident integration has a line of singularities $t = s$ in the parameter space $\{t, s\}$, this is actually a somewhat simpler situation than the adjacent integration with its single singular point. We therefore begin the journey through the singular integral evaluations with this easier task.

Let the element $E_P = E_Q$ be defined by the nodes $Q_1 = P_1 = (x_1, y_1)$ and $Q_2 = P_2 = (x_2, y_2)$. The exterior boundary limit is implemented by replacing P with

$$
P_\varepsilon = P + \varepsilon\mathbf{N}
\tag{3.10}
$$

where $\varepsilon > 0$ and the outward normal times the Jacobian is $J\mathbf{N} = J\mathbf{n} = (y_2 - y_1, x_1 - x_2)$. The distance $r^2 = \|R\|^2 = \|Q - P\|^2$ is therefore

$$r^2(s,t) = \|Q(s) - P(t)\|^2 = a^2(s-t)^2 + \varepsilon^2 , \tag{3.11}$$

where a is equal to the Jacobian,

$$a = J = \left((x_2 - x_1)^2 + (y_2 - y_1)^2 \right)^{1/2} . \tag{3.12}$$

As noted above, the key aspect of the linear element is that $r^2(s,t)$, appearing in the denominators of the kernel functions, is a quadratic function of the parameters.

It has been emphasized that the integration procedures are the same for all kernel functions. It would therefore suffice to treat the most difficult integral, the hypersingular second derivative. However, as this is our first look at singular integration, it is worthwhile to examine all three integrals in detail, starting with the simplest and working up in complexity.

Integral of G

Substituting Eq.(3.11) and Eq.(3.2) into Eq.(3.1), the integrals to be evaluated are

$$-\frac{a^2}{4\pi} \int_0^1 \psi_k(t) \int_0^1 \psi_j(s) \log \left(a^2 (s-t)^2 + \varepsilon^2 \right) ds \, dt . \tag{3.13}$$

Note that, at $\varepsilon = 0$, the integrand is only weakly singular along the line $s = t$, and we could therefore immediately set $\varepsilon = 0$. However, as we wish to emphasize that we always work with nonsingular integrands, we keep $\varepsilon > 0$ until after the integration. The integrals with respect to s and t are easily evaluated, the final result seen to be

$$\begin{array}{ll} -\frac{a^2 [-7 + 4\log(a)]}{32\pi} & k = j \\[2ex] -\frac{a^2 [-5 + 4\log(a)]}{32\pi} & k \neq j . \end{array} \tag{3.14}$$

The simple Maple script that was used to obtain this result will be discussed below.

Integral of $\partial G / \partial \mathbf{n}$ and $\partial G / \partial \mathbf{N}$

As the normal vector is constant, these two integrals differ only by a sign (resulting from differentiating with respect to P rather than Q), and thus it suffices to treat only the former. The integral takes the form

$$-\frac{a}{2\pi} \int_0^1 \psi_k(t) \int_0^1 \psi_j(s) \frac{J_q \mathbf{n} \cdot R}{r^2} ds \, dt , \tag{3.15}$$

where the factor a in front is from the P Jacobian. It is convenient in this case to attach the Q Jacobian J to the kernel function, as $J\mathbf{n} = (y_1 - y_0, x_0 - x_1)$ is simpler to compute than the unit normal; this is not critical here, as everything is constant, but this observation will be useful when working with higher order interpolations. Moreover, $Q - P$ is a vector that is parallel to $Q_1 - Q_0$, and thus

$$J\mathbf{n} \cdot R = J\mathbf{n} \cdot (-\varepsilon \mathbf{N}) = -a\varepsilon . \tag{3.16}$$

50 TWO DIMENSIONAL ANALYSIS

Thus, for a linear element, the only contribution to this integral is from the limit, and this is therefore, in boundary limit terms, the replacement for the CPV evaluation process. The inner s integration is

$$\int_0^1 \psi_j(s) \frac{\varepsilon}{a^2(s-t)^2 + \varepsilon^2} \, ds \, , \tag{3.17}$$

and thus the value of this integral will involve an inverse tangent, $\tan^{-1}(a/\varepsilon)$. The limiting value of this expression is $\pi/2$, and the result for Eq.(3.15) is then simply

$$\begin{aligned} \frac{a}{6} \, , & \quad k = j \\ \frac{a}{12} \, , & \quad k \neq j \, . \end{aligned} \tag{3.18}$$

As advertised, the Cauchy Principal Value integral has been evaluated in a straightforward manner, without requiring symmetric surgery of the surface. Most importantly, the same boundary limit procedure will also work for analyzing the hypersingular integral, as will be demonstrated next.

Integral of $\partial^2 G/\partial n \partial N$

The $\partial G/\partial n$ integral, which is somewhat contentious for collocation, is well behaved for Galerkin, and thus nothing very interesting has emerged as yet from the analytic integrations and limit process. However, based upon the simple limit calculation in the previous section, the hypersingular integral should prove less boring. This integral can be written as

$$\begin{aligned} -\frac{1}{2\pi} \int_0^1 \psi_k(t) \int_0^1 \psi_j(s) &\left\{ -\frac{(J_p\mathbf{N}) \bullet (J_q\mathbf{n})}{r^2} + 2\frac{(J_q\mathbf{n} \bullet R)(J_p\mathbf{N} \bullet R)}{r^4} \right\} ds \, dt \\ &= -\frac{1}{2\pi} \int_0^1 \psi_k(t) \int_0^1 \psi_j(s) \left\{ -\frac{a^2}{r^2} + 2\frac{a^2\varepsilon^2}{r^4} \right\} ds \, dt \, , \end{aligned} \tag{3.19}$$

and the singularity in the kernel functions is of order $(s-t)^{-2}$. Note that in this case it is convenient to absorb both Jacobians into the normal vectors.

As shown by the simple Galerkin example discussed in Chapter 2, the coincident integral *by itself* cannot have a finite limit. The result that will eventually be established is that the coincident and adjacent integrals are separately divergent in the limit $\varepsilon \to 0$, but their sum is finite. Thus, the complete boundary integral exists, even though its component pieces do not. Moreover, the cancellation of divergent terms will occur assuming only continuity of ϕ on the boundary.

The first point of interest in the evaluation of Eq.(3.19) is that, after integration with respect to s, the r^{-2} and r^{-4} expressions each produce a term ε^{-1}. Fortunately, this divergent quantity appears with opposite signs in the two terms, and cancels by itself.

Carrying out the complete integration, but keeping $\varepsilon > 0$, we find

$$\begin{aligned} \frac{1}{4\pi} \left(1 - \log(a^2)\right) + \frac{1}{4\pi} \log(\varepsilon^2) \, , & \quad k = j \\ -\frac{1}{4\pi} \, , & \quad k \neq j \, . \end{aligned} \tag{3.20}$$

Note that only the diagonal terms are divergent. In other words, in the equation for, say, P_1, the divergent term in the coincident integral is

$$\frac{\phi(P_1)}{4\pi} \log(\varepsilon^2) , \tag{3.21}$$

and only involves $\phi(P_1)$, not $\phi(P_2)$. Thus, the divergences can be associated with the ends of the element; as discussed previously, divergences arise at the element edges because they lack a 'complete neighborhood' of the singular point.

This situation should be contrasted with the divergent term from the collocation calculation in Chapter 2. There the $\log(\varepsilon^2)$ term originated with the term from $\phi(Q(s))$ that was linear in the parameter s, and therefore related to the derivative of ϕ at the singular point. With Galerkin, it is the constant term, and thus all that will be required is that this constant term be the same for the adjacent integral, $i.e.$, continuity. It will be established below that this $\log(\varepsilon^2)$ term does indeed cancel with a corresponding term from the adjacent integration.

A significant advantage of the direct limit approach is that the canceling divergent terms can be identified and removed from the calculation. The cancellation is therefore achieved *exactly*, and each individual integral is a well defined quantity. The alternative, approximate quadrature of divergent integrals and effecting the cancellation numerically, is not an attractive proposition.

3.2.2 Coincident: Symbolic Computation

The Maple script employed to derive the above results is presented below. As mentioned above, symbolic computation is not really needed for handling the simple 2D Laplace integrals, and this material certainly does not make for exciting reading. The justification for including this section is that for more complicated kernel functions, $e.g.$, three dimensional elasticity, symbolic computation does become, if not absolutely essential, extremely useful for obtaining the necessary analytic formulas. Moreover, this section will also serve to reinforce the mechanics of how the approximations and the boundary limit are executed.

On the other hand, this section is certainly not a tutorial on how to write elegant and efficient Maple code, there are surely better ways to accomplish the necessary tasks. We have instead attempted to present things in as simple and straightforward manner as possible.

The script begins by defining the shape functions (called mp and mq to distinguish between the inner and outer integrations), the vector R, and the kernel functions. The boundary limit parameter is the variable eps, a stand-in for ε employed above. In the discussion we use eps and ε interchangeably.

```
# Shape function
      mq[1]  := 1-s; mq[2]  := s;
      mp[1]  := 1-t; mp[2]  := t;

# Coordinates of Q
      xq     := x[1]*mq[1] + x[2]*mq[2];
      yq     := y[1]*mq[1] + y[2]*mq[2];

# Coordinates of P
```

52 TWO DIMENSIONAL ANALYSIS

```
        xp    := x[1]*mp[1] + x[2]*mp[2] + eps*N[1];
        yp    := y[1]*mp[1] + y[2]*mp[2] + eps*N[2];

# R = Q - P

        Rx    := xq-xp;  Ry := yq-yp;
        rsq   := Rx^2 + Ry^2;
        rsq   := expand(rsq);
        rsq   := subs(N[1]^2=1-N[2]^2,rsq);
        rsq   := collect(expand(rsq),s);
        rsq   := a^2*(s-t)^2 + eps^2;

# Kernels

        ker0  := -a^2*ln(rsq)/(4*pi);
        ker1  := a^2*eps/rsq/(2*pi);
        ker2  := (a^2/rsq - 2*a^2*eps^2/rsq/rsq)/(2*pi);
```

Note that rather than use the form for rsq that Maple calculates, this quantity is redefined in terms of the (for Maple, undefined) coefficient a^2 (the square of the length of the element). The value of this constant is immaterial for the subsequent integrations, and keeping a undefined results in much simpler formulas. Similarly, Maple could calculate the inner products $\mathbf{n} \cdot R$ and $\mathbf{N} \cdot R$ needed to define the Green's functions quantities ker1 and ker2, but in this case it is simpler to put in the known form. The next section of code does the integrations, with two loops to accommodate the two different shape functions $\psi_j(s)$ and $\psi_k(t)$ for the inner and outer integrations.

```
# Analytical Integration

# Q Integration

for j from 1 to 2 do
 G0 := int(mq[j]*ker0,s=0..1):
 G1 := int(mq[j]*ker1,s=0..1):
 G2 := int(mq[j]*ker2,s=0..1):

# P Integration

 for k from 1 to 2 do
  GG0[k,j] := int(mp[k]*G0,t=0..1):
  GG1[k,j] := int(mp[k]*G1,t=0..1):
  GG2[k,j] := int(mp[k]*G2,t=0..1):

# Boundary limit: eps = 0
   GG0[k,j] := subs(ln(eps^2)=2*loge,ln(eps)=loge,arctan(a/eps)=pi/2,
                  ln((a^2+eps^2)/eps^2)=ln(a^2)-2*loge,eps=0,
                  expand(GG0[k,j]));
   GG1[k,j] := subs(ln(eps^2)=2*loge,ln(eps)=loge,arctan(a/eps)=pi/2,
                  ln((a^2+eps^2)/eps^2)=ln(a^2)-2*loge,eps=0,
                  expand(GG1[k,j]));
 od;
od;
```

The desired goal, evaluation of the boundary limit, was to set eps equal to zero. For G and $\partial G/\partial \mathbf{n}$, this substitution is permissible once Maple's arctan(a/eps) is replaced by $\pi/2$, and ln(eps) replaced by loge. The divergent loge quantities are multiplied by a positive power of ε, and thus when eps is set to zero, the correct limiting value for these terms, namely 0, is obtained.

For the hypersingular kernel, a bit more work is required before the substitution $\varepsilon = 0$ can be executed. It is necessary to extricate the ε^2 factor from the complicated logarithmic expressions so as to identify all the $\log(\varepsilon^2)$ terms. This is not difficult, and the details are not particularly instructive, so we spare the reader this piece of code. The most important comment is that, as shown in Eq.(3.20), a loge term will survive the substitution $\varepsilon = 0$.

3.2.3 Adjacent Integration

The coincident integrations were relatively trivial, and complete analytical evaluation of the double integral was possible. Moreover, the resulting expressions for the integrals were also quite simple. For the adjacent singular integration, the singularities are necessarily weaker, being at a single point in the two-dimensional $\{t, s\}$ parameter space, but the integrands are more difficult to work with. The complication is that the expression for r^2 is now a full quadratic having a first order term, e.g.,

$$r^2 = b_2(1 - t + s)^2 + b_1\varepsilon(1 - t + s) + \varepsilon^2 . \tag{3.22}$$

In this expression we have assumed that, in the orientation of the boundary curve, the Q element E_Q follows E_P, so that the common point for the two elements is given by $t = 1$ and $s = 0$. For $b_1 \neq 0$, the double integration with this function in the denominators, even if possible, would result in very lengthy expressions.

As a consequence, we adopt a hybrid method, the analytical integration limited to one dimension in parameter space, and one dimension left to numerical means. However, having only one analytic integration at our disposal, it is essential to make the most of it. Most importantly, for the hypersingular integral, this single integration must be successful in producing the divergent $\log(\varepsilon^2)$ that will cancel with the coincident integral. Analytically integrating either s or t will not work, it is necessary to include both conditions defining the singular point, $t = 1$ and $s = 0$.

A procedure that is successful, and one that will be repeatedly exploited in Chapter 4 for 3D singular integrations, is to introduce a polar coordinate transformation, $\{t, s\} \rightarrow \{\rho, \theta\}$, centered at the singular point. This transformation will have three noteworthy aspects:

- the two conditions for singularity, $t = 1$ and $s = 0$, are equivalent to the one condition $\rho = 0$;

- as the singular part of the kernel functions is associated only with ρ, the analytic integration is with respect to this variable; the resulting function of θ is nonsingular.

- the jacobian of the transformation, $\rho d\rho d\theta$, contributes to reducing the strength of the singularity at $\rho = 0$.

Thus, the analytic integration with respect to ρ is, in effect, somewhat more than just integrating one parameter dimension, and this is why one analytic integration will succeed in producing the desired $\log(\varepsilon^2)$ term.

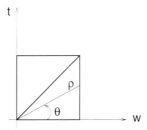

Figure 3.2 Polar coordinates for the adjacent singular integrations.

For an element $E_P = (P_1, P_2)$, there are *normally* two cases to consider, one in which the adjacent Q element E_Q follows E_P, i.e., the common point is P_2 (*right adjacent*), and the *left adjacent* case where the common point is P_1. In the following only the *right adjacent* case, $E_Q = (P_2, P_3)$, $P_k = (x_k, y_k)$, will be treated, the left adjacent can be treated analogously. Or, possibly more simply, the left adjacent becomes right adjacent by reversing the direction of the elements; thus, right adjacent formulas can be employed, provided appropriate sign changes are incorporated to account for the wrong orientation.

We said *normally* because there are important situations when an element can have more than two adjacent neighbors. In particular, for a *surface crack*, a crack surface that intersects the domain boundary, the crack element at the juncture will have three adjacent neighbors. This situation will be discussed in detail in Chapter 9. In addition, for a complicated 'self-intersecting' crack geometry, the number of adjacent elements can be more than three.

Referring to Fig. 3.1, the right adjacent case is $E_P = E_0$, $E_Q = E_1$, and the common point P_1 in each element occurs for $t = 1$, $s = 0$. Changing variables, the singular point is at the origin of the parameter space $\{w, s\}$ where $w = 1 - t$, and polar coordinates is then defined by

$$w = \rho \cos(\theta) \qquad (3.23)$$
$$s = \rho \sin(\theta)$$

As indicated in Fig. 3.2, this will necessitate splitting the θ integral into two parts, $0 < \theta < \pi/4$ and $\pi/4 < \theta < \pi/2$, as the expression for the upper limit on ρ is either $\sec(\theta)$ or $\csc(\theta)$, depending upon the sub-triangle. Thus, ignoring the specific integrands for the moment, the integrals to be evaluated are of the form

$$\int_0^{\frac{\pi}{4}} \int_0^{\sec(\theta)} \cdots \rho\, d\rho\, d\theta + \int_{\frac{\pi}{4}}^{\frac{\pi}{2}} \int_0^{\csc(\theta)} \cdots \rho\, d\rho\, d\theta\ . \qquad (3.24)$$

Transferring to polar coordinates, Eq.(3.22) becomes

$$r^2 = a_2 \rho^2 + a_1 \rho \varepsilon + \varepsilon^2 \qquad (3.25)$$

where the coefficients $\{a_2, a_1\}$ are now functions of the nodal coordinates, and $\cos(\theta)$ and $\sin(\theta)$. To be specific,

$$R = (a_x \rho - \varepsilon N_1, a_y \rho - \varepsilon N_2) \qquad (3.26)$$

where

$$
\begin{aligned}
a_x &= (x_2 - x_1)\cos(\theta) + (x_3 - x_2)\sin(\theta) \\
a_y &= (y_2 - y_1)\cos(\theta) + (y_3 - y_2)\sin(\theta) \\
a_2 &= a_x^2 + a_y^2 \\
a_1 &= -2\,(a_x N_1 + a_y N_2)
\end{aligned}
\tag{3.27}
$$

Thus, incorporating the jacobians (note that the Q and P jacobians and normals are in general different), the expressions required for the Green's functions are

$$
\begin{aligned}
J_p \mathbf{N} \bullet R &= j_p^1 \rho - J_p \varepsilon \\
j_p^1 &= J_p\,(N_1 a_x + N_2 a_y) \\
J_q \mathbf{n} \bullet R &= j_q^1 \rho - J_q \mathbf{N} \bullet \mathbf{n}\varepsilon \\
j_q^1 &= J_q\,(n_1 a_x + n_2 a_y)\ .
\end{aligned}
\tag{3.28}
$$

These formulas will be needed for establishing the cancellation of the $\log(\varepsilon^2)$ term appearing in the hypersingular integral.

Integral of G and $\partial G/\partial \mathbf{n}$

As the singularity now resides at one point in a two-dimensional parameter space, these two kernels, and the related $\partial G/\partial \mathbf{N}$, are now weakly singular. Thus, their integration is straightforward, they are discussed together. Denoting by L_ρ the upper limit on ρ, $0 < \rho < L_\rho$ ($L_\rho = \sec(\theta)$ or $L_\rho = \csc(\theta)$), the integrals over ρ become

$$
\mathcal{I}_{kj}^0 = -\frac{J_p J_q}{4\pi}\int_0^{L_\rho} \psi_k(\rho,\theta)\psi_j(\rho,\theta)\log(r^2)\rho\,d\rho
\tag{3.29}
$$

$$
\mathcal{I}_{kj}^1 = -\frac{J_p J_q}{2\pi}\int_0^{L_\rho} \psi_k(\rho,\theta)\psi_j(\rho,\theta)\frac{J_q \mathbf{n}\bullet R}{r^2}\rho\,d\rho
\tag{3.30}
$$

where r^2 is given in Eq.(3.25). As a final observation concerning the polar coordinate transformation, note that both P and Q shape functions are functions of ρ. Thus, Q and P are equally involved in the ρ integration, the change of variables effectively 'mixes' both inner and outer integrations. The integrals of the Green's function turn out to be

$$
\begin{aligned}
\mathcal{I}_{11}^0 &= -\frac{J_p\,J_q}{144}\cos(\theta)\,L_\rho^3\,\big(6\,(3\,L_\rho\sin(\theta)-4)\log\big(a_2\,L_\rho^2\big)-9\,L_\rho\sin(\theta)+16\big) \\
\mathcal{I}_{12}^0 &= \frac{J_p\,J_q}{16}\cos(\theta)\sin(\theta)\,L_\rho^4\,\big(2\log\big(a_2\,L_\rho^2\big)-1\big) \\
\mathcal{I}_{21}^0 &= \frac{J_p\,J_q}{144}\,L_\rho^2\,\big(-9\,L_\rho^2\cos(\theta)\sin(\theta)+16\,L_\rho\,(\cos(\theta)+\sin(\theta))-36 \\
&\quad \big(18\,L_\rho^2\cos(\theta)\sin(\theta)-24\,L_\rho(\cos(\theta)+\sin(\theta))+36\big)\log\big(a_2\,L_\rho^2\big)\big) \\
\mathcal{I}_{22}^0 &= -\frac{J_p\,J_q}{144}\sin(\theta)\,L_\rho^3\,\big(6\,(3\,L_\rho\cos(\theta)-4)\log\big(a_2\,L_\rho^2\big)-9\,L_\rho\cos(\theta)+16\big)
\end{aligned}
\tag{3.31}
$$

The main observation is that these formulas are smooth functions of θ. The $\log(x)$ function is potentially dangerous at $x = 0$, but in this case the variable $a_2\,L_\rho^2$

56   TWO DIMENSIONAL ANALYSIS

does not come close to zero. Integrating the derivative of the Green's function is even easier,

$$
\begin{aligned}
\mathcal{I}_{11}^1 &= -\frac{J_p \alpha}{6 a_2} L_\rho^2 \cos(\theta)\,(2\,L_\rho \sin(\theta) - 3) \\
\mathcal{I}_{12}^1 &= \frac{J_p \alpha}{3 a_2} L_\rho^3 \cos(\theta) \sin(\theta) \\
\mathcal{I}_{21}^1 &= -\frac{J_p \alpha}{6 a_2} L_\rho \left(6 - 3\,L_\rho \sin(\theta) - 3 \cos(\theta) L_\rho + 2\,L_\rho^2 \cos(\theta) \sin(\theta)\right) \\
\mathcal{I}_{22}^1 &= -\frac{J_p \alpha}{6 a_2} L_\rho^2 \sin(\theta)\,(2 \cos(\theta) L_\rho - 3)\ ,
\end{aligned}
\tag{3.32}
$$

where

$$
\alpha = \mathbf{n}_1 a_x + \mathbf{n}_2 a_y = \cos(\theta)\,\{(y_3 - y_2)(x_2 - x_1) + (x_2 - x_3)(y_2 - y_1)\}
\tag{3.33}
$$

Integral of $\partial^2 G / \partial \mathbf{N} \partial \mathbf{n}$

The ρ integral for the hypersingular kernel takes the form

$$
\mathcal{I}_{kj}^2 = \frac{1}{2\pi} \int_0^{L_\rho} \psi_k(\rho,\theta)\psi_j(\rho,\theta) \left[\frac{J_p \mathbf{N} \cdot J_q \mathbf{n}}{r^2} - 2\frac{(J_q \mathbf{n} \cdot R)(J_q \mathbf{n} \cdot R)}{r^4} \right] \rho\,d\rho
\tag{3.34}
$$

Note that now the integrands behave as ρ^{-1}, and thus a $\log(\varepsilon^2)$ term can be expected from this integral. However, with the exception of $k = 2$, $j = 1$, the products $\psi_k(\rho,\theta)\psi_j(\rho,\theta)$ contain a factor of ρ; as a consequence these integrals are also well behaved. However, for consistency, we will nevertheless integrate these terms with $\varepsilon > 0$. The results are

$$
\begin{aligned}
\mathcal{I}_{11}^2 &= \frac{1}{2 a_2^2} \cos(\theta)\,L_\rho \left(a_2 \mathbf{n} \cdot \mathbf{N}(2 - L_\rho \sin(\theta)) + j_q^1 j_p^1 (2\,L_\rho \sin(\theta) - 4)\right) \\
\mathcal{I}_{12}^2 &= -\frac{1}{2 a_2^2} \cos(\theta) \sin(\theta) L_\rho^2 \left(-a_2\,\mathbf{n} \cdot \mathbf{N} + 2\,j_q^1 j_p^1\right) \\
\mathcal{I}_{22}^2 &= \frac{1}{2 a_2^2} \sin(\theta)\,L_\rho \left(a_2 \mathbf{n} \cdot \mathbf{N}(2 - \cos(\theta) L_\rho) + j_q^1 j_p^1 (2 \cos(\theta) L_\rho - 4)\right)
\end{aligned}
\tag{3.35}
$$

We now focus on the integration for $k = 2$, $j = 1$: of all the calculations, this last integral for two dimensional analysis is the most complicated. The formula for $\mathcal{I}_{21}^2$, while important, is rather long and therefore omitted. Most importantly, we find a divergent term of the form

$$
\frac{\left(2\,j_q^1 j_p^1 - a_2\,\mathbf{n} \cdot \mathbf{N}\right)}{4\pi\,a_2^2} \log(\varepsilon^2)
\tag{3.36}
$$

3.2.4   Cancellation of $\log(\varepsilon^2)$

The remaining, and vital, task is to show that the complete hypersingular integral is finite, *i.e.*, Eq.(3.36) cancels with the divergent contribution from the coincident integral Eq.(3.20). This at first does not look promising, as other than the logarithm itself, the divergent term for the adjacent integral looks nothing like the simple $\log(\varepsilon^2)/4\pi$ found for the coincident. Remember, however, that the adjacent calculation is not complete, there remains the integration over θ. As Eq.(3.36) is independent of the upper limit L_ρ – as it should be, as the divergence should arise solely from $\rho = 0$ – it is not necessary to split this integral at $\pi/4$ and the actual divergent term is

$$\phi(P_2)\frac{\log(\varepsilon^2)}{4\pi} \int_0^{\pi/2} \left[\frac{2\,j_q^1\,j_p^1}{a_2^2} - \frac{\mathbf{n} \cdot \mathbf{N}}{a_2} \right] d\theta \ . \tag{3.37}$$

It will turn out that this integral can be evaluated exactly, and yields the desired result. This term cancels the divergence from the coincident integral multiplying $\phi(P_2)$; the remaining divergence from the coincident integral is removed when the left adjacent integral is considered.

Ignoring many details, which can in fact be left to Maple, we indicate only the main steps in the evaluation of Eq.(3.37). Note that the denominator is of the form

$$a_2^m = (a_x^2 + a_y^2)^m = \left(d_{cc}\cos^2(\theta) + d_{cs}\cos(\theta)\sin(\theta) + d_{ss}\sin^2(\theta) \right)^m \tag{3.38}$$

where $\{a_x, a_y\}$ are defined in Eq.(3.27) and $m = 1, 2$. Rewrite this expression by removing a factor of $d_{ss}\sin^2(\theta)$ and making the substitution

$$q = \cotan(\theta) \ , \tag{3.39}$$

resulting in

$$a_2^m = d_{ss}^m \sin^{2m}(\theta) \left(d_2 q^2 + d_1 q + 1 \right)^m \ . \tag{3.40}$$

The new coefficients are defined in an obvious manner. As $dq = -d\theta/\sin^2(\theta)$, it is easily seen that the integrand with $m = 1$ becomes

$$\frac{\mathbf{n} \cdot \mathbf{N}}{d_{ss}} \int_0^\infty \frac{dq}{d_2 q^2 + d_1 q + 1} \ . \tag{3.41}$$

whereas for $m = 2$ we obtain, using Eq.(3.27) and Eq.(3.28),

$$\frac{1}{d_{ss}^2} \int_0^\infty \frac{\gamma_2 q^2 + \gamma_1 q + \gamma_0}{(d_2 q^2 + d_1 q + 1)^2} \, dq \ . \tag{3.42}$$

These integrals can be evaluated analytically, and the desired result is achieved, all hypersingular divergent terms cancel. Thus, the $\log(\varepsilon^2)$ contributions can be legitimately discarded, and it has been established that the hypersingular Galerkin integral is finite, assuming only continuity of the potential on the boundary.

Finally, as noted above, if the domain contains a surface crack, the adjacency configuration is more complicated, *i.e.*, two left or right adjacent elements. The hypersingular integral is nevertheless finite at a crack/outer boundary junction; this will be discussed in Chapter 9.

3.2.5 Adjacent: shape function expansion

For the analysis of the two-dimensional Laplace equation, the exact formulas for the adjacent integration are relatively small expressions which do not require much

58   TWO DIMENSIONAL ANALYSIS

computational effort. However, for other Green's functions, such as elasticity, higher order interpolations, and in three dimensions, this will not be the case. Thus, it is useful to note here that there is a useful alternative to the complete evaluation of every term $\mathcal{I}_{kj}^l$.

What we have in mind is to first recognize that the shape function products can be written as polynomials in ρ,

$$\psi_k(\rho, \theta)\psi_j(\rho, \theta) = \sum_{l=0}^{m_l} c_{kj}^l \rho^l \,, \tag{3.43}$$

with the coefficients $c_{kj}^l(\theta)$ easily obtainable. Thus, it is only necessary to evaluate the basic component integrals

$$\int_0^{L_\rho} \rho^{1+l} \mathcal{K}(\rho, \theta) \, d\rho \, d\theta \,, \tag{3.44}$$

for the powers of ρ, $0 < l < m_l$. With the two dimensional linear approximation, the shape function products are quadratic, $m_l = 2$, and there are three integrals to do. The complete integral can for any $\{k, j\}$ can then be pieced together using the coefficients $c_{kj}^l(\theta)$.

By replacing the shape function product by a simple monomial, the size of the Maple formulas can be significantly reduced. Moreover, this will also reduce computational effort: the lengthier expressions contain much duplicated effort that is avoided by using the shape function expansions.

Thus, for example, the three integrals $\mathcal{I}_0^l$ for G turn out to be

$$\begin{aligned}
\mathcal{I}_0^0 &= \frac{1}{4} J_p J_q L_\rho^2 \left(\log \left(a_2 L_\rho^2 \right) - 1 \right) \\
\mathcal{I}_0^1 &= \frac{1}{18} J_p J_q L_\rho^3 \left(3 \log \left(a_2 L_\rho^2 \right) - 2 \right) \\
\mathcal{I}_0^2 &= \frac{1}{16} J_p J_q L_\rho^4 \left(2 \log \left(a_2 L_\rho^2 \right) - 1 \right),
\end{aligned} \tag{3.45}$$

and for $\partial G/\partial \mathbf{n}$

$$\begin{aligned}
\mathcal{I}_0^0 &= \frac{J_p j_q^1 L_\rho}{a_2} \\
\mathcal{I}_0^1 &= \frac{J_p j_q^1 L_\rho^2}{2a_2} \\
\mathcal{I}_0^2 &= \frac{J_p j_q^1 L_\rho^3}{3a_2}.
\end{aligned} \tag{3.46}$$

Even in this simple situation, the formulas are much more compact.

3.2.6   Numerical Tests

As a validation of the above procedures, and to demonstrate the level of accuracy one can expect, the results from some simple test calculations will now be presented. Consider first a Dirichlet problem on the unit disk, with boundary values $\phi = x^2 - y^2$. As this is a harmonic function, the solution – namely $\phi(x, y) = x^2 - y^2$

everywhere on the disk – is known, and thus the errors can be computed. Two measures of the error in the computed surface flux are examined: the maximum error over the boundary, and the pointwise L^2 error defined as

$$\left(\frac{1}{M}\sum_{k=1}^{M} e_k^2\right)^{1/2}. \tag{3.47}$$

Here M denotes the number of nodes and e_k is the error at node k. Table 3.1 compares the errors for the Symmetric-Galerkin and collocation methods, both employing a linear element approximation. The integrations for both methods were equivalent: all numerical quadratures were computed using a 4-point Gauss rule, and the singular integrals in the collocation method were also calculated analytically. Results for four uniform discretizations of the circle are reported, $M = 30,\ 60,\ 150,\ 300$. In this case, the symmetric Galerkin results are significantly more accurate than collocation.

Table 3.1 A comparison of symmetric Galerkin and collocation nodal errors for a Dirichlet problem on the unit disk.

M	Symmetric-Galerkin		Collocation	
	Max	L^2	Max	L^2
30	.00326354	.00230768	.01518792	.01073948
60	.00086900	.00061446	.00417751	.00295394
150	.00014453	.00010220	.00070568	.00049900
300	.00003703	.00002619	.00017957	.00012698

For the Dirichlet boundary conditions, the symmetric Galerkin method employs the potential equation, and thus the above example does not exercise the hypersingular flux equation. The collocation solution was also obtained using the potential equation, not only to compare the methods as closely as possible, but also because the hypersingular equation cannot be used with the C^0 linear element approximation. To examine the performance of Galerkin with the hypersingular equation, a Neumann problem is solved exterior to the unit disk. The chosen exact solution is the harmonic function $\phi = x/(x^2 + y^2)$, which also turns out to be the form of the normal flux boundary condition. The errors in the computed potential are shown in Table 3.2, and once again symmetric Galerkin is more accurate than collocation.

Table 3.2 A comparison of symmetric Galerkin and collocation nodal errors for a Neumann problem on the domain exterior to the unit disk.

M	Symmetric-Galerkin		Collocation	
	Max	L^2	Max	L^2
30	.00191174	.00135180	.00941169	.00665506
60	.00047500	.00033588	.00231805	.00163911
150	.00009619	.00006802	.00036755	.00025990
300	.00006431	.00004547	.00009158	.00006476

60   TWO DIMENSIONAL ANALYSIS

That Galerkin is more accurate than collocation is not surprising. Similar behavior was observed for two dimensional elasticity (using a non-symmetric Galerkin formulation) by de Paula and Telles [71].

3.3 HIGHER ORDER INTERPOLATION

As analytic integration is necessary to evaluate the boundary limit, it may appear that the techniques described above are inherently limited to a linear element. The purpose of this section is to demonstrate that, with an additional step, higher order approximations can be accommodated.

Linear interpolation works nicely primarily because the distance function r^2 is a quadratic polynomial. (It also helps that the jacobian is a constant, rather than a square root of a polynomial, but this is comparatively minor.) With a higher order interpolation, the denominators in the kernel functions become sufficiently complicated that exact integration is not *immediately* possible.

The strategy is to split the singular integrals into two pieces. One part is a singular term that will look essentially like a linear element integral, and will therefore be amenable to analytic integration and limit evaluation. The second part, a remainder term, will be completely non-singular and thus ε can be set to zero and this term can be integrated numerically. Moreover, this remainder will usually be relatively small in value compared to the singular part, and can therefore be treated efficiently by low order numerical quadrature. This section will show how to accomplish this decomposition.

The splitting techniques are generally applicable, but it is simplest to describe the procedures using a specific curved interpolation. We choose a three-noded quadratic element, defined by the shape functions, $0 \leq t \leq 1$

$$
\begin{aligned}
\psi_1(t) &= (1-t)(1-2t) \\
\psi_2(t) &= 4t(1-t) \\
\psi_3(t) &= t(2t-1) \ .
\end{aligned}
\tag{3.48}
$$

Thus, for the boundary interpolation of the three boundary nodes $Q_j = (x_j, y_j)$ we have

$$
Q(s) = \left(\sum_{j=1}^{3} x_j \psi_k(s), \sum_{k=1}^{3} y_j \psi_k(s) \right) \ ,
\tag{3.49}
$$

and isoparametric approximation for ϕ,

$$
\phi(Q(s)) = \sum_{j=1}^{3} \phi(P_j)\psi_j(s) \ ,
\tag{3.50}
$$

and a similar expression for the flux.

To illustrate the splitting into singular and non-singular parts, it suffices to consider the coincident integration, the corresponding treatment of the adjacent case being straightforward. Including the limit to the boundary term, $\mathbf{R} = Q(s) -$

$(P(t) + \epsilon \mathbf{N}) = (r_1(t,s), r_2(t,s))$ and using the above equations,

$$
\begin{aligned}
r_1(t,s) &= a_2(s-t)^2 + a_1(s-t) + \epsilon N_1 \\
r_2(t,s) &= b_2(s-t)^2 + b_1(s-t) + \epsilon N_2 \\
r^2 &= \epsilon^2 + \sum_{l=2}^{4} \alpha_l (s-t)^l ,
\end{aligned}
\tag{3.51}
$$

where the coefficients $\{a_i, b_i\}$ are given in terms of the nodal coordinates. In addition, the jacobian is given by

$$
J_q(s) = \left\| \frac{d}{ds} Q(s) \right\| = (j_0 + j_1 s)^{\frac{1}{2}} .
\tag{3.52}
$$

3.3.1 Integral of G

The splitting for $G(P, Q)$ is accomplished by appropriately rewriting each factor in the product $J_q(s) \log(r^2)$. The logarithm is straightforward:

$$
\begin{aligned}
\log(r^2) &= \log \left(\sum_{l=2}^{4} \alpha_l (s-t)^l + \varepsilon^2 \right) \\
&= \log \left(\alpha_2 (s-t)^2 + \varepsilon^2 \right) + \log \left(1 + \sum_{l=1}^{2} \beta_l (s-t)^l \right) ,
\end{aligned}
\tag{3.53}
$$

where, dropping the ε^2 in the second term, $\beta_1 = \alpha_3/\alpha_2$ and $\beta_2 = \alpha_4/\alpha_2$. The jacobian is handled by writing

$$
J_q(s) = \hat{J}(s,t) + \left(J_q(s) - \hat{J}(s,t) \right)
\tag{3.54}
$$

where $\hat{J}(s,t)$ is a Taylor expansion of $J_q(s)$ expanded around $s = t$. The partitioning is therefore given by

$$
\begin{aligned}
J_q(s) \log(r^2) &= \hat{J}(s,t) \log \left(\alpha_2 (s-t)^2 + \varepsilon^2 \right) + J_q(s) \log \left(1 + \sum_{l=1}^{2} \beta_l (s-t)^l \right) \\
&\quad + \left(J_q(s) - \hat{J}(s,t) \right) \log \left(\alpha(s-t)^2 \right) .
\end{aligned}
\tag{3.55}
$$

The first term on the right is the 'linear element' contribution containing all of the singularity: it simple enough to be integrated exactly with respect to s, and the two remainder terms are sufficiently well behaved to be integrated numerically. The analytic integration results in $\log(t)$ and $\log(1-t)$ terms, and a second Taylor expansion of $J_p(t)$ with respect to t will allow an accurate numerical treatment of these weakly singular integrals.

There remains the question of the order of the Taylor expansion $\hat{J}(s,t)$. The main concern is the integration of the last term in Eq.(3.55), the remainder in the Taylor series, which is $\mathcal{O}((s-t)^\gamma)$, multiplied by the log singularity. It is easy to include any number of terms in $\hat{J}(s,t)$, but for efficiency it is still desirable to limit the complexity of the analytic formulas as much as possible. A three term expansion has been found to be satisfactory if standard Gauss quadrature is used to

62   TWO DIMENSIONAL ANALYSIS

evaluate this last term. However, a special quadrature technique that specifically accounts for the logarithmic singularity can also be employed [17, 67, 110]. This special Gauss rule is based on observing that

$$\int_0^1 f(t) \log(t)\, dt = -\int_0^1 \int_0^1 f(st)\, ds\, dt \ . \tag{3.56}$$

For well behaved $f(t)$, the non-singular double integral can be evaluated with a simple Gauss rule.

3.3.2 Integral of $\partial G/\partial \mathbf{n}$ and $\partial G/\partial \mathbf{N}$

For $\partial G/\partial \mathbf{n}$, $J_q(s)$ can be absorbed into the normal $\mathbf{n}(Q)$ that appears in this first derivative of G, leaving only $J_p(t)$ in the outer integration. For this reason, when dealing with $\partial G/\partial \mathbf{N}$ it is convenient to reverse the order of integration and execute P first. The numerator in this kernel function, at least for the first analytic integration, is polynomial, and does not present a problem, the main stumbling block is the presence of rational functions with denominators containing fourth degree polynomials for r^2, Eq.(3.51). As with the logarithm, the goal will be to find a modified form of the rational function which is exactly integrable but which nevertheless retains all of the singularity at $t = s$. This is achieved by replacing the r^{-2} terms with the simpler expression $\hat{r}^{-2}$,

$$\frac{1}{\hat{r}^2} = \frac{1}{(\alpha_2(s-t)^2 + \epsilon^2)} \ . \tag{3.57}$$

consisting of the first two non-zero terms in r^2, Eq.(3.51). These are the most important terms as $s \to t$, and contain the essence of the singularity. Some algebraic manipulation results in

$$
\begin{aligned}
\frac{1}{r^2} &= \frac{1}{\hat{r}^2} + \left(\frac{1}{r^2} - \frac{1}{\hat{r}^2} \right) \\
&= \frac{1}{\hat{r}^2} + \left(\frac{\hat{r}^2 - r^2}{r^2 \hat{r}^2} \right)
\end{aligned}
\tag{3.58}
$$

For the term in parenthesis, the lowest order term appearing in the numerator is $(s - t)^3$, and the denominator is essentially $(s - t)^4$. The singularity in this remainder is therefore one degree less than r^{-2}, which behaves as $(s - t)^{-2}$. Even though the remainder is still singular, the above expansion suffices in this case, as the numerator in the kernel is also of order $(s - t)$. However, it is also possible to obtain a completely nonsingular remainder by iterating this process. Replacing the r^{-2} on the right in Eq.(3.58) by this same expression results in

$$
\begin{aligned}
\frac{1}{r^2} &= \frac{1}{\hat{r}^2} + \frac{\hat{r}^2 - r^2}{\hat{r}^4} + \frac{(\hat{r}^2 - r^2)^2}{r^2 \hat{r}^4} \\
&= \frac{1}{\hat{r}^2} - \frac{\beta(s-t)^3}{\hat{r}^4} + I_0(t, s) \ ,
\end{aligned}
\tag{3.59}
$$

where $I_0(t, s)$ is a rational function which is nonsingular at $t = s$. Thus, one can set $\epsilon = 0$ in I_0 and integrate numerically. As with the linear element, the first two terms in Eq.(3.59) can be integrated analytically without any difficulty, and the limit $\epsilon \to 0$ can then be computed.

3.3.3 Integral of $\partial^2 G/\partial N \partial n$

For this kernel, analytic integration of both s and t is essential to carry out the limit, and it is therefore helpful that both jacobians can be incorporated into the kernel function. Note too that in this case the r^{-2} component in $\partial^2 G/\partial N \partial n$ has a numerator of $\mathbf{n} \cdot \mathbf{N}$. Consequently, the two-term expansion Eq.(3.59) is definitely required.

The r^{-4} term in the hypersingular kernel can be manipulated in the same manner described above. Thus,

$$
\begin{aligned}
\frac{1}{r^4} &= \frac{1}{\hat{r}^4} + \left(\frac{\hat{r}^4 - r^4}{r^4 \hat{r}^4} \right) \\
&= \frac{1}{\hat{r}^4} + \frac{\hat{r}^4 - r^4}{\hat{r}^8} + \frac{(\hat{r}^4 - r^4)^2}{r^4 \hat{r}^8} \\
&= \frac{1}{\hat{r}^4} - 2\frac{\beta(s-t)^3}{\hat{r}^6} + I_1(t,s) \ ,
\end{aligned}
\tag{3.60}
$$

In this case I_1 is still singular, of the form $(s - t)^{-2}$, but this will be taken care of by the zero in the numerator of the kernel function at $t = s$. As with the choice of order in the Taylor expansion for the jacobian, the process represented by Eq.(3.59) or Eq.(3.60) can be repeated as often as desired. More iterations will make the numerical part weaker and possibly improve accuracy of the integration. However, it also increases the complexity (and hence the eventual computational cost) of the analytic formulas. Finally note that for the adjacent integration, the Taylor expansions are naturally employed around $\rho = 0$.

These simple techniques allow curved elements to be treated *almost* as easily as the linear element. The two main differences are that numerical integration will always be required, where before complete analytic integration was often possible, and that the analytic integration is more involved (*e.g.*, the $\hat{r}^{-4}$ term in Eq.(3.59)). Symbolic computation can alleviate much of the drudgery involved in carrying out this program.

3.4 OTHER GREEN'S FUNCTIONS

The basic idea of partitioning integrals into singular and non-singular parts can also be applied to boundary integral formulations involving more complicated fundamental solutions [108]. For example, the Green's function kernels for elastic wave scattering (frequency domain) are defined in terms of Hankel functions $H_m(r)$, or exponentials e^{ikr}, in two [130] and three dimensions [236], respectively. For the Hankel function, the singular integral analysis analysis can be carried out by invoking the asymptotic form of the Hankel function as $z \to 0$ [204],

$$
H_m^{(1)}(z) \approx -\frac{2i}{\pi} \left\{ \frac{(m-1)!}{z^m} - J_m(z) \log \left(\frac{z}{2} \right) \right\} \ ,
\tag{3.61}
$$

where J_m is the Bessel function of order m. Only the leading order term in this expansion is shown, the lower order contributions, $z^{-k}, k < m$, will become regular after multiplication by the appropriate coefficient from the kernel function. However, the integrable logarithm term is displayed, to indicate that this term should be pulled out and handled via the special quadrature discussed above.

With the techniques presented in the previous section, the limit definition and analytic integration process can likely be applied to any boundary integral formulation. An important example of a more complicated Green's function is that for three dimensional Laplace equation in axisymmetric geometries. This leads to a two dimensional formulation wherein the Green's functions are given in terms of complete elliptic integrals. Due to the usefulness of this formulation, this topic is covered separately in Chapter 6.

3.5 CORNERS

It has been shown that the Galerkin form of the hypersingular integral is well defined, and can be evaluated, without the C^1 constraint required with collocation. While this is the most important distinction between the two approximations, a second important advantage of Galerkin is in the treatment of boundary corners, and this section will describe the Galerkin corner/edge procedures.

In two dimensions, the corner is represented by the usual technique of a 'double node' pair, one node for each side of the corner, the coordinates of the two nodes being the same. The basic idea is to now exploit the flexibility provided by the choice of weight function in the outer P integration [71]. Each node in the pair has its own weight function, which is non-zero only on its side of the corner. Thus, corner nodes have effectively 'half' the weight function compared to a node where the surface is smooth. Obviously the situation in three dimensions – corners and edges – can be more complicated, but the basic idea will be the same.

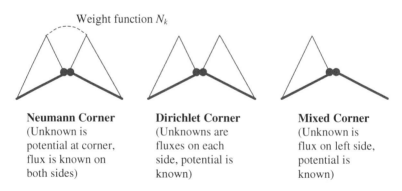

Figure 3.3 Corner treatment in the Galerkin BEM

With the weight functions defined as above, the Galerkin equations are first assembled as usual, and then adjusted appropriately depending upon the boundary conditions. First, if the potential is specified on both sides, then there are two flux values to be determined, and no further manipulation is required. Galerkin therefore provides two independent equations for solving for the two (different) unknown normal derivatives.

If flux is specified on one side and potential on the other, then the only unknown is the flux at the Dirichlet node, as the potential must be continuous at the corner. In this case the equation at the Neumann node can be ignored. Finally, if flux is specified on both sides, then the single unknown is the potential, which is the same on both sides. Hence, the two equations should be added, and now the weight function spans both sides of the corner. The three possibilities are illustrated in Figure 3.3.

3.6 NONLINEAR BOUNDARY CONDITIONS

An area where boundary integral methods have proven to be quite useful, for both scientific research and industrial applications, is electrochemistry [27, 292]. Part of the reason is that the boundary conditions on electrode surfaces are neither specified potential nor specified flux; rather a complicated nonlinear relationship between surface potential and flux (equivalently, current density) is given,

$$\mathcal{I} = \mathcal{F}(\phi) \,, \tag{3.62}$$

where the current density $\mathcal{I}$ is

$$\mathcal{I} = -\kappa \frac{\partial \phi}{\partial n} \,, \tag{3.63}$$

κ is the electrolyte conductivity, There are a number of immediate advantages of a boundary formulation for this type of system. First of all, the integral equation explicitly (*without* a numerical differentiation of the potential function) involves the normal derivative, and thus it is easier to couple with the boundary condition. Second, it is only necessary for the nonlinear iteration to converge to a solution on the boundary, rather than throughout the volume. Finally, as will be discussed in detail below, the highly local ('near' diagonal dominance) nature of the boundary integral matrices can be exploited to develop an effective iteration algorithm.

The electrode condition Eq.(3.62) is often referred to as *polarization*, and represents the details of the electrochemical reactions occurring near the electrode surfaces. Simultaneously solving the Laplace equation together with the nonlinear boundary conditions can be quite challenging. Although Eq.(3.62) is 'local' (point by point), all polarization values are very much coupled through the solution of the Laplace equation. Depending upon the geometry of the problem, this can be a very large coupled system, making each iteration computationally expensive. Moreover, depending upon the parameters of the system, the voltage/current relationship can be highly nonlinear. Thus, restricting the nonlinear iteration to just the boundary can be quite useful.

The electrochemical applications [21,198] that have been studied with boundary integral equations include the modeling and design of cathode protection systems [1,8,74,298], modeling of industrial electroforming operations [32,73,102], simulation of Scanning Electrochemical Microscopy [10,96,97,103] and the analysis of an electrospray emitter tube [281,282]. This emitter tube converts material in solution to gas phase ions that can be analyzed by a mass spectrometer, the physical set-up is shown in Fig. 3.4.

In the remainder of this section we present a nonlinear algorithm for a generic polarization function $\mathcal{F}$. The explicit definition of the polarization for each of the applications can be found in the cited papers.

Nonlinear algorithm. To describe the algorithm, label the electrode surface where Eq.(3.62) holds with a subscript E. We can then write the boundary integral matrix

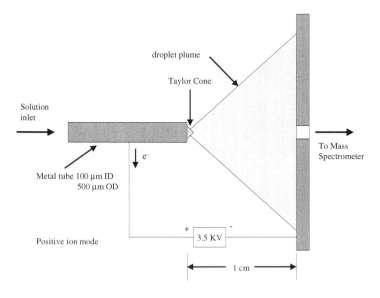

Figure 3.4 Electrospray mass spectrometry configuration.

equation with the electrode nodes listed first,

$$H \begin{bmatrix} \phi_E \\ \phi \end{bmatrix} = G \begin{bmatrix} \mathcal{I}_E \\ \mathcal{I} \end{bmatrix}, \tag{3.64}$$

where ϕ and $\mathcal{I}$ denote the potential and current density on the remainder of the boundary. Now rearrange the columns of this system depending upon which values of ϕ and $\mathcal{I}$ are known,

$$\tilde{H} \begin{bmatrix} \phi_E \\ b \end{bmatrix} = \tilde{G} \begin{bmatrix} \mathcal{I}_E \\ x \end{bmatrix}, \tag{3.65}$$

where b denotes the known boundary data and x the unknowns on the non-electrode part of the surface. Multiplying both sides by $\tilde{G}^{-1}$ results in the block matrix equation

$$\begin{bmatrix} \mathcal{I}_E \\ x \end{bmatrix} = \begin{bmatrix} A^{11} & A^{12} \\ A^{21} & A^{22} \end{bmatrix} \begin{bmatrix} \phi_E \\ b \end{bmatrix}, \tag{3.66}$$

where $A = \tilde{G}^{-1}\tilde{H}$. (Of course A is not computed in this fashion: the kth column of A, A_k, is obtained by solving $\tilde{G}A_k = \tilde{H}_k$, the right hand side being the kth column of $\tilde{H}$.) The rationale for reformulating the equations in this way is that it is now possible to write an equation for the current at anode node j

$$\mathcal{I}_j = \sum_k A^{11}_{jk} \phi_k + \sum_l A^{12}_{jl} b_l . \tag{3.67}$$

which does not involve any other anode current values. Moreover, note that the boundary integral matrices are, roughly speaking, 'local', in that influences from nodes away from node j are relatively unimportant. (This is a consequence of the Green's function dying off with distance.) It is therefore likely that A will share this property, and this can be exploited by subtracting out the diagonal term in

Eq.(3.67),

$$\mathcal{I}_j - A_{jj}^{11} \, \phi_j = \sum_{k \neq j} A_{jk}^{11} \, \phi_k + \sum_l A_{jl} \, b_l \, . \tag{3.68}$$

The potential ϕ_j can now be replaced by its equivalent in terms of current,

$$\phi_j = \mathcal{F}(\mathcal{I}_j) \, , \tag{3.69}$$

and we obtain

$$\mathcal{I}_j - A_{jj}^{11} \, \mathcal{F}(\mathcal{I}_j) = \sum_{k \neq j} A_{jk}^{11} \, \phi_k + \sum_l A_{jl} \, b_l \, . \tag{3.70}$$

Note that the vector

$$\sum_l A_{jl}^{12} \, b_l \, . \tag{3.71}$$

that appears in Eq.(3.70) is obviously independent of the iteration, and is therefore computed once and saved. The virtue of this form is that, treating all electrode potentials except at node j as known quantities, this single variable *nonlinear* equation can be solved for $\mathcal{I}_j$ and ϕ_j (for example, by a simple bisection algorithm). Cycling through the anode nodes, continually updating the right hand side with the new potentials, yields a Gauss-Seidel type iteration in which the latest information for ϕ^A is employed in solving the next nonlinear equation. A second benefit of the reformulation in Eq.(3.66) is that the iteration can proceed by dealing solely with the electrode nodes: with Eq.(3.66), the solution for x (if needed) automatically falls into place once the values on the electrode have converged. The solution is considered converged when the maximum change in ϕ_j, $1 \leq j \leq M_A$, from one iteration to the next is less than a specified tolerance.

3.7 CONCLUDING REMARKS

The primary focus of this chapter has been the development of algorithms for evaluating two-dimensional singular Galerkin integrals. Hybrid analytical/numerical procedures based upon the boundary limit definition have been presented, the limit providing a mathematically sound and physically sensible definition of *all* singular integrals. The algorithms are, moreover, completely general, valid for arbitrary interpolation or Green's function. The task of carrying out the analytic integrations and the limit process, admittedly onerous by hand, is conveniently automated with symbolic computation.

CHAPTER 4

THREE DIMENSIONAL ANALYSIS

Synopsis: The primary endeavor, as in the previous chapter, is to develop effective algorithms for the evaluation of singular integrals. The singular cases now come in three flavors – when the inner and outer elements are the same, or when they share a common edge or a common vertex. The boundary limit analysis for surface integrals is necessarily lengthier, but nevertheless involves nothing more complicated than polar coordinate transformations and interchanging orders of integration. The coincident and adjacent edge hypersingular integrals, as can be expected from the previous chapter, are not separately finite, and the key task will again be to identify the divergent terms and to show that they cancel. One new feature of the algorithms is that, after the limit $\varepsilon \to 0$ has been achieved, it will be necessary to execute another analytic integration to obtain an expression that can be easily computed numerically. The chapter concludes with a discussion of the boundary integral formulation for anisotropic elasticity. Despite the fact that the anisotropic Green's function contains an angular component which is not known in simple closed form, it is shown that the boundary limit process, suitably modified, can still be applied.

4.1 PRELIMINARIES

As in the discussion for two dimensions, the methods for singular integration form the core of this chapter. The algorithms in three dimensions will be applicable to most, if not all, boundary integral formulations, and thus the techniques are first presented for the simplest possible setting, the solution of the Laplace equation based upon a linear element approximation. The extension to higher order interpolation then follows along the same lines as in Chapter 3, namely splitting of the integrand into singular and non-singular components. The additional complication in three dimensions is that the power of the distance r is now a half-integer instead of an integer, and as a consequence, the splitting will require a bit more algebra. To justify the above statement about the general applicability of the boundary limit approach, we then consider the opposite extreme, the integration of the Green's function for anisotropic elasticity.

We begin with a recapitulation of the three dimensional formulation presented in Chapter 2. The exterior limit boundary integral equations for surface potential and flux take the form

$$\mathcal{P}(P) \equiv \lim_{\varepsilon \to 0} \int_\Sigma \left[\phi(Q) \frac{\partial G}{\partial \mathbf{n}}(P_\varepsilon, Q) - G(P_\varepsilon, Q) \frac{\partial \phi}{\partial \mathbf{n}}(Q) \right] dQ = 0 , \qquad (4.1)$$

and

$$\mathcal{F}(P) \equiv \lim_{\varepsilon \to 0} \int_\Sigma \left[\phi(Q) \frac{\partial^2 G}{\partial \mathbf{N} \partial \mathbf{n}}(P_\varepsilon, Q) - \frac{\partial G}{\partial \mathbf{N}}(P_\varepsilon, Q) \frac{\partial \phi}{\partial \mathbf{n}}(Q) \right] dQ = 0 , \qquad (4.2)$$

where Σ is the boundary surface of $\mathcal{V}$ and $P_\varepsilon = P + \varepsilon \mathbf{N}(P)$. As in two dimensions, $\mathbf{n}(Q)$ and $\mathbf{N}(P)$ are the unit outward normals, and P_ε are exterior points. The decision to work with an exterior limit is simply a matter of convenience, inasmuch as it avoids the 'free term' outside the integral that is present in the interior limit. The geometry of the limit process is illustrated in Figure 4.1.

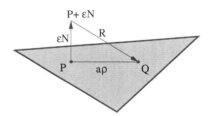

Figure 4.1 The three points Q, P, and $P_\varepsilon = P + \varepsilon \mathbf{N}$ that enter into the boundary limit analysis. The distance $a\rho$ will appear shortly in the coincident analysis.

To complete the specification of these two equations, the fundamental solution $G(P, Q)$ and its derivatives are

$$\begin{aligned}
G(P, Q) &= \frac{1}{4\pi} \frac{1}{r} \\
\frac{\partial G}{\partial \mathbf{n}}(P, Q) &= -\frac{1}{4\pi} \frac{\mathbf{n} \bullet R}{r^3} \\
\frac{\partial G}{\partial \mathbf{N}}(P, Q) &= \frac{1}{4\pi} \frac{\mathbf{N} \bullet R}{r^3} \\
\frac{\partial^2 G}{\partial \mathbf{N} \partial \mathbf{n}}(P, Q) &= \frac{1}{4\pi} \left(\frac{\mathbf{n} \bullet \mathbf{N}}{r^3} - 3 \frac{(\mathbf{n} \bullet R)(\mathbf{N} \bullet R)}{r^5} \right) ,
\end{aligned} \qquad (4.3)$$

where $R = Q - P$, and $r = \|R\|$ is the distance between P and Q.

For the approximation of Eq.(4.1) and Eq.(4.2) via a linear triangular element, the interpolations are conveniently expressed in terms of the shape functions

$$\psi_1(\eta,\xi) = \frac{\sqrt{3}(1-\eta) - \xi}{2\sqrt{3}}$$
$$\psi_2(\eta,\xi) = \frac{\sqrt{3}(1+\eta) - \xi}{2\sqrt{3}} \quad (4.4)$$
$$\psi_3(\eta,\xi) = \frac{\xi}{\sqrt{3}},$$

where the parameter space $\{\eta,\xi\}$ is the equilateral triangle $\{\eta,\xi\}$, $-1 \leq \eta \leq 1$, $0 \leq \xi \leq \sqrt{3}(1-|\eta|)$ (Fig. 4.2). The shape functions are defined by the property that $\psi_j(v_k) = \delta_{kj}$, $1 \leq j,k \leq 3$, where $\{v_k\}$ are the three vertices of the equilateral triangle, $v_1 = (-1,0)$, $v_2 = (1,0)$, and $v_3 = (0,\sqrt{3})$.

The choice of parameter space is of course more or less arbitrary. This equilateral triangle has been selected for one reason, it turns out to be very convenient for the coincident integration, Section 4.2.2.

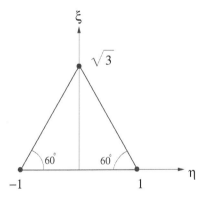

Figure 4.2 Isoparametric equilateral triangular linear element in $\{\eta,\xi\}$ space, where $-1 \leq \eta \leq 1$, $0 \leq \xi \leq \sqrt{3}(1-|\eta|)$.

For the Galerkin approximation, the Green's function is integrated over outer variable P and inner variable Q, and thus two separate parameter spaces are required. The parametric variables for the outer P integration will be denoted by (η,ξ), and that for Q by (η^*,ξ^*). Thus, for a triangular patch on the surface defined by the three nodal points $\{Q_j\}$, $Q_j = (x_j, y_j, z_j)$, the linear element interpolation of the surface is given by

$$Q(\eta^*,\xi^*) = \sum_{j=1}^{3} \psi_j(\eta^*,\xi^*)(x_j,y_j,z_j) \quad (4.5)$$
$$= \left(\sum_{j=1}^{3} x_j \psi_j(\eta^*,\xi^*), \sum_{j=1}^{3} y_j \psi_j(\eta^*,\xi^*), \sum_{j=1}^{3} z_j \psi_j(\eta^*,\xi^*) \right)$$

with Jacobian

$$J(\eta^*,\xi^*) = \left\| \frac{\partial Q(\eta^*,\xi^*)}{\partial \eta^*} \times \frac{\partial Q(\eta^*,\xi^*)}{\partial \xi^*} \right\|, \quad (4.6)$$

the cross product being the un-normalized outward normal vector, $J(\eta^*, \xi^*)\mathbf{n}(Q)$. Similar expressions hold for $P = P(\eta, \xi)$. For the linear element defined above, this Jacobian is, as in two dimensions, a constant.

The approximations of the boundary functions, potential and flux, are also conveniently accomplished as parametric mappings using the shape functions,

$$\phi(\eta^*, \xi^*) \equiv \phi(Q(\eta^*, \xi^*)) = \sum_{j=1}^{3} \psi_j(\eta^*, \xi^*)\phi(Q_j)$$

$$\frac{\partial \phi}{\partial \mathbf{n}}(\eta^*, \xi^*) \equiv \frac{\partial \phi}{\partial \mathbf{n}}(Q(\eta^*, \xi^*)) = \sum_{j=1}^{3} \psi_j(\eta^*, \xi^*)\frac{\partial \phi}{\partial \mathbf{n}}(Q_j) , \qquad (4.7)$$

where $\phi(Q_j)$ and $\partial \phi/\partial \mathbf{n}(Q_j)$ are the values of the functions at the nodal points.

In a Galerkin approximation, the reduction to a finite system of equations is accomplished by using the shape functions as weight functions, enforcing the integral equations Eq.(4.1) and Eq.(4.2) 'on average'. More precisely, the potential and flux equations are solved numerically in the form

$$\int_{\Sigma} \hat{\psi}_k(P)\mathcal{P}(P)\,dP = 0$$
$$\int_{\Sigma} \hat{\psi}_k(P)\mathcal{F}(P)\,dP = 0. \qquad (4.8)$$

If there are M nodes defining the boundary, the needed M equations are generated by selecting M different weight functions. For the equation associated with a particular boundary node P_k, the weight function $\hat{\psi}_k(P)$ consists of all shape functions $\psi_l(P)$ that are nonzero at the node P_k. The weight function $\hat{\psi}_k(P)$ therefore has limited support, being non-zero only on the elements containing P_k. This is illustrated in Figure 4.3.

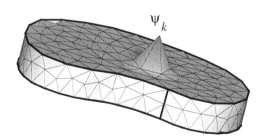

Figure 4.3   Galerkin weight functions for 3D analysis.

To obtain an explicit computational form of the boundary integral equation, first use the decomposition of the boundary as a sum of elements to write the potential equation as

$$\sum_{E_P, E_Q} \int_{E_P} \hat{\psi}_k(P) \int_{E_Q} \left[\phi(Q) \frac{\partial G}{\partial \mathbf{n}}(P_\varepsilon, Q) - G(P_\varepsilon, Q) \frac{\partial \phi}{\partial \mathbf{n}}(Q) \right] \mathrm{d}Q \, \mathrm{d}P = 0 , \quad (4.9)$$

the limit $\varepsilon \to 0$ being understood. Incorporating next the parametric interpolations and the attendant jacobians, the Galerkin equation becomes

$$\sum_{E_P, E_Q} \sum_{k,j} \phi(Q_j) \int \psi_k(\eta, \xi) J_P \int \psi_j(\eta^*, \xi^*) J_Q \frac{\partial G}{\partial \mathbf{n}}(P_\varepsilon, Q) \, \mathrm{d}\xi^* \mathrm{d}\eta^* \, \mathrm{d}\xi \mathrm{d}\eta \quad (4.10)$$

$$- \frac{\partial \phi}{\partial \mathbf{n}}(Q_j) \int \psi_k(\eta, \xi) J_P \int \psi_j(\eta^*, \xi^*) J_Q G(P_\varepsilon, Q) \, \mathrm{d}\xi^* \mathrm{d}\eta^* \, \mathrm{d}\xi \mathrm{d}\eta = 0 ,$$

where, for brevity, indications of the parametric surface mappings $Q = Q(\eta^*, \xi^*)$ and $P = P(\eta, \xi)$ have been omitted. The jacobians will in general also be a function of position, but as noted above, are constant for a linear element. The flux equation can obviously be written similarly.

Incorporating all of the boundary approximations, this is a somewhat lengthy expression. However, it does clearly show how the approximate solution is computed. The parameter space integrals involve only known quantities, the Green's function and the shape functions, and can therefore be evaluated. These numbers depend only on the geometry Σ, and contribute to the system matrix equation, multiplying either the potential or the flux vector, as shown in Eq.(4.10). Once the matrices are assembled, the problem can be solved by applying the boundary conditions and solving the linear equations for the unknown quantities.

This process would of course be utterly trivial if the Green's function and its derivatives were well behaved functions, Gauss quadrature rules could be applied to evaluate all integrals. However, the parameter space integrals have to be computed for every pair of elements $\{E_P, E_Q\}$ and singular integrals will arise whenever Q can be equal to P. For three dimensional problems, there are four possible configurations for the two elements containing source point P and field point Q, as shown in Figure 4.4:

- *Non-singular case*: the source element E_P and field element E_Q have null intersection;

- *Coincident case*: $E_P = E_Q$;

- *Edge-adjacent case*: E_P and E_Q share a common edge;

- *Vertex-adjacent case*: E_P and E_Q share only a common vertex.

We now begin the discussion of the singular cases.

4.2 LINEAR ELEMENT ANALYSIS

Based upon the two dimensional analysis, the singular integrals of the various kernel functions can be expected to behave in the following manner. The Green's function itself, although singular, is an integrable function and can be evaluated without a boundary limit process. In the integration of the first derivative (Cauchy

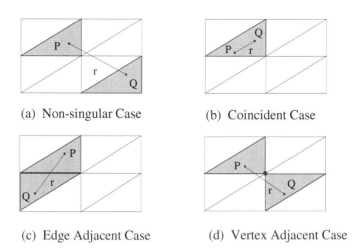

Figure 4.4 Four different cases considered for integration: (a) non-singular; (b) coincident; (c) edge adjacent; and (d) vertex adjacent.

Principal Value) integral, the limit analysis of the coincident integral will, at least initially, produce a divergent $\log(\varepsilon^2)$ term. However, this term will self-cancel after integrating, with respect to an angle variable, completely around the singular point. (Essentially the same thing happens in two dimensions, but this gets lost in the very simple complete analytic integration of this term.) Thus, this integral is finite, as are the adjacent edge and vertex integrals, these latter integrals again not requiring a limit definition. For the hypersingular kernel, neither the coincident nor the adjacent edge is finite, $\log(\varepsilon^2)$ contributions appearing in both, and it will be necessary to first exhibit the divergent terms, and then to demonstrate that these divergences cancel when the complete integral is assembled. The hypersingular vertex integral will be seen to be finite.

As has been repeatedly emphasized, one theme of this book is that the process for evaluating the singular integrals is the same for all kernel functions. Having already seen this in detail in two dimensions, here it suffices to focus on the most difficult case, the integration of the hypersingular kernel, the techniques apply equally well to G and its first derivatives. However, a discussion of the coincident CPV integral, and the disappearance of the $\log(\varepsilon^2)$ term, is also warranted, as this is somewhat different from two dimensions. First, we very briefly discuss the nonsingular integration.

4.2.1 Nonsingular Integration

For the non-singular integration, a two-dimensional Gauss quadrature rule is simple and convenient, for both inner and outer integrals. Thus, the integral over the equilateral parameter space is accomplished as

$$\int_{-1}^{1} \int_{0}^{\sqrt{3}(1-|\eta|)} f(\eta^*, \xi^*) \, d\xi^* \, d\eta^* \approx \sum_{l=1}^{n_g} \omega_l f(\eta^*_l, \xi^*_l) \,, \qquad (4.11)$$

where $\{(\eta^*_l, \xi^*_l)\}$ and ω_l are the Gauss points and weights for an n_g-point rule. Tables for $n_g = 3, 7, 12$ are given in the Appendix.

4.2.2 Coincident Integration

For $E_P = E_Q = E$, the coincident hypersingular integral is

$$\int_E \psi_k(P) \int_E \phi(Q) \frac{\partial^2 G}{\partial \mathbf{N} \partial \mathbf{n}}(P_\varepsilon, Q) \, \mathrm{d}Q \, \mathrm{d}P = \tag{4.12}$$

$$\sum_{j=1}^3 \phi(Q_j) \int_E \psi_k(P) \int_E \psi_j(Q) \frac{\partial^2 G}{\partial \mathbf{N} \partial \mathbf{n}}(P_\varepsilon, Q) \, \mathrm{d}Q \, \mathrm{d}P \,,$$

where E is defined by nodes P_k, $1 \le k \le 3$. Transferring the integral to parameter space requires including the (constant) Jacobian J_P ($J_Q = J_P$), conveniently incorporated into the hypersingular kernel,

$$J_P^2 \frac{\partial^2 G}{\partial \mathbf{N} \partial \mathbf{n}}(P, Q) = \frac{1}{4\pi} \left(\frac{J_P^2}{r^3} - 3 \frac{(J_P N \cdot R)^2}{r^5} \right). \tag{4.13}$$

In parameter space Eq.(4.12) is a four dimensional integral. The ultimate goal is to use analytic integration to reduce this to a *one-dimensional nonsingular* integral that can be evaluated numerically. The first two analytic integrations will produce the divergent $\log(\varepsilon^2)$ term, plus a finite integral. This finite integral however will contain a logarithmic singularity, and as a result accurate numerical evaluation is difficult. The final analytic integration is employed to remove this weak singularity and simplify the numerical evaluation. Maple codes for performing these operations are provided in Appendix C.

4.2.2.1 1st Analytic Integration As mentioned above, the parametric variables for the outer P integration will be denoted by (η, ξ), and that for Q by (η^*, ξ^*). For the inner Q integration, considering for the moment P to be fixed, the first step is to define a polar coordinate system $\{\rho, \theta\}$ centered at the singular point $P = (\eta, \xi)$,

$$\begin{aligned} \eta^* - \eta &= \rho \cos(\theta) \\ \xi^* - \xi &= \rho \sin(\theta) \end{aligned} \tag{4.14}$$

as illustrated in Fig. 4.5. Polar coordinate transformations centered at the singular point are highly effective: the jacobian of the transformation $\rho \, \mathrm{d}\rho$ reduces the order of the singularily, and it is generally possible to integrate ρ analytically. These transformations will therefore be repeatedly exploited for all integrations in three dimensions.

The ρ integration is over $0 \le \rho \le \rho_L(\theta)$, and as the expression for the upper limit is different as θ traverses each edge of the parametric triangle, the (ρ, θ) integration must be split into the three sub-triangles shown in Figure 4.5. In the following we carry out the calculation for the lower (shaded) subtriangle associated with the edge $\xi^* = 0$, defined via $\Theta_1 \le \theta \le \Theta_2$,

$$\begin{aligned} \Theta_1 &= -\frac{\pi}{2} + \tan^{-1}\left(\frac{1 - \eta}{\xi} \right) \\ \Theta_2 &= -\frac{\pi}{2} - \tan^{-1}\left(\frac{1 + \eta}{\xi} \right) \end{aligned} \tag{4.15}$$

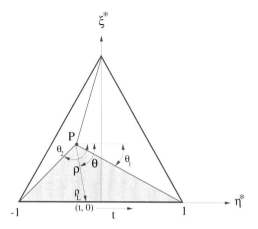

Figure 4.5 First polar coordinate transformation, $\{\eta^*, \xi^*\} \to \{\rho, \theta\}$, for the coincident integration. The variable t eventually replaces θ.

and for this sub-triangle

$$\rho_L(\theta) = -\frac{\xi}{\sin(\theta)} \ . \tag{4.16}$$

Although the remaining two cases could be handled in exactly the same manner, this would require repeating the analysis and generating different analytic formulas for each sub-triangle. An alternate procedure, which has the benefit of simplifying the implementation, is to exploit the symmetry of the equilateral parameter space – the reason for choosing this parameter domain. The remaining two sub-triangles are handled by rotating the Q element and employing the formulas for the lower subtriangle. Compared to computing the integrals over all three sub-triangles at the same time, this does create some small additional computational overhead. However, overall the coincident integrals are not a major contributor to the computational cost, and the benefits of simplicity outweigh this consideration.

With the polar coordinate transformation the vector $R = Q - P$ takes the form

$$R = (a_1 \rho - \varepsilon N_1, a_2 \rho - \varepsilon N_2, a_3 \rho - \varepsilon N_3) \ . \tag{4.17}$$

The vector of coefficients $\mathbf{a} = (a_1, a_2, a_3)$ is a tangent vector, $\mathbf{a} \cdot \mathbf{N} = 0$, and moreover

$$a_l(\theta) = a_l^c \cos(\theta) + a_l^s \sin(\theta) \tag{4.18}$$

where the coefficients $\{a_l^c, a_l^s\}$ depend solely on the nodal coordinates $\{x_j, y_j, z_j\}$. The distance $r = \|R\|$ therefore takes the simple form

$$r^2(\rho, \theta) = \varepsilon^2 + a^2(\theta)\rho^2 \ , \tag{4.19}$$

where $a^2 = \mathbf{a} \cdot \mathbf{a} > 0$.

With Eq.(4.14), the shape function $\psi_j(Q)$ becomes a linear function of ρ,

$$\psi_j(\rho, \theta) = c_{j,0}(\eta, \xi) + c_{j,1}(\eta, \xi, \theta)\rho \ , \tag{4.20}$$

where $c_{j,0}(\eta, \xi) = \psi_j(P)$ and

$$c_{j,1}(\eta, \xi, \theta) = c_j^c(\eta, \xi) \cos(\theta) + c_j^s(\eta, \xi) \sin(\theta) \ , \tag{4.21}$$

LINEAR ELEMENT ANALYSIS **77**

and again the coefficients $\{c_j^c(\eta, \xi), c_j^s(\eta, \xi)\}$ depend upon the nodal coordinates. To simplify the expressions that follow, the arguments will be dropped and the coefficients denoted simply as a^2, $c_{j,0}$ and $c_{j,1}$. Thus, employing the boundary limit procedure and expressing the kernel function in polar coordinates, Eq.(4.12) becomes, for $m = 0, 1$,

$$\frac{J_P^2}{4\pi} \sum_{m=0}^{1} \int_{-1}^{1} d\eta \int_{0}^{\sqrt{3}(1-|\eta|)} \psi_k(\eta, \xi) d\xi \int_{\Theta_1}^{\Theta_2} c_{j,m} d\theta$$

$$\int_{0}^{\rho_L} \rho^{m+1} \left(\frac{1}{(a^2\rho^2 + \varepsilon^2)^{3/2}} - 3\frac{\varepsilon^2}{(a^2\rho^2 + \varepsilon^2)^{5/2}} \right) d\rho , \qquad (4.22)$$

the extra factor of ρ coming from the jacobian of polar coordinates. The innermost ρ integrals are easily evaluated analytically. The more singular term, $m = 0$ is

$$F_0 = -\frac{\rho_L^2}{(\varepsilon^2 + a^2 \rho_L^2)^{3/2}} , \qquad (4.23)$$

and for the moment, nothing more need be said about this expression other than it is not yet permissible to set $\varepsilon = 0$.

For $m = 1$ the result is

$$F_1 = -\frac{1}{a^3} \left[\log(\varepsilon^2)/2 - \log\left(a\rho_L + \sqrt{\varepsilon^2 + a^2\rho_L^2} \right) + \frac{2a^3\rho_L^3 + \varepsilon^2 a\rho_L}{(\varepsilon^2 + a^2\rho_L^2)^{3/2}} \right] , \qquad (4.24)$$

and we now proceed to simplify this expression.

In the first place, the $\log(\varepsilon^2)$ term that appears in F_1 is *not* the divergent term that is being sought, it will in fact vanish in the integration over θ. To see this, first note that this term is independent of ρ_L, and thus the θ integration of this term can be carried out for $[0, 2\pi]$. Incorporating the $(m = 1)$ coefficient from the shape function the θ integral is

$$-\frac{\log(\varepsilon^2)}{2} \int_{0}^{2\pi} \frac{c_{j,1}(\eta, \xi, \theta)}{a^3} d\theta . \qquad (4.25)$$

However, the coeffecent $c_{j,1}(\eta, \xi, \theta)$ is linear in $\cos(\theta)$ and $\sin(\theta)$, and therefore satisfies $c_{j,1}(\eta, \xi, \pi + \theta) = -c_{j,1}(\eta, \xi, \theta)$. On the other hand, from Eq.(4.18), and recalling that a is always positive, $a(\pi + \theta) = a(\theta)$. As a consequence, the integral in Eq.(4.25) is zero, and the $\log(\varepsilon^2)$ term in Eq.(4.24) can therefore be removed from F_1.

The same argument shows that the contribution from the last term

$$-\frac{1}{a^3}\frac{2a^3\rho_L^3}{(\varepsilon^2 + a^2\rho_L^2)^{3/2}} = -\frac{2}{a^3} + \left(\frac{2}{a^3} - \frac{1}{a^3}\frac{2a^3\rho_L^3}{(\varepsilon^2 + a^2\rho_L^2)^{3/2}} \right) \qquad (4.26)$$

also vanishes in the integration over θ. The term $2/a^3$ is likewise independent of ρ_L and goes away for the same reasons as the $\log(\varepsilon^2)$ term, while the remainder disappears in the limit $\varepsilon \to 0$.

The final formula for the $m = 1$, ρ integration is then

$$F_1 = \frac{\log(2a\rho_L)}{a^3} - \frac{1}{a^3}\frac{\varepsilon^2 a\rho_L}{(\varepsilon^2 + a^2\rho_L^2)^{3/2}} . \qquad (4.27)$$

78 THREE DIMENSIONAL ANALYSIS

Roughly speaking this term behaves as ε^2/ρ_L^2 and it is not immediately clear what happens in the limit, as $\rho_L \approx 0$ occurs in the P integration. It will turn out that the ε^2 factor dominates and this term will eventually vanish, but there is no harm in keeping it for the subsequent integrations.

It is worth noting that, if instead of integrating the entire hypersingular kernel, the r^{-3} and r^{-5} fractions were treated separately, it would be seen that ε^{-1} terms appear. This is not surprising, as both fractions behave as ρ^{-2}. The self-cancellation of ε^{-1} contributions is typical for hypersingular integration.

4.2.2.2 2nd *Analytic Integration*

The first analytic integration has failed to display the expected divergent term. However, when $\xi \approx 0$, it follows that $\rho_L \approx 0$, and the expressions for F_0 and F_1 are both singular at $\rho_L = 0$, with of course F_0 being the more dangerous. (This is for the lower subtriangle in the Q integration; for the other two cases, singularities would appear as P approaches the respective edge). The weak (integrable) singularity in F_1 will obviously not produce a divergent term, but for numerical implementation it is clearly beneficial to integrate $\log(\rho_L)$ analytically. For F_0 however, the behavior is ρ_L^{-1}, and this will produce a $\log(\varepsilon^2)$ contribution upon integration; in the following we therefore consider only $m = 0$, as $m = 1$ is handled similarly.

The remaining integral is over the parameters $\{\eta, \xi, \theta\}$. As just noted, the only source of trouble is $\xi = 0$, the dependence of the integrand on θ being complicated but harmless. As θ cannot be disposed of by an analytic integration, it must be moved out of the way to allow an integration of ξ. However, the needed interchange in the order of integrations is impeded by the fact that the limits of the θ integral, Θ_1 and Θ_2, Eq.(4.15), depend on η and ξ.

To manuever around this, we change variables, θ replaced by t defined as

$$\theta = -\frac{\pi}{2} + \tan^{-1}(\frac{t - \eta}{\xi}) ,$$

$$\frac{d\theta}{dt} = \frac{\xi}{\xi^2 + (t - \eta)^2} . \tag{4.28}$$

As indicated in Figure 4.5, for a value of θ, t is the 'end-point' $(t, 0)$ of ρ on the ξ^*-axis; the virtue of t is that $-1 \le t \le 1$, independent of η and ξ. Making this change and then interchanging the order of integration, Eqs. (4.22) and (4.23) for $m = 0$ now becomes $(c_{j,0} = \psi_j(\eta, \xi))$

$$-\frac{J_P^2}{4\pi} \int_{-1}^{1} d\eta \int_{-1}^{1} dt \int_{0}^{\sqrt{3}(1-|\eta|)} \psi_k(\eta, \xi) \, \psi_j(\eta, \xi) \frac{\rho_L^2}{(\varepsilon^2 + a^2 \rho_L^2)^{3/2}} \, d\xi . \tag{4.29}$$

The singularity is at $\rho_L = 0$, and with the change of variables in Eq.(4.16), $\rho_L = (\xi^2 + (t - \eta)^2)^{1/2}$. Thus, the singularity is at $t = \eta$, $\xi = 0$, and this once again suggests polar coordinates $\{\Lambda, \Psi\}$ to replace $\{t, \xi\}$,

$$t = \Lambda \cos(\Psi) + \eta$$

$$\xi = \Lambda \sin(\Psi) , \tag{4.30}$$

the singular point becoming simply $\Lambda = 0$. The next analytic integration will obviously be with respect to Λ. The combined changes of variables, $\theta \to t$ and $\{t, \xi\} \to \{\Lambda, \Psi\}$ is remarkably simple: $d\theta/dt$ becomes $\sin(\Psi)/\Lambda$, and the Λ factor cancels with the jacobian from the polar coordinate transformation. The total jacobian is simply $\sin(\Psi)$. Moreover, $\cos(\theta)$ becomes $\cos(\Psi)$ and $\sin(\theta)$ becomes $-\sin(\Psi)$. Thus, $a(\theta)$, Eq.(4.18), is simply $a(\Psi)$ (with an appropriate change of sign) and is a constant as far as the Λ integration is concerned.

As shown in Fig. 4.6, the $\{t,\xi\}$ domain is a rectangle, $-1 < t < 1$, $0 < \xi < \sqrt{3}(1-|\eta|)$, and integrating over $\{\Lambda,\Psi\}$ will necessitate a decomposition into three subdomains $0 \le \Psi \le \Psi_1$, $\Psi_1 \le \Psi \le \Psi_2$, and $\Psi_2 \le \Psi \le \pi$, where

$$\Psi_1 = \tan^{-1}\left(\frac{\sqrt{3}(1-|\eta|)}{1-\eta}\right)$$

$$\Psi_2 = \pi - \tan^{-1}\left(\frac{\sqrt{3}(1-|\eta|)}{1+\eta}\right). \quad (4.31)$$

In addition, $0 < \Lambda < \Lambda_L$, where Λ_L depends upon the particular subtriangle.

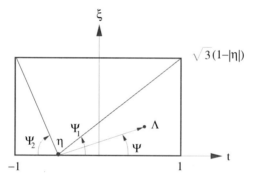

Figure 4.6 Geometry of the second polar coordinate transformation, $\{t,\xi\} \to \{\Lambda,\Psi\}$, for the coincident integration.

Finally, with $\xi = \Lambda\sin(\Psi)$, the shape functions are linear in Λ and the product is quadratic,

$$\psi_k(\eta,\xi)\psi_j(\eta,\xi) = \sum_{s=0}^{2} d_{kjs}(\eta,\Psi)\Lambda^s \quad (4.32)$$

and the integrals to be evaluated are of the form

$$-\frac{J_P^2}{4\pi}\int_{-1}^{1} d\eta \int_{\Psi_i}^{\Psi_{i+1}} d_{kjs}(\eta,\Psi)\sin(\Psi)\,d\Psi \int_0^{\Lambda_L} \Lambda^s \frac{\Lambda^2}{(\varepsilon^2 + a^2\Lambda^2)^{3/2}}\,d\Lambda \quad (4.33)$$

where $\Psi_0 = 0$, $\Psi_3 = \pi$, $i = 0,1,2$ and $s = 0,1,2$. For $s > 0$ it is permissible to immediately set $\varepsilon = 0$, and the Λ integration is clearly trivial. Thus, we now focus solely on $s = 0$.

Note that the coefficients in this case are the products of shape functions evaluated at $\xi = 0$, $d_{kj0} = \psi_k(\eta,0)\psi_j(\eta,0)$, and

$$\psi_1(\eta,0) = \frac{1-\eta}{2} \qquad \psi_2(\eta,0) = \frac{1+\eta}{2} \qquad \psi_3(\eta,0) = 0 . \quad (4.34)$$

Thus, as is appropriate for the lower subtriangle in Fig. 4.5, there will be no singular term associated with the node not on the edge $\eta^* = 0$ (this will of course

80   THREE DIMENSIONAL ANALYSIS

cycle appropriately when the Q element is rotated to consider the two remaining subtriangles). Carrying out the Λ integration, we find a finite contribution

$$J_P^2 \sin(\Psi) \frac{1 - \log(2a\Lambda_L)}{a^3} \tag{4.35}$$

plus divergent terms

$$L_{kj}^c = \log(\varepsilon^2) \frac{J_P^2}{8\pi} \int_{-1}^{1} \psi_k(\eta, 0)\, \psi_j(\eta, 0)\, \mathrm{d}\eta \int_0^{\pi} \frac{\sin(\Psi)}{a^3}\, \mathrm{d}\Psi \ . \tag{4.36}$$

Note that the divergent term is, appropriately, independent of Λ_L (the singularity being at $\Lambda = 0$), and thus the Ψ integral can be written over the full domain $[0, \pi]$ without subdivision. Moreover, as $a = a(\Psi)$ is independent of η, this expression simplifies to

$$L_{kj}^c = \log(\varepsilon^2) \frac{J_P^2}{8\pi} \frac{1 + \delta_{kj}}{3} \int_0^{\pi} \frac{\sin(\Psi)}{a^3}\, \mathrm{d}\Psi \ , \tag{4.37}$$

where δ_{kj} is the usual Kronecker delta function and $1 \leq k, j \leq 2$.

Note that a mixing of both inner and outer integrations was required to expose the $\log(\varepsilon^2)$ contribution, *i.e.*, simply executing the Q-integral completely would not work. It is therefore expected that the same will be true for the edge adjacent integral. We must postpone a further discussion of this divergent term until section 4.2.6, following the analysis of the edge integrals.

At first look the reader may have found this to be a long and arduous journey, so it should be noted that quite a bit has been accomplished: a complicated, four dimensional, *divergent* integral has been reduced to a finite, computable, two dimensional ($\{\eta, \Psi\}$) integral, together with an explicit divergent term. Once it is shown that the $\log(\varepsilon^2)$ contribution cancels with the edge calculation, the complete hypersingular integral will have been shown to be a mathematically well defined quantity, and appropriate formulas for the evaluation of its constituent parts will have been developed.

4.2.2.3 3rd Analytic Integration

Although the primary goal, explicit separation of the divergent term $\log(\varepsilon^2)$ term, has been achieved, a small difficulty remains *for the hypersingular integral.*

For $\eta \to \pm 1$ the height of the rectangle in Fig. 4.6 will shrink to zero, and as a consequence, Λ_L can also become small. Thus, the part of Eq.(4.35) that depends upon $\log(\Lambda_L)$ is weakly singular at $\eta = \pm 1$. Direct evaluation via Gauss quadrature is inefficient, as a large number of Gauss points would be required to get an accurate value. Moreover, the use of specialized Gauss rules for a logarithmic singularity are in this case difficult to apply, as the integrands are not in the explicit form $\log(|\eta|)$.

A solution is to proceed further with analytic evaluation, integrating the $\log(\Lambda_L)$ term exactly. However, as the difficulties are now associated with $\eta = \pm 1$, it will once again be necessary to interchange the order of integration to move η from second to first. In this case however there is nothing especially complicated about this process, it is just elementary calculus, and thus we present the steps as briefly as possible.

It should also be noted that the re-ordering and analytic integration of η is not absolutely essential for G, $\partial G/\partial \mathbf{n}$, or for that matter the parts of the hypersingular integral that do not depend upon $\log(\Lambda_L)$. Thus, whether or not to carry out this

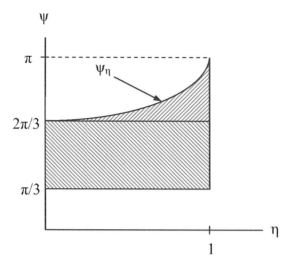

Figure 4.7 The $\{\eta, \Psi\}$ domain for $\eta \geq 0$.

last analytic step for these components is a matter of choice. The main argument in favor is that it reduces computational effort, leaving a one dimensional integral for the numerical quadrature instead of a double integral. As long as this process is being implemented for $\log(\Lambda_L)$, it is just as easy to then include all integrals. The argument against is that, for the non-weakly singular terms, it is just simpler to leave them as a double integral.

The interchange of order requires the separate consideration of $\eta > 0$ and $\eta < 0$. For $\eta \geq 0$, $\Psi_1 = \pi/3$ and the subdivision of the Ψ integral and the expressions for Λ_L are

$$\begin{array}{ll} 0 \leq \Psi \leq \frac{\pi}{3} & \Lambda_L = (1-\eta)/\cos(\Psi) \\ \frac{\pi}{3} \leq \Psi \leq \Psi_2 & \Lambda_L = \sqrt{3}(1-\eta)/\sin(\Psi) \\ \Psi_2 \leq \Psi \leq \pi & \Lambda_L = -(1+\eta)/\cos(\Psi) \end{array} \quad (4.38)$$

where $\Psi_2(\eta) = \pi - \tan^{-1}(\sqrt{3}(1-\eta)/(1+\eta))$. Consulting Fig. 4.7, the interchanges for the three Ψ integrals are easily seen to be

$$\int_0^1 d\eta \int_0^{\pi/3} d\Psi = \int_0^{\pi/3} d\Psi \int_0^1 d\eta$$

$$\int_0^1 d\eta \int_{\pi/3}^{\Psi_2} d\Psi = \int_{\pi/3}^{2\pi/3} d\Psi \int_0^1 d\eta + \int_{2\pi/3}^{\pi} d\Psi \int_\alpha^1 d\eta \quad (4.39)$$

$$\int_0^1 d\eta \int_{\Psi_2}^{\pi} d\Psi = \int_{2\pi/3}^{\pi} d\Psi \int_0^\alpha d\eta$$

where $\alpha = \alpha(\Psi)$ is obtained by inverting the formula for Ψ_2,

$$\alpha(\Psi) = \frac{1-\beta}{1+\beta} \qquad \beta = \tan(\pi - \Psi)/\sqrt{3} \quad (4.40)$$

The η integrals are all easily evaluated analytically, leaving the Ψ to be computed numerically.

The case $\eta \leq 0$ is similar. In this case $\Psi_2 = 2\pi/3$ and the subdivision of the Ψ integral is

$$\begin{array}{ll} 0 \leq \Psi \leq \Psi_1 & \Lambda_L = (1-\eta)/\cos(\Psi) \\ \Psi_1 \leq \Psi \leq 2\pi/3 & \Lambda_L = \sqrt{3}(1-\eta)/\sin(\Psi) \\ 2\pi/3 \leq \Psi \leq \pi & \Lambda_L = -(1+\eta)/\cos(\Psi) \end{array} \quad (4.41)$$

where $\Psi_1 = \tan^{-1}(\sqrt{3}(1+\eta)/(1-\eta))$.

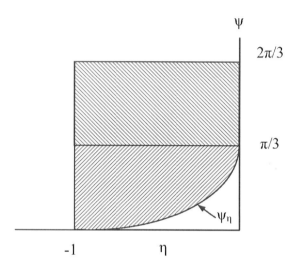

Figure 4.8 The $\{\eta, \Psi\}$ domain for $\eta < 0$.

The different subregions of the parameter domain are shown in Fig. 4.8, and the three Ψ integrals become

$$\int_{-1}^{0} d\eta \int_{0}^{\Psi_1} d\Psi = \int_{0}^{\pi/3} d\Psi \int_{\alpha}^{0} d\eta \qquad (4.42)$$

$$\int_{-1}^{0} d\eta \int_{\Psi_1}^{2\pi/3} d\Psi = \int_{0}^{\pi/3} d\Psi \int_{-1}^{\alpha} d\eta + \int_{\pi/3}^{2\pi/3} d\psi \int_{-1}^{0} d\eta \qquad (4.43)$$

$$\int_{-1}^{0} d\eta \int_{2\pi/3}^{\pi} d\Psi = \int_{2\pi/3}^{\pi} d\Psi \int_{-1}^{0} d\eta ,$$

where now

$$\alpha(\Psi) = -\frac{1-\beta}{1+\beta} \qquad \beta = \tan(\Psi)/\sqrt{3} . \qquad (4.44)$$

This concludes the discussion of the coincident integral.

4.2.3 Coincident CPV integral

Before proceeding to the edge integration we discuss, as promised, the coincident integration for the first derivative (with respect to P or Q) of the Green's function. This kernel is one order less singular than the second derivative, and thus in the initial ρ integration only the constant term, $m = 0$, from the shape functions is singular. Although this term might therefore be expected to behave analogously to $m = 1$ for the hypersingular integral, this is not the case, and it is instructive to see why.

Specifically, for $\partial G/\partial \mathbf{n}$ and $m = 0$, the ρ integrand is of the form

$$\frac{\varepsilon\rho}{(\varepsilon^2 + a^2\rho_L^2)^{3/2}} \cdot \tag{4.45}$$

Note that the difference between this and the corresponding r^{-3} integral from the hypersingular kernel is that the numerator is $\varepsilon\rho$ instead of ρ^2. Integration with respect to ρ yields

$$\frac{1}{a^2} - \frac{\varepsilon}{(\varepsilon^2 + a^2\rho_L^2)^{1/2}} \cdot \tag{4.46}$$

and no divergent terms are present. The integration of the second term with respect to Λ will produce a $\log(\varepsilon^2)$ quantity, but the factor of ε then kills off this entire contribution in the limit. Again, in the limit process, the 'Cauchy Principal Value' contribution has been neatly captured in the first term, without the need for any kind of symmetric cancelling of divergences. It can also be seen from Eq.(4.45) that this term changes sign if an interior limit is employed, compensating for the 'free term' in the interior integral equation.

4.2.4 Edge Adjacent Integration

As it is often instructive to first point out what *does not work*, the discussion of the adjacent edge case will begin by describing a method that will *not* be employed. This will not be a completely wasted digression however, as the early part of this discussion is necessary to describe the method that will be successful.

For the edge adjacent integral, it is convenient to orient the elements so that the shared edge is defined by $\xi = 0$ in E_P, $\xi^* = 0$ for E_Q, and the singularity occurs when $\eta = -\eta^*$, Fig. 4.9(a). Attempting to mimic the first two analytic integrations for the coincident integral, a seemingly reasonable approach would be to begin with polar coordinates (as always, centered at the singular point) for the Q integration,

$$\begin{aligned} \eta^* &= \rho\cos(\theta) - \eta \\ \xi^* &= \rho\sin(\theta) \ . \end{aligned} \tag{4.47}$$

Thus, $\rho = 0$ is equivalent to two of the three conditions for $P = Q$, namely $\eta^* = -\eta$ and $\xi^* = 0$. Integrating ρ, and subsequently ξ, analytically would encompass all of the singularity conditions, and should therefore produce the desired $\log(\varepsilon^2)$ term.

The difficulty with this approach is that the distance function takes the form

$$r^2 = \varepsilon^2 + b_{00}\xi^2 + (b_{10}\varepsilon + b_{11}\xi)\rho + b_{22}\rho^2 \ , \tag{4.48}$$

and it is the (unavoidable) presence of the first order term in ρ that creates serious problems. The expressions obtained by integrating ρ will have very complicated denominators, involving two quadratic factors, one having an integer exponent and one having a half-integer exponent. The proposed second analytic integration with respect to ξ may be technically possible, but would result in horrendous formulas;

the numerical implementation would therefore be, at best, inefficient and cumbersome. Moreover, if the Laplace analysis produces ugly expressions, those for the elasticity Green's function and other fundamental solutions would likely be unbearable.

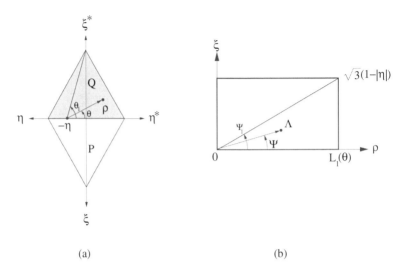

Figure 4.9 (a) Polar coordinate transformation employed in the Q element, $\{\eta^*, \xi^*\} \to \{\rho, \theta\}$; (b) Second polar coordinate transformation $\{\rho, \xi\} \to \{\Lambda, \Psi\}$ for the edge-adjacent integration.

As two separate analytic integrations is not feasible, we seek to produce the divergent term using just one. However, for this to be successful, the variable of integration must represent all three conditions for singularity, $\xi = \xi^* = 0$ and $\eta = -\eta^*$. The procedure for accomplishing this is described below. Moreover, as with the coincident integral, after the divergent term is identified, an additional exact integration will be useful for handling a weakly singular log integral. Thus, the edge adjacent integral will eventually be reduced to a two parameter integral that can be computed numerically.

4.2.4.1 1st Analytic Integration A simpler path for the edge integration begins with the polar coordinate transformation introduced above, Eq.(4.47). As shown in Fig. 4.9(a), the θ integration must be split into two pieces (for simplicity the integrands are omitted, but it will be useful to retain the jacobians of the transformations)

$$\int_{-1}^{1} d\eta \int_{0}^{\sqrt{3}(1-|\eta|)} d\xi \left[\int_{0}^{\Theta_1} d\theta \int_{0}^{L_1(\theta)} \rho \, d\rho + \int_{\Theta_1}^{\pi} d\theta \int_{0}^{L_2(\theta)} \rho \, d\rho \right], \quad (4.49)$$

where

$$L_1(\theta) = \frac{\sqrt{3}(1+\eta)}{\sin(\theta) + \sqrt{3}\cos(\theta)}$$

$$L_2(\theta) = \frac{\sqrt{3}(1-\eta)}{\sin(\theta) - \sqrt{3}\cos(\theta)}, \quad (4.50)$$

The key observation is that the break-point in θ,

$$\Theta_1 = \Theta_1(\eta) = \frac{\pi}{2} - \tan^{-1}\left(\frac{\eta}{\sqrt{3}}\right) \tag{4.51}$$

is only a function of η. It is therefore an easy matter to rearrange the integrals,

$$\int_{-1}^{1} d\eta \int_{0}^{\Theta_1} d\theta \int_{0}^{\sqrt{3}(1-|\eta|)} d\xi \int_{0}^{L_1(\theta)} \rho\, d\rho\; + \tag{4.52}$$

$$\int_{-1}^{1} d\eta \int_{\Theta_1}^{\pi} d\theta \int_{0}^{\sqrt{3}(1-|\eta|)} d\xi \int_{0}^{L_2(\theta)} \rho\, d\rho\; .$$

As the singularity occurs when $\rho = \xi = 0$, it makes sense to now introduce a second polar coordinate transformation

$$\rho = \Lambda \cos(\Psi) \tag{4.53}$$
$$\xi = \Lambda \sin(\Psi)$$

The ρ, ξ domain is a rectangle, and thus the Ψ integration must also be taken in two parts, as illustrated in Fig. 4.9(b). The resulting four integrals are

$$\int_{0}^{\Theta_1} d\theta \left[\int_{0}^{\Psi_1} d\Psi \int_{0}^{\lambda_{11}} \cos(\Psi)\Lambda^2\, d\Lambda + \int_{\Psi_1}^{\pi/2} d\Psi \int_{0}^{\lambda_{12}} \cos(\Psi)\Lambda^2\, d\Lambda \right] +$$

$$\int_{\Theta_1}^{\pi} d\theta \left[\int_{0}^{\Psi_2} d\Psi \int_{0}^{\lambda_{21}} \cos(\Psi)\Lambda^2\, d\Lambda + \int_{\Psi_2}^{\pi/2} d\Psi \int_{0}^{\lambda_{22}} \cos(\Psi)\Lambda^2\, d\Lambda \right] \tag{4.54}$$

followed by integrating η. The formulas for the Λ limits are simply

$$\lambda_{n1} = L_n(\theta)/\cos(\Psi)$$
$$\lambda_{n2} = L_n(\theta)/\sin(\Psi) \tag{4.55}$$

for $n = 1, 2$. Most importantly, the distance takes the form

$$r^2 = \varepsilon^2 - \varepsilon b_1 \Lambda + b_2 \Lambda^2\; , \tag{4.56}$$

but now the $\varepsilon b_1 \Lambda$ term does not present a major problem: there is only one integration and moreover the presence of the ε factor means that the resulting expressions will eventually simplify when $\varepsilon \to 0$. Observe that, as desired, $\Lambda = 0$ encapsulates all three conditions for $r = 0$, $\xi = \xi^* = 0$, $\eta = -\eta^*$, and that one analytic integration suffices because the total jacobian contains a Λ^2 factor. The hypersingular kernel will therefore behave as Λ^{-1}, and a $\log(\varepsilon^2)$ term will appear. For the hypersingular integral, carrying out the Λ integration results in a finite quantity

$$-\frac{J_P J_Q T_0}{b_2^{3/2}} + \log\left(\frac{4b_2^{3/2}\Lambda_L}{2b_2 + b_2^{1/2}b_1}\right)\frac{3J_P j_{1q} b_1 + 2\mathbf{n}\cdot\mathbf{N}b_2}{2b_2^{5/2}} \tag{4.57}$$

plus a divergent contribution

$$L_{kj}^e = \log(\varepsilon^2)\frac{1}{2\pi}\int_{-1}^{1} \hat{\psi}_k^0\, \hat{\psi}_j^0\, d\eta \int_{0}^{\pi} d\theta \int_{0}^{\pi/2} \cos(\Psi)\left(\frac{3j_{1p}j_{1q}}{b_2^{5/2}} - \frac{\mathbf{n}\cdot\mathbf{N}}{b_2^{3/2}}\right)\, d\Psi\; . \tag{4.58}$$

86 THREE DIMENSIONAL ANALYSIS

Here j_{1p} and j_{1q} are the coefficients of Λ in $J_Q\mathbf{n}\cdot\mathbf{R}$ and $J_P\mathbf{N}\cdot\mathbf{R}$,

$$
\begin{aligned}
J_P\mathbf{N}\cdot\mathbf{R} &= j_{1p}\Lambda - J_P\varepsilon \\
J_Q\mathbf{n}\cdot\mathbf{R} &= j_{1q}\Lambda - J_Q\mathbf{n}\cdot\mathbf{N} .
\end{aligned}
\tag{4.59}
$$

The expression for T_0 in Eq.(4.57) is not overly complicated, but the precise expression need not concern us here; what is of interest is that the finite term once again contains a weakly singular log term for $\Lambda_L \to 0$ plus the divergent contribution L_{kj}^e. As with the coincident integral, Eq.(4.36), the divergent term is independent of Λ_L and thus the complete θ integral can be included. Also as before, $\hat{\psi}_k^0$ and $\hat{\psi}_j^0$ denote the shape functions evaluated at $\Lambda = 0$. Here we have a slight problem with notation in matching the edge L_{kj}^e terms with the coincident L_{kj}^c. In the edge ordering, node 1 in P is node 2 in Q and vice versa. Thus, L_{11}^e corresponds to the off-diagonal contribution L_{12}^c and L_{12}^e should cancel with L_{11}^c. To simplify the discussion of the cancellation, we adopt, for Eq.(4.58), the convention that the subscripts refer to the coincident integral (*i.e.*, the numbering of the P element). In this case, $\hat{\psi}_l^0$ are the same for P and Q, and moreover, given by Eq.(4.34). It is important to note that b_2 is a function of the nodal coordinates and the angles θ, Ψ, but not of η, and indeed all other quantities in the integrand are independent of η. Thus, Eq.(4.58) reduces to

$$
L_{kj}^e = \log(\varepsilon)\,\frac{1}{4\pi}\,\frac{1+\delta_{kj}}{3}\int_0^\pi d\theta \int_0^{\pi/2}\cos(\Psi)\left(\frac{3j_{1p}j_{1q}}{b_2^{5/2}} - \frac{\mathbf{n}\cdot\mathbf{N}}{b_2^{3/2}}\right)d\Psi ,
\tag{4.60}
$$

again for $1 \le k,j \le 2$. This expression and that for L_{kj}^c, Eq.(4.37), do not, at first sight, appear to cancel. The proof that they do will be given following a discussion of the vertex integration.

4.2.4.2 2nd Analytic Integration The discussion thus far has reduced the edge adjacent integral to the divergent term, plus four 3 dimensional integrals over the parameters (from outermost to innermost) $\{\eta, \theta, \Psi\}$. The four integrals are a result of having to split both the θ and ψ integrals into two pieces.

As in the coincident integral, the integrands resulting from the hypersingular kernel contain a log function that is weakly singular at $\eta = \pm 1$. For reasons of accuracy and computational efficiency it is therefore desirable to once again integrate η analytically. The process in this case is a bit longer – η must be interchanged with both θ and Ψ – but still a simple calculus exercise. As before, the problem is at $\eta \pm 1$, and we seek to move the η integral to the front.

In the following we consider $\eta \ge 0$, $\eta \le 0$ is handled similarly. In addition, recall that the θ integral is divided at $\theta_\eta = \tan^{-1}(\sqrt{3}/\eta)$, and we must examine $0 \le \theta \le \theta_\eta$ and $\theta_\eta \le \theta \le \pi$ separately. As the latter is the more dangerous integral, in that the upper limit on ρ can go to zero, we begin with it. The two terms are

$$
\int_0^1 d\eta \int_{\theta_\eta}^\pi d\theta \int_0^{\Psi_1} d\Psi + \int_0^1 d\eta \int_{\theta_\eta}^\pi d\theta \int_{\Psi_1}^\pi d\Psi
\tag{4.61}
$$

where

$$
\Psi_1 = \Psi_1(\theta) = \tan^{-1}\left(\sin(\theta) - \sqrt{3}\cos(\theta)\right) .
\tag{4.62}
$$

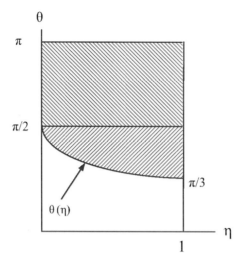

Figure 4.10 The $\{\eta, \theta\}$ domain for the first shift of the integral, $\theta_\eta \leq \theta \leq \pi$.

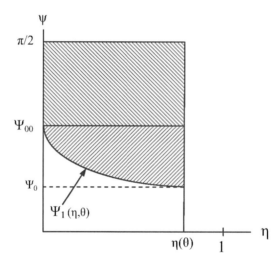

Figure 4.11 The domain for interchanging the integrals $\{\eta, \Psi\}$, for a fixed value of θ and $\hat{\eta}_\theta < 1$.

Interchanging the η and θ integrals, see Fig. 4.10, yields

$$\int_{\pi/2}^{\pi} d\theta \int_0^1 d\eta \int_0^{\Psi_1} d\Psi + \int_{\pi/3}^{\pi/2} d\theta \int_{\hat{\eta}_\theta}^1 d\eta \int_0^{\Psi_1} d\Psi + \qquad (4.63)$$

$$\int_{\pi/2}^{\pi} d\theta \int_0^1 d\eta \int_{\Psi_1}^{\pi/2} d\Psi + \int_{\pi/3}^{\pi/2} d\theta \int_{\hat{\eta}_\theta}^1 d\eta \int_{\Psi_1}^{\pi/2} d\Psi$$

where

$$\hat{\eta}_\theta = \frac{\sqrt{3}}{\tan(\theta)} \qquad (4.64)$$

88 THREE DIMENSIONAL ANALYSIS

As ψ_1 is not a function of η, the η integral slides to the front

$$\int_{\pi/2}^{\pi} d\theta \int_0^{\Psi_1} d\Psi \int_0^1 d\eta + \int_{\pi/3}^{\pi/2} d\theta \int_0^{\Psi_1} d\Psi \int_{\hat{\eta}_\theta}^1 d\eta \; + \qquad (4.65)$$

$$\int_{\pi/2}^{\pi} d\theta \int_{\Psi_1}^{\pi/2} d\Psi \int_0^1 d\eta \; + \; \int_{\pi/3}^{\pi/2} d\theta \int_{\Psi_1}^{\pi/2} d\Psi \int_{\hat{\eta}_\theta}^1 d\eta$$

and can be integrated analytically.

For $0 < \theta \le \theta_e$ the integrals are better behaved, but the interchange is more involved. The two starting integrals in this case are

$$\int_0^1 d\eta \int_0^{\theta_\eta} d\theta \int_0^{\Psi_1} d\Psi + \int_0^1 d\eta \int_0^{\theta_\eta} d\theta \int_{\Psi_1}^{\pi/2} d\Psi \qquad (4.66)$$

where now

$$\Psi_1 = \Psi_1(\eta, \theta) = \tan^{-1}\left(\frac{1-\eta}{1+\eta}(\sin(\theta) - \sqrt{3}\cos(\theta)) \right). \qquad (4.67)$$

Interchanging the θ and η integrals, Fig. 4.11, results in

$$\int_0^{\pi/3} d\theta \int_0^1 d\eta \int_0^{\Psi_1} d\Psi + \int_{\pi/3}^{\pi/2} d\theta \int_0^{\hat{\eta}_\theta} d\eta \int_0^{\Psi_1} d\Psi \; + \qquad (4.68)$$

$$\int_0^{\pi/3} d\theta \int_0^1 d\eta \int_{\Psi_1}^{\pi/2} d\Psi \; + \; \int_{\pi/3}^{\pi} d\theta \int_0^{\hat{\eta}_\theta} d\eta \int_{\Psi_1}^{\pi/2} d\Psi$$

where $\hat{\eta}_\theta$ is as above. From Fig. 4.12, moving the η integral to the front in the first two integrals $(0 < \Psi < \Psi_1)$ results in

$$\int_0^{\pi/3} d\theta \int_0^{\Psi_1^0} d\Psi \int_0^{\hat{\eta}_\psi} d\eta + \int_{\pi/3}^{\pi/2} d\theta \int_0^{\psi_\alpha} d\Psi \int_0^{\hat{\eta}_\theta} d\eta + \int_{\pi/3}^{\pi/2} d\theta \int_{\psi_\alpha}^{\Psi_1^0} d\Psi \int_0^{\hat{\eta}_\psi} d\eta$$
$$(4.69)$$

where

$$\Psi_1^0 \;=\; \Psi_1(\eta = 0) = \tan^{-1}\left((\sin(\theta) - \sqrt{3}\cos(\theta)) \right) \qquad (4.70)$$

$$\psi_\alpha \;=\; \Psi_1(\eta = \hat{\eta}_\theta) = \tan^{-1}\left(\frac{1-\hat{\eta}_\theta}{1+\hat{\eta}_\theta}(\sin(\theta) - \sqrt{3}\cos(\theta)) \right)$$

Similarly, the second two integrals $(\Psi_1 < \Psi < \pi/2)$ become

$$\int_0^{\pi/3} d\theta \int_0^{\Psi_1^0} d\Psi \int_{\hat{\eta}_\psi}^1 d\eta + \int_0^{\pi/3} d\theta \int_{\Psi_1^0}^{\pi/2} d\Psi \int_0^1 d\eta \; + \qquad (4.71)$$

$$\int_{\pi/3}^{\pi} d\theta \int_{\psi_\alpha}^{\Psi_1^0} d\Psi \int_{\hat{\eta}_\psi}^1 d\eta \; + \; \int_{\pi/3}^{\pi} d\theta \int_{\Psi_1^0}^{\pi/2} d\Psi \int_0^{\hat{\eta}_\theta} d\eta$$

4.2.5 Vertex Adjacent Integration

As mentioned above, the vertex adjacent integrals are separately finite. The singularity is in this case limited to a single point, as opposed to a line, in the four

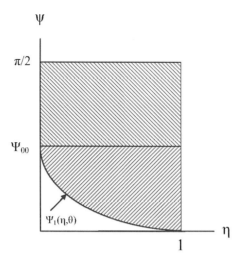

Figure 4.12 The domain for interchanging the integrals $\{\eta, \Psi\}$, for a fixed value of θ, $\hat{\eta}_\theta = 1$.

dimensional integration. Thus, any number of different algorithms can be used to evaluate the vextex adjacent pairs. However, the procedures described above, when suitably modified for the vertex situation, will integrate the singularity as completely as possible using one analytic integration, and would therefore appear to be an effective and efficient approach. The discussion that follows will simply outline the basic procedure, highlighting only the differences from the edge adjacent case.

Orient the P and Q elements so that in the parametric mappings the singular point corresponds to $\eta = -1$ and $\eta^* = -1$. Separate polar coordinates can then be introduced in each element,

$$\begin{aligned} \eta^* &= \rho_q \cos(\theta_q) - 1 & \xi^* &= \rho_q \sin(\theta_q) \\ \eta &= \rho_p \cos(\theta_p) - 1 & \xi &= \rho_p \sin(\theta_p) \end{aligned} \quad (4.72)$$

This results in an integral of the form (again omitting the kernel function and just keeping track of the jacobians)

$$\int_0^{\pi/3} d\theta_p \int_0^{L_p(\theta_p)} \rho_p \, d\rho_p \int_0^{\pi/3} d\theta_q \int_0^{L_q(\theta_q)} \rho_q \, d\rho_q = \quad (4.73)$$
$$\int_0^{\pi/3} d\theta_p \int_0^{\pi/3} d\theta_q \int_0^{L_p(\theta_p)} \rho_p \, d\rho_p \int_0^{L_q(\theta_q)} \rho_q \, d\rho_q$$

where $L_p(\theta_p) = 2\sqrt{3}/\left[\sin(\theta_p) + \sqrt{3}\cos(\theta_p)\right]$ and similarly for L_q. The lone singularity is at the common vertex $\rho_p = \rho_q = 0$, and thus one further polar coordinate transformation is warranted

$$\begin{aligned} \rho_p &= \Lambda \cos(\Psi) \\ \rho_q &= \Lambda \sin(\Psi). \end{aligned} \quad (4.74)$$

As indicated in Fig. 4.13, the $\{\rho_p, \rho_q\}$ domain is a rectangle, and thus the Ψ integration must be taken in two pieces. The combined jacobian in this case is

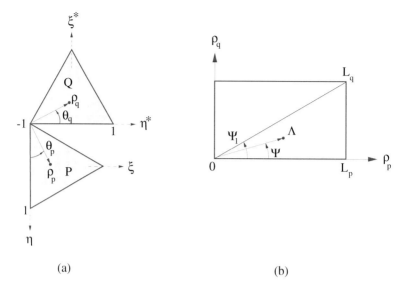

Figure 4.13 (a) Initial polar coordinate transformation employed in both P and Q elements; (b) Final polar coordinate transformation $\{\rho_p, \rho_q\} \to \{\Lambda, \Psi\}$ for the vertex adjacent integration.

$\cos(\Psi)\sin(\Psi)\Lambda^3$, and thus Eq.(4.73) becomes

$$\int_0^{\pi/3} d\theta_p \int_0^{\pi/3} d\theta_q \left[\int_0^{\Psi_1} \cos(\Psi)\sin(\Psi)\, d\Psi \int_0^{L_1(\Psi)} \Lambda^3 \, d\Lambda + \right. \tag{4.75}$$

$$\left. \int_{\Psi_1}^{\pi/2} \cos(\Psi)\sin(\Psi)\, d\Psi \int_0^{L_2(\Psi)} \Lambda^3 \, d\Lambda \right],$$

where $L_1(\Psi) = L_P(\theta_p)/\cos(\Psi)$ and $L_2(\Psi) = L_Q(\theta_q)/\sin(\Psi)$. With the Λ^3 factor the kernel function simplifies, as it is possible to immediately set $\varepsilon = 0$, and the distance is then $r^2 = b^2 \Lambda^2$ (the coefficient b^2 being a function of all three angles and nodal coordinates). Thus, it is then apparent that this integral is finite, and it is a simple matter to execute the analytic integrations.

4.2.6 Proof of Cancellation

Recapitulating the above results, the coincident and edge adjacent integrations give rise to divergent $\log(\varepsilon^2)$ terms of the form

$$L_{kj}^c = \frac{J_P^2}{4\pi} \frac{1+\delta_{kj}}{3} \int_0^\pi \frac{\sin(\Psi)}{a^3} d\Psi \tag{4.76}$$

$$L_{kj}^e = \frac{1}{4\pi} \frac{1+\delta_{kj}}{3} \int_0^\pi d\theta \int_0^{\pi/2} \cos(\Psi) \left(\frac{3j_{1p}j_{1q}}{b_2^{5/2}} - \frac{\mathbf{n}\cdot\mathbf{N}}{b_2^{3/2}} \right) d\Psi ,$$

where $k, j = 1, 2$ refer to the two nodes P_1 and P_2 along the common edge. It is therefore necessary to establish that

$$J_P^2 \int_0^\pi \frac{\sin(\Psi)}{a^3} \, d\Psi = -\int_0^\pi d\theta \int_0^{\pi/2} \cos(\Psi) \left(\frac{3 j_{1p} j_{1q}}{b_2^{5/2}} - \frac{\mathbf{n} \cdot \mathbf{N}}{b_2^{3/2}} \right) d\Psi \, , \qquad (4.77)$$

and this will be accomplished by brute force, evaluating the integrals. This is most easily carried out using a symbolic computation program.

To simplify matters, it is convenient (and permissible) to shift and rotate the elements so that $P_1 = (0, 0, 0)$, $P_2 = (x_2, 0, 0)$ and $P_3 = (x_3, y_3, 0)$, and thus $\mathbf{N} = (0, 0, 1)$ and $J_P = x_2 y_3 / 2\sqrt{3}$. Note that for the edge adjacent Q-element, the convention is that $Q_1 = P_2$ and $Q_2 = P_1$. Setting $Q_3 = (x_3^*, y_3^*, z_3^*)$, $J_Q \mathbf{n} = (0, z_3^* x_2, -y_3^* x_2)$.

From Eq.(4.18) and the comment below Eq.(4.30), the coincident integral takes the form

$$J_P^2 \int_0^\pi \frac{\sin(\Psi)}{\left(a_{cc} \cos(\Psi)^2 - a_{cs} \cos(\Psi) \sin(\Psi) + a_{ss} \sin(\Psi)^2 \right)^{3/2}} \, d\Psi \, , \qquad (4.78)$$

and for the shifted geometry

$$
\begin{aligned}
a_{cc} &= \frac{1}{4} x_2^2 \\
a_{cs} &= \sqrt{3} \, x_2 \, (2 x_3 - x_2) / 6 & (4.79) \\
a_{ss} &= \left(x_2^2 + 4 x_3^2 + 4 y_3^2 - 4 x_3 x_2 \right) / 12 & (4.80)
\end{aligned}
$$

After substituting $q = \cotan(\Psi)$, Eq.(4.78) becomes

$$-J_P^2 \int_{-\infty}^\infty \frac{1}{\left(a_{cc} q^2 - a_{cs} q + a_{ss} \right)^{3/2}} \, dq \, , \qquad (4.81)$$

and carrying out the integration we find that the coincident divergent term becomes simply

$$J_P^2 \int_0^\pi \frac{\sin(\Psi)}{a^3} \, d\Psi = x_2 \, . \qquad (4.82)$$

Thus, as expected, the divergent term does not depend upon P_3.

As a consequence of the double integration, the evaluation of the edge integral is considerably more involved. Although symbolic computation will eventually execute all of the required calculus and algebra, manipulation is required to modify the forms of the expressions, and care is required to keep the size of the expressions from exceeding the available memory. The discussion below will therefore only outline the procedure. As a function of Ψ, the coefficient b_2 defined in Eq.(4.56) takes the form

$$b_2 = c_2 \cos^2(\Psi) + c_1 \cos(\Psi) \sin(\Psi) + c_0 \sin^2(\Psi) \qquad (4.83)$$

where the c_j are functions of $\cos(\theta)$ and $\sin(\theta)$. Thus, as with L_{kj}^c, substituting $q = \cotan(\Psi)$ is convenient, resulting in an integral of the form

$$\int_0^\infty \left[\alpha_1 \frac{q^2}{\left(c_2 q^2 + c_1 q + c_0 \right)^{5/2}} + \alpha_2 \frac{q}{\left(c_2 q^2 + c_1 q + c_0 \right)^{3/2}} \right] dq \, . \qquad (4.84)$$

92 THREE DIMENSIONAL ANALYSIS

The function of θ that results from this integration once again benefits from the substitution $p = \cot(\theta)$, and the θ integral becomes

$$\int_{-\infty}^{\infty} \frac{h_1(p)}{\left(s_2p^2 + s_1p + s_0\right)^2}\, \mathrm{d}p+ \tag{4.85}$$
$$\int_{-\infty}^{\infty} \frac{h_2(p)}{\sqrt{(t_2p^2 + t_1p + t_0)}\left(s_2p^2 + s_1p + s_0\right)^2}\, \mathrm{d}p\ ,$$

where $h_1(p)$ and $h_2(p)$ are quadratic and cubic polynomials, respectively. The coefficients $\{s_j\}$ and $\{t_j\}$ are now just functions of the nodal coordinates. The first integral is found to be 0, while the second is, as desired, $-x_2$.

4.3   HIGHER ORDER INTERPOLATION

In direct limit evaluation it is essential to perform basic operations – integrating and taking of the boundary limit – analytically. A linear element is sufficiently simple that it is immediately possible to carry out these calculations, either by hand or using symbolic computation. A linear approximation will suffice for many situations, but higher order is essential for at least one important application, namely fracture analysis. This is due to the availability of the quadratic 'quarter-point' element [15, 120, 134] which accurately and conveniently captures the critical singular behavior at the crack front. Moreover, fracture, discussed in detail in Chapter 9 is a key application for hypersingular boundary integral equations [64]. Thus, for direct evaluation to be viable, it is necessary to extend the techniques in to higher order interpolation.

Going beyond a linear approximation requires one additional aspect to the algorithm, a splitting of the integrands into singular and nonsingular components. The nonsingular term must be tame enough that numerical quadrature can be used, while the singular term must be simple enough that the analytic manipulations can be carried out. This decomposition is relatively simple for the adjacent edge integral (wherein the outer and inner integration elements share a common edge), as only a single analytic integration is required. However, as detailed above, the coincident integration demands a minimum of two separate analytic integrations to produce the divergent boundary limit term, and thus the splitting here is somewhat more involved.

The decomposition of the integrand is not unique, as one can choose to include some nonsingular integrals into the analytic component. While including as much as possible into the singular component would possibly provide some additional accuracy, the tradeoff is that the exact integration formulas can become quite lengthy. For the discussion below it suffices to employ the simplest possible Green's function, that for the Laplace equation, but in general the kernel functions are considerably more involved, *e.g.*, three dimensional elasticity. Thus, the approach taken herein will be to keep the singular component as small as possible, whatever can be safely evaluated numerically will be relegated to the nonsingular term.

For the purposes of discussion it also suffices to consider a six-noded quadratic triangular element, applied to the hypersingular Laplace kernel function. The application to other elements, and less singular integrands, is relatively immediate.

4.4 HYPERSINGULAR BOUNDARY INTEGRAL: QUADRATIC ELEMENT

The quadratic interpolation will be based upon the shape functions

$$\psi_1(\eta, \xi) = \frac{(\xi + \sqrt{3}\eta - \sqrt{3})(\xi + \sqrt{3}\eta)}{6}$$

$$\psi_2(\eta, \xi) = \frac{(\xi - \sqrt{3}\eta - \sqrt{3})(\xi - \sqrt{3}\eta)}{6}$$

$$\psi_3(\eta, \xi) = \frac{(2\xi - \sqrt{3})\xi}{3} \qquad (4.86)$$

$$\psi_4(\eta, \xi) = \frac{(\xi + \sqrt{3}\eta - \sqrt{3})(\xi - \sqrt{3}\eta - \sqrt{3})}{3}$$

$$\psi_5(\eta, \xi) = -2\frac{(\xi - \sqrt{3}\eta - \sqrt{3})\xi}{3}$$

$$\psi_6(\eta, \xi) = -2\frac{(\xi + \sqrt{3}\eta - \sqrt{3})\xi}{3} ,$$

once again defined on the equilateral triangle parameter space $\{\eta, \xi\}$, $-1 \leq \eta \leq 1$, $0 \leq \xi \leq \sqrt{3}(1 - |\eta|)$. The nodes $\{v_j\}$ on the $\{\eta, \xi\}$ triangle are the vertices plus the mid-side nodes, as shown in Fig. 4.14, and the shape functions are constructed so that $\psi_l(v_j) = \delta_{lj}$, $1 < l, j < 6$.

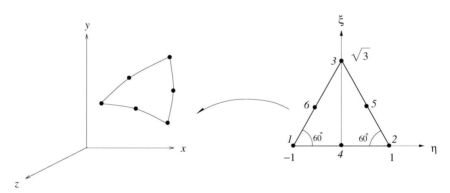

Figure 4.14 A triangle in the 3D space is mapped to an equilateral triangular quadratic element in $\{\eta, \xi\}$ space, where $-1 \leq \eta \leq 1$, $0 \leq \xi \leq \sqrt{3}(1 - |\eta|)$.

For an element consisting of the nodal points $\{Q_j = (x_j, y_j, z_j)\}$, $1 < j < 6$ the approximate boundary surface is given by

$$\Sigma(\eta, \xi) = \sum_{j=1}^{6}(x_j, y_j, z_j)\psi_j(\eta, \xi) , \qquad (4.87)$$

while the boundary potential and flux are similarly defined in terms of their nodal values *i.e.* ,

$$\phi(Q) = \sum_j \phi(Q_j)\psi_j(Q)$$

$$\frac{\partial \phi}{\partial \mathbf{n}}(Q) = \sum_j \frac{\partial \phi}{\partial \mathbf{n}}(Q_j)\psi_j(Q) . \qquad (4.88)$$

94   THREE DIMENSIONAL ANALYSIS

4.4.1 Coincident Integration

As above we take $E_P = E_Q = E$, and thus the coincident integral is now

$$\int_E \psi_k(P) \int_E \phi(Q) \frac{\partial^2 G}{\partial \mathbf{N} \partial \mathbf{n}}(P,Q)\, dQ\, dP = \tag{4.89}$$

$$\sum_{j=1}^{6} \phi(Q_j) \int_E \psi_k(P) \int_E \psi_j(Q) \frac{\partial^2 G}{\partial \mathbf{N} \partial \mathbf{n}}(P,Q)\, dQ\, dP \; ,$$

where E is defined by nodes P_k, $1 \le k \le 6$. Transferring the integral to parameter space, denoted by $\{\eta, \xi\}$ for the outer P integral and $\{\eta^*, \xi^*\}$ for Q, requires including the Jacobians J_P and $= J_Q$,

$$J_Q(\eta^*, \xi^*) = \left\| \frac{\partial Q(\eta^*, \xi^*)}{\partial \eta^*} \times \frac{\partial Q(\eta^*, \xi^*)}{\partial \xi^*} \right\| , \tag{4.90}$$

where $Q(\eta^* \xi^*) = \Xi_Q(\eta^*, \xi^*)$ is defined as in Eq.(4.87). This results in

$$\frac{1}{4\pi} \sum_{k,j} \phi(Q_j) \int \psi_k(\eta, \xi) \int \psi_j(\eta^*, \xi^*) \left(\frac{J_Q \mathbf{n} \cdot J_P \mathbf{N}}{r^3} - 3 \frac{(J_Q \mathbf{n} \cdot R)(J_P \mathbf{N} \cdot R)}{r^5} \right) d\xi^* d\eta^*\, d\xi d\eta,$$

$$\tag{4.91}$$

as the Jacobians are conveniently incorporated into the normal vectors $\mathbf{n}(Q)$ and $\mathbf{N}(P)$, and thus the square root in Eq.(4.90) will not appear in the integrand. However, for integrands in which the Jacobian does appear explicitly, a Taylor series expanded around the singular point can be used to obtain a function that can be integrated analytically.

4.4.2 r Expansion

As detailed above for the linear element, the first step in the integration is to consider $P = (\eta, \xi)$ fixed and define a polar coordinate system centered at this point,

$$\eta^* - \eta = \rho \cos(\theta) \tag{4.92}$$
$$\xi^* - \xi = \rho \sin(\theta) \; . \tag{4.93}$$

With the quadratic interpolation, the components of $\mathbf{R} = \{R_j\}$ take the form

$$R_j = c_2 \rho^2 + c_1 \rho - \varepsilon N_j \tag{4.94}$$

and the expression for r^2 is therefore a fourth order polynomial

$$r^2 = a_4 \rho^4 + a_3 \rho^3 + a_2 \rho^2 + \varepsilon^2 \; , \tag{4.95}$$

where $a_l = a_l(\eta, \xi, \theta)$. The important change to note is the dependence of a_2 on ξ; recall that for the linear approximation it was solely a function θ.

As in two dimensions, Chapter 3, the key step is to introduce the simpler, linear-type expression,

$$\hat{r}^2 = a_2(\eta, \xi, \theta) \rho^2 + \varepsilon^2 \; , \tag{4.96}$$

into the integrand by writing, for example,

$$
\frac{1}{r^3} = \frac{1}{\hat{r}^3} + \left(\frac{1}{r^3} - \frac{1}{\hat{r}^3}\right)
$$

$$
= \frac{1}{\hat{r}^3} + \frac{\hat{r}^3 - r^3}{r^3\hat{r}^3} \ . \tag{4.97}
$$

$(r^{-1}$ and r^{-5} can be handled similarly). Moreover, rationalizing the numerator,

$$
\frac{\hat{r}^3 - r^3}{r^3\hat{r}^3} = \frac{\hat{r}^6 - r^6}{r^3\hat{r}^3(r^3 + \hat{r}^3)} \ , \tag{4.98}
$$

it can be seen that $\hat{r}^6 - r^6$ is $\mathcal{O}(\rho^7)$ (taking both ρ and ε into account). Thus the remainder term is $\mathcal{O}(\rho^{-2})$, one degree less singular than the original r^{-3}. A less singular remainder is required for the hypersingular analysis, and this can be achieved by repeating the above process,

$$
\frac{1}{r^3} = \frac{1}{\hat{r}^3} + \frac{\hat{r}^3 - r^3}{r^3\hat{r}^3} \tag{4.99}
$$

$$
= \frac{1}{\hat{r}^3} + \frac{\hat{r}^3 - r^3}{\hat{r}^6} + \frac{(\hat{r}^3 - r^3)^2}{r^3\hat{r}^6} \ , \tag{4.100}
$$

and now the last term is $\mathcal{O}(\rho^{-1})$. Note that the numerator of $\hat{r}^{-6}$ is not polynomial. Converting this term into something that can be integrated analytically is more work than in two dimensions, a consequence of the half-integer exponents of r^2, but can be accomplished as in Eq.(4.98), followed by

$$
\frac{1}{r^3 + \hat{r}^3} = \frac{1}{2\hat{r}^3} + \frac{\hat{r}^3 - r^3}{2\hat{r}^3(r^3 + \hat{r}^3)} \ . \tag{4.101}
$$

Thus, the final result of the expansion process is to produce a singular term

$$
\frac{1}{r^3} = \frac{1}{\hat{r}^3} + \frac{\hat{r}^6 - r^6}{2\hat{r}^9} \tag{4.102}
$$

that can be integrated analytically, plus a remainder that can safely be dealt with numerically. Note that the polynomial $\hat{r}^6 - r^6$ will contain terms that are nonsingular after divsion by $\hat{r}^9$; these can be left as part of the singular component or integrated numerically.

4.4.3 First Integration

The goal in this section is to discuss the extra steps needed for the quadratic interpolation, and thus most of the details of the analysis will be omitted. They are essentially the same as for the linear interpolation earlier in this chapter.

As above, it suffices to carry out the ρ integration for the lower subtriangle associated with the edge $\xi^* = 0$, and the integration limits $0 \le \rho \le \rho_L$ and $\Theta_1 \le \theta \le \Theta_2$ are as before, Eq.(4.16) and Eq.(4.15). The shape functions $\psi_j(Q)$ are quadratic polynomials in ρ, but for the purpposes herein it suffices to consider only the most singular (constant) term, $\psi_j(\rho = 0) = \psi_j(\eta, \xi)$. The hypersingular

96 THREE DIMENSIONAL ANALYSIS

integral in Eq.(4.89) therefore becomes

$$
\frac{1}{4\pi} \int_{-1}^{1} d\eta \int_{0}^{\sqrt{3}(1-|\eta|)} \psi_k(\eta,\xi)\psi_j(\eta,\xi)\,d\xi \int_{\Theta_1}^{\Theta_2} d\theta
$$
$$
\int_{0}^{\rho_L} \rho \left(\frac{J_Q \mathbf{n} \cdot J_P \mathbf{N}}{(a^2\rho^2 + \varepsilon^2)^{3/2}} - 3\frac{(J_Q \mathbf{n} \cdot R)(J_P \mathbf{N} \cdot R)}{(a^2\rho^2 + \varepsilon^2)^{5/2}} \right) d\rho, \tag{4.103}
$$

where $a^2 = a_2(\eta,\xi,\theta)$. Again simplifying matters down to what is necessary for this discussion, we consider only the lowest order terms in the numerators, $J_Q\mathbf{n} \cdot J_P\mathbf{N} \approx J_P^2$ and $(J_Q\mathbf{n} \cdot R)(J_P\mathbf{N} \cdot R) \approx J_P^2\varepsilon^2$.

Thus, the analytic evaluation of the ρ integral results, as before, in

$$
-\frac{J_P^2}{4\pi}\frac{\rho_L^2}{(\varepsilon^2 + a^2\rho_L^2)^{3/2}}, \tag{4.104}
$$

where J_P^2 is not necessarily constant, but it is a polynomial. The second integration is essential for treating this term, and it is here that the ξ dependence of a^2 comes into play.

Getting to the second integration, requires the two changes of variables, θ replaced by t, Eq.(4.28), followed by $\{t,\xi\}$ are relaced by the second polar coordinate transformation, $\{\Lambda,\Psi\}$, Eq.(4.30).

The intention is of course to integrate

$$
\int \frac{\Lambda^2}{(\varepsilon^2 + a^2\Lambda^2)^{3/2}}\,d\Lambda. \tag{4.105}
$$

with respect to Λ. However, with the two changes of variables, $a_2(\eta,\xi,\theta)$ is (unlike for the linear element) a function of Λ

$$
a^2 = c_2\Lambda^2 + c_1\Lambda + c^2 \tag{4.106}
$$

where c^2 is the coefficient that would appear in the linear analysis. The procedure for handling this more complicated denominator is as in the r expansion, namely

$$
\frac{\Lambda^2}{(\varepsilon^2 + a^2\Lambda^2)^{3/2}} = \frac{\Lambda^2}{(\varepsilon^2 + c^2\Lambda^2)^{3/2}} + \Lambda^2 \left(\frac{1}{(\varepsilon^2 + a^2\Lambda^2)^{3/2}} - \frac{1}{(\varepsilon^2 + c^2\Lambda^2)^{3/2}} \right). \tag{4.107}
$$

Again rationalizing the numerator, it can be seen that the remainder term on the right is finite at $\varepsilon = 0$, and that the singular term is easily integrated. Moreover, as the singular integration is of the same form as in the linear interpolation (and this will be true in the edge case as well), the proof of cancellation of the divergent $\log(\varepsilon^2)$ term will follow as in the linear analysis.

4.4.4 Edge Integration

The adjacent edge integral is less singular, requiring a single analytic integration to produce the canceling $\log(\varepsilon^2)$ contribution. Thus, in this case, the modifications required for the quadratic interpolation are much simpler. The expression for r^2 that results from the two polar coordinate transformations in Section 4.2.4 can once again be simplified to a 'linear form' by a suitable expansion as in Eq.(4.97). Again due to the lesser singularity, this process need only be applied once.

4.5 CORNERS

The treatment of corners in the Galerkin BEM is simple and elegant due to the flexibility in choosing the weight function for the Galerkin approximation. The corner treatment in 3D is similar to 2D as explained in Section 3.5 in Chapter 3. Corners are represented by multiple nodes, and on each side different weight functions are used (see Figure 4.15). Figure 4.15 shows an assembly of six planes with different orientation of the normals where, each plane has been prescribed with either Dirichlet or Neumann boundary conditions. Consequently at the intersection of two planes double nodes are applied. For a mixed corner (flux is unknown in one side of the corner, potential is known), a non zero weight function is assigned only on the side where flux is unknown. For a Neumann corner (flux specified on both sides of the corner, potential is the unknown), the weight functions are combined together. On a Dirichlet corner (unknowns are flux on each sides, potential is known) the usual weight functions are assigned on both sides of the corners.

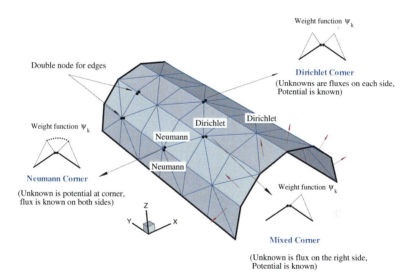

Figure 4.15 Corner treatment in the Galerkin BEM. Notice that the 6 normal vectors in the figure define the 6 planes that compose the semi-cylindrical geometry (with axis along the y-direction).

4.6 ANISOTROPIC ELASTICITY

Although the above discussion of singular integration has considered the simplest possible Green's function, other differential equations can be handled by precisely the same techniques. The ability to carry out the necessary analytic integrations is the key requirement, and therefore the limit methods are directly applicable to isotropic elasticity, even though the fundamental solution expressions are much more complicated. Moreover, Green's functions such as

98 THREE DIMENSIONAL ANALYSIS

$$G(P,Q) = \frac{1}{4\pi} \frac{e^{ikr}}{r} \ . \tag{4.108}$$

for the scalar Helmholtz equation [16] (acoustic wave propagation)

$$\left(\nabla^2 + \kappa^2\right) u = 0 \tag{4.109}$$

also present no difficulties. Taylor series expansions of the exponential, centered at the singular point, can be employed to once again (as with higher order interpolations) split the integrals into two parts: a 'rational function' singular part and nonsingular part that is safely integrated numerically.

Thus, together with the asymptotic expansions discussed in Chapter 3, singular integration for most any Green's function is amenable to the boundary limit procedures. Even if the expressions for the Green's function and its derivatives are lengthy, as is the case, for example, in elastic wave scattering (frequency domain), the work in generating the exact integration formulas and evaluating the limits can be eased by symbolic computation.

However, for anisotropic elasticity, arguably the most complicated fundamental solution, it is not immediately clear that direct limit evaluation can be applied. Moreover, as the ability to treat a general linear elastic solid is essential for many important applications, this topic warrants a separate discussion.

The anisotropic Green's function is given in terms of a function $\tilde{\mathcal{U}}(P,Q)$ which can only be known through numerical evaluation. This in itself does not rule out analytic methods: the Green's function for the two-dimensional Helmholtz equation is the Bessel function $H_0^1(kr)$, and as discussed in Chapter 3, the singular integration can be based upon asymptotic expansions at $r = 0$. Thus, having a Green's function which is not expressed in simple algebraic form is not necessarily a problem. However, what is different for anisotropic elasticity is that the nonsingular $\tilde{\mathcal{U}}(P,Q)$ contributes an *angular variation* to the Green's function at the singular point. Thus, asymptotic expansions or Taylor series are not appropriate, and it is not immediately clear how to execute the necessary analytic integrations *completely around the singular point.*

Despite the complications introduced by $\tilde{\mathcal{U}}(P,Q)$, it will be shown that splitting of the integrands into appropriate singular and non-singular components is still possible. The necessary analytic integrations will be possible because $\tilde{\mathcal{U}}$ will not be involved. The integrand for the non-singular component will depend upon $\tilde{\mathcal{U}}$, but as this integral will be evaluated numerically, this is not a problem. The primary task of this section therefore is to show how to accomplish this splitting.

A secondary, but important, problem is the numerical evaluation of $\tilde{\mathcal{U}}(P,Q)$. The definition in terms of a contour integral, Eq.(4.113) below, is not suitable for computations due to the expense, and a number of methods have been proposed for overcoming this problem [14,72,291]. The calculations presented below will employ the method based upon residue evaluations [239].

The discussion for the Laplace equation focused on the hypersingular integral, as the techniques filtered down to the less singular integrals in a more or less obvious way. This section will discuss the boundary integral equation for surface displacement, involving the Green's function and its first derivative. At this point, the limit procedures have not been explicitly implemented for the anisotropic traction equation, though this the techniques presented here are expected to carry over directly to the hypersingular integral.

4.6.1 Anisotropic Elasticity Boundary Integral Formulation

The exterior limit boundary integral equation for surface displacement $\mathbf{u}$, analogous to the potential equation Eq.(4.1), can be written as

$$\mathcal{P}(P) = \lim_{\varepsilon \to 0} \int_{\Sigma} [\mathcal{T}(P_\varepsilon, Q)\mathbf{u}(Q) - \mathcal{U}(P_\varepsilon, Q)\boldsymbol{\tau}(Q)] \, dQ = 0, \qquad (4.110)$$

where, as above, P_ε are points exterior to the domain V converging to the point $P \in \Sigma$. The corresponding Galerkin form is

$$\int_{\Sigma} \hat{\psi}_k(P) \int_{\Sigma} \mathcal{P}(P) \, dP = 0 \ . \qquad (4.111)$$

The Green's function is

$$\mathcal{U}(P,Q) = \frac{1}{8\pi^2 r}\tilde{\mathcal{U}}(P,Q) = \frac{1}{8\pi^2 r}\tilde{\mathcal{U}}(\mathbf{R}/r) = \frac{1}{8\pi^2 r}\tilde{\mathcal{U}}(\zeta,\varphi) \ , \qquad (4.112)$$

where $\{\zeta,\varphi\}$ are spherical coordinate angles defining the unit direction vector $\mathbf{R}/r$, φ being the polar or azimuthal angle. In the following the function $\tilde{\mathcal{U}}$ will be written either as a function of P and Q, or $\mathbf{R}$, or the spherical angles, whichever is the most convenient at the moment. To complete the specification of the Green's function, the 3×3 matrrix $\tilde{\mathcal{U}}$ can be expressed as

$$\tilde{\mathcal{U}}(P,Q) = \oint_{S^1} \mathbf{K}^{-1}(\boldsymbol{\xi}) \, ds(\boldsymbol{\xi}) \ , \qquad (4.113)$$

and the Christoffel matrix $\mathbf{K}(\boldsymbol{\xi})$, $\boldsymbol{\xi} \in \mathcal{R}^3$, is defined by $K_{ij} = C_{iljm}\xi_l\xi_m$, $\mathcal{C}$ the elastic constants. The line integral contour S^1 is the unit circle in the plane having normal $\mathbf{R}/r$,

$$S^1 = S^1(P,Q) = \{\boldsymbol{\xi} \in \mathcal{R}^3 \ | \ \|\boldsymbol{\xi}\| = 1, \ \boldsymbol{\xi} \cdot \mathbf{R} = 0\} \ . \qquad (4.114)$$

The function $\tilde{\mathcal{U}}$ can be conveniently computed as a function of the spherical coordinate angles $\{\zeta,\varphi\}$, as can its partial derivatives with respect to these angles [239]. These angles can be referenced to any choice of spherical coordinate system, and the flexibility to select the coordinate system will be important. In what follows, we will employ three spherical coordinate sysytems, the ones for which a cartesian axis, X, Y, or Z, is the polar axis.

The traction kernel $\mathcal{T}(P,Q)$ is obtained from $\mathcal{U}$ in the same manner as the traction $\boldsymbol{\tau}$ is obtained from displacement. Thus,

$$\mathcal{T} = \frac{1}{2}\sum_{m,l} C_{imjl} \left(\mathcal{U}_{ij,l} + \mathcal{U}_{il,j}\right) \qquad (4.115)$$

where $\mathcal{U}_{ij,l}$ is the derivative with respect to the lth component of Q

$$\begin{aligned}
\mathcal{U}_{,l}(P,Q) &= \frac{1}{8\pi^2}\left(-\frac{R_l}{r^3}\tilde{\mathcal{U}}(P,Q) + \frac{1}{r}\frac{\partial}{\partial q_l}\tilde{\mathcal{U}}(P,Q)\right) & (4.116)\\
&= \frac{1}{8\pi^2}\left(-\frac{R_l}{r^3}\tilde{\mathcal{U}} + \frac{1}{r}\left\{\frac{\partial\tilde{\mathcal{U}}}{\partial\zeta}\frac{\partial\zeta}{\partial q_l} + \frac{\partial\tilde{\mathcal{U}}}{\partial\psi}\frac{\partial\psi}{\partial q_l}\right\}\right) .
\end{aligned}$$

100 THREE DIMENSIONAL ANALYSIS

As $\mathcal{T}$ is a linear combination of $\mathcal{U}_{,l}(P,Q)$, it will suffice to examine the integration of these individual derivatives. To simplify notation, the component indices $\{i,j\}$ for the Green's function will be dropped.

We first tackle the evaluation of the singular integrals involving the traction kernel $\mathcal{T}$. Integrals for the simpler displacement kernel are only weakly singular, and therefore exist with P directly on the boundary, *i.e.*, a limit process is not required. The same techniques will apply for this integral, only much simplified.

4.6.2 $\mathcal{T}$ Kernel: Coincident Integration

From the above discussion, the coincident integral to be evaluated is

$$\frac{1}{8\pi^2} \int_E \hat{\psi}_k(P) \int_E \left(-\frac{R_l}{r^3} \tilde{\mathcal{U}}(P_\varepsilon, Q) + \frac{1}{r}\frac{\partial}{\partial q_l}\tilde{\mathcal{U}}(P_\varepsilon, Q) \right) \mathbf{u}(Q)\,\mathrm{d}Q\,\mathrm{d}P \,, \tag{4.117}$$

where, as earlier in this chapter, the element E is defined by linear interpolation of the vertices (x_m, y_m, z_m), $1 \le m \le 3$. Formulated as a parameter space integral this becomes

$$\frac{J_P^2}{8\pi^2} \int_{-1}^1 \int_0^{e(\eta)} \psi_k(\eta, \xi)\mathrm{d}\eta\mathrm{d}\xi \tag{4.118}$$

$$\int_{-1}^1 \int_0^{e(\eta^*)} \psi_j(\eta^*, \xi^*) \left(-\frac{R_l}{r^3} \tilde{\mathcal{U}}(P_\varepsilon, Q) + \frac{1}{r}\frac{\partial}{\partial q_l}\tilde{\mathcal{U}}(P_\varepsilon, Q) \right) \mathrm{d}\eta^*\mathrm{d}\xi^* \,,$$

where $e(\eta) = \sqrt{3}(1 - |\eta|)$ and J_P the (constant) Jacobian. It will be convenient to explicitly identify the boundary point as P_0, *i.e.* , $P_\varepsilon = P_0 + \varepsilon\mathbf{N}$, as illustrated in Figure 4.1.

The initial steps proceed as in Section 4.2.2: the polar coordinate transformation Eq.(4.14), illustrated in 4.5), results in

$$\begin{aligned}
\mathbf{R} &= (R_1, R_2, R_3) = (a_1\rho - \varepsilon N_1, a_2\rho - \varepsilon N_2, a_3\rho - \varepsilon N_3) \\
a_l &= a_l^c \cos(\theta) + a_l^s \sin(\theta) \\
r^2 &= a^2\rho^2 + \varepsilon^2 \,,
\end{aligned} \tag{4.119}$$

and as before it suffices to examine just the lower subtriangle associated with the edge $\xi^* = 0$ in the $\{\eta^*, \xi^*\}$ parameter space. At this point, it is convenient to consider the two constituents of $\mathcal{T}(P,Q)$, Eq.(4.117), separately.

4.6.2.1 $\tilde{\mathcal{U}}$ With the polar coordinate transformation, the $Q-$shape function is

$$\psi_j(\eta^*, \xi^*) = \psi_j(\eta, \xi) + c_j(\theta)\rho \,, \tag{4.120}$$

and as the integration with the $c_j(\theta)\rho$ term is relatively easy, the constant term integral

$$-\frac{J_P^2}{8\pi^2} \lim_{\varepsilon \to 0} \int_{-1}^1 \int_0^{e(\eta)} \psi_k(\eta, \xi)\psi_j(\eta, \xi)\mathrm{d}\eta\mathrm{d}\xi \tag{4.121}$$

$$\int_{\Theta_1}^{\Theta_2} \int_0^{\rho_L} \rho \frac{a_l\rho - \varepsilon N_l}{(a^2\rho^2 + \varepsilon^2)^{3/2}} \tilde{\mathcal{U}}(P_\varepsilon, Q)\mathrm{d}\rho\mathrm{d}\theta$$

ANISOTROPIC ELASTICITY **101**

will be examined first. The obvious roadblock to integrating ρ is that $\tilde{\mathcal{U}}(P_\varepsilon, Q)$ is a function of ρ and a closed form for this function is lacking. The first step in getting around this difficulty is to rewrite this function as (see Fig. 4.1)

$$\tilde{\mathcal{U}}(P_\varepsilon, Q) = \tilde{\mathcal{U}}(P_0, Q) + \left[\tilde{\mathcal{U}}(P_\varepsilon, Q) - \tilde{\mathcal{U}}(P_0, Q)\right] , \tag{4.122}$$

and to note that for the linear element $(Q - P_0)/\|Q - P_0\| = \mathbf{a}/a$. Thus, $\tilde{\mathcal{U}}(P_0, Q)$ is independent of ρ (as it should be, being the angular dependence of the Green's function), and executing the ρ integration for the first term on the right hand side of Eq.(4.122) is straightforward,

$$\tilde{\mathcal{U}}(P_0, Q) \int_0^{\rho_L} \rho \frac{a_l \rho - \varepsilon N_l}{(a^2 \rho^2 + \varepsilon^2)^{3/2}} \, d\rho = \tag{4.123}$$

$$\tilde{\mathcal{U}}(P_0, Q) \left(-\frac{a_l}{a^3} \log(2a\rho_L) + \frac{N_l}{a^2} + \frac{a_l \left(2 + \log(\varepsilon^2)\right)}{2a^3}\right) .$$

As discussed for the Laplace kernel, the $\log(\varepsilon^2)$ term appearing in this expression is precisely the term that the Cauchy Principal Value [127] procedure eliminates by removing a symmetric neighborhood of the singular point. As before, this term simply cancels on its own,

$$\frac{1 + \log(\varepsilon^2)}{2} \int_0^{2\pi} \tilde{\mathcal{U}}(P_0, Q) \frac{a_l}{a^3} \, d\theta = 0 \tag{4.124}$$

as $a_l(\pi + \theta) = -a_l(\theta)$ and $\tilde{\mathcal{U}}(P, Q) = \tilde{\mathcal{U}}(Q, P)$ implies that $\tilde{\mathcal{U}}(\mathbf{a}/a) = \tilde{\mathcal{U}}(-\mathbf{a}/a)$. This term can therefore be removed from Eq.(4.123), leaving a finite expression for the limit. Note however that this expression contains an (integrable) logarithmic singularity, as ρ_L approaches zero for ξ approaching zero. We return to this point in Section 4.6.4.

A complete analytic integration of the 'limit term' from Eq.(4.122),

$$\int_0^{\rho_L} \frac{\rho \left(a_l \rho - \varepsilon N_l\right)}{(a^2 \rho^2 + \varepsilon^2)^{3/2}} \left[\tilde{\mathcal{U}}(P_\varepsilon, Q) - \tilde{\mathcal{U}}(P_0, Q)\right] d\rho . \tag{4.125}$$

is clearly out of the question. However, the purpose of the exact integration is to isolate and remove (in this case via the θ integration) the potentially divergent $\log(\varepsilon^2)$ term. This then allows the computation of the boundary limit. Fortunately, an analytic evaluation of Eq.(4.125) is *not* essential: it will now be shown that the limiting behavior of this integral is not a problem.

The key observation needed to justify this last statement is that $\tilde{\mathcal{U}}(P_\varepsilon, Q)$ is, by simple geometry, a function of ρ/ε. Thus, making the change of variables

$$s = \rho/\varepsilon , \tag{4.126}$$

the above integral becomes

$$\lim_{\varepsilon \to 0} \int_0^{\rho_L/\varepsilon} s \frac{\left(a_l s - N_l\right)}{(1 + a^2 s^2)^{3/2}} f_0(s) \, ds \tag{4.127}$$

102 THREE DIMENSIONAL ANALYSIS

where, with a slight abuse of notation, $f_0(s) = \tilde{\mathcal{U}}(s) - \tilde{\mathcal{U}}(\mathbf{a})$ (the dependence of f_0 on θ is suppressed). Moreover, note that

$$f_0(s) \int_0^{2\pi} \frac{a_l s^2}{(1 + a^2 s^2)^{3/2}} \, \mathrm{d}\theta = 0 \tag{4.128}$$

for the same reason that the $\log(\varepsilon^2)$ contribution in Eq.(4.123) disappears.

Thus, ignoring the fact that ρ_L is a function of θ (this can be made rigorous), Eq.(4.127) becomes

$$\int_0^\infty s \frac{(a_l s - N_l)}{(1 + a^2 s^2)^{3/2}} f_0(s) \, \mathrm{d}s . \tag{4.129}$$

To justify taking the limit, *i.e.*, replacing the upper limit by infinity, note that $f_0(s)$ goes to zero at infinity, and the other part of the integrand is $\mathcal{O}(s^{-1})$. For numerical computation it is more convenient to work with a finite length interval, and introducing the change of variables $s = \tan(q)/a$, the integral is instead over $[0, \pi/2]$. Finally, as the above expression is independent of ρ_L, it holds for all three sub-triangles, and thus the θ integral is over $[0, 2\pi]$. Moreover, the inner integration $\{\theta, s\}$ is independent of $\{\eta, \xi\}$, and thus the complete integral for this limit term is of the form

$$\frac{J_P^2}{4\pi^2} \alpha_{kj} \int_0^\pi \frac{1}{a^2} \, \mathrm{d}\theta \int_0^{\pi/2} \beta(q) \, f_0(\tan(q)/a) \, \mathrm{d}q , \tag{4.130}$$

where

$$\alpha_{kj} = \int_{-1}^1 \int_0^{e(\eta)} \psi_k(\eta, \xi) \psi_j(\eta, \xi) \, \mathrm{d}\eta \mathrm{d}\xi . \tag{4.131}$$

As an example, Figure 4.16 plots $f_0(q)$ for the diagonal components of the Green's function; the elastic constants are for silicon ($C_{11} = 1.657$, $C_{12} = 0.639$, $C_{44} = 0.796$, using the reduced notation for $\mathcal{C}$), $\mathbf{N} = (1, 0, 0)$ and $\mathbf{a} = (0, 1, 0)$.

Recall that thus far only the constant term from the shape function $\psi_j(\eta^*, \xi^*)$, Eq.(4.120), has been considered. However, for the linear term (and obviously any higher power if a curved interpolation is employed) this integral is well defined and finite *without* the limit process. Specifically, for the linear term $c_j(\theta)\rho$, the ρ integration is trivial and Eq.(4.121) becomes

$$-\frac{J_P^2}{8\pi^2} \int_{-1}^1 \int_0^{e(\eta)} \psi_k(\eta, \xi) \mathrm{d}\eta \mathrm{d}\xi \int_{\Theta_1}^{\Theta_2} c_j(\theta)\rho_L \frac{a_l}{a^3} \tilde{\mathcal{U}}(P_0, Q) \, \mathrm{d}\theta . \tag{4.132}$$

The fact that only the constant component of the shape function contributes a 'limit term', and moreover that this limit term eventually reduces to an innocuous integral, should not come as a surprise. What is being computed (in a Galerkin form) with the constant term is essentially the CPV 'free term' which is simply $\mathbf{u}(P)/2$.

Recapping, the boundary limit algorithm for anisotropic elasticity first splits the angle dependent part of the kernel function as in Eq.(4.122). The 'non-limit' $\varepsilon = 0$ term from this split, namely $\tilde{\mathcal{U}}(P_0, Q)$, will incorporate all of the singularity of the integral, and as it is independent of ρ, it can be treated as with the Laplace kernels. The remaining 'limit' $\varepsilon \neq 0$ term is expressed as a well behaved, computable integral. This process will now be applied to the second term in Eq.(4.117). The linear term from the shape function is, as above, relatively trivial, and thus only the constant term will be investigated.

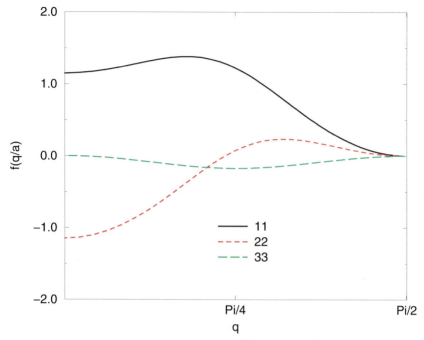

Figure 4.16 The diagonal components $f_0(q) = \tilde{\mathcal{U}}_{kk}(P_\varepsilon, Q) - \tilde{\mathcal{U}}_{kk}(P_0, Q)$ for silicon elastic constants.

4.6.2.2 Derivative of $\tilde{\mathcal{U}}$
Corresponding to Eq.(4.121), the second integral is

$$-\frac{J_P^2}{8\pi^2} \lim_{\varepsilon \to 0} \int_{-1}^{1} \int_{0}^{e(\eta)} \psi_k(\eta, \xi) \psi_j(\eta, \xi) \mathrm{d}\eta \mathrm{d}\xi \qquad (4.133)$$
$$\int_{\Theta_1}^{\Theta_2} \int_{0}^{\rho_L} \frac{\rho}{(a^2 \rho^2 + \varepsilon^2)^{1/2}} \frac{\partial}{\partial q_l} \tilde{\mathcal{U}}(P_\varepsilon, Q) \mathrm{d}\rho \mathrm{d}\theta \; ,$$

and analogously to Eq.(4.122) the derivative of $\tilde{\mathcal{U}}$ is split as

$$\frac{\partial}{\partial q_l} \tilde{\mathcal{U}}(P_\varepsilon, Q) = \frac{\partial}{\partial q_l} \tilde{\mathcal{U}}(P_0, Q) + \frac{\partial}{\partial q_l} \left[\tilde{\mathcal{U}}(P_\varepsilon, Q) - \tilde{\mathcal{U}}(P_0, Q) \right] \; . \qquad (4.134)$$

The q_l derivative can be expressed as

$$\frac{\partial}{\partial q_l} \tilde{\mathcal{U}} = \frac{\partial \tilde{\mathcal{U}}}{\partial \zeta} \frac{\partial \zeta}{\partial q_l} + \frac{\partial \tilde{\mathcal{U}}}{\partial \varphi} \frac{\partial \varphi}{\partial q_l} \qquad (4.135)$$

and thus the ρ integration for the first term is

$$\frac{\partial \tilde{\mathcal{U}}}{\partial \zeta}(P_0, Q) \int_{0}^{\rho_L} \frac{\rho}{(a^2 \rho^2 + \varepsilon^2)^{1/2}} \frac{\partial \zeta}{\partial q_l}(P_0, Q) \, \mathrm{d}\rho + \qquad (4.136)$$
$$\frac{\partial \tilde{\mathcal{U}}}{\partial \varphi}(P_0, Q) \int_{0}^{\rho_L} \frac{\rho}{(a^2 \rho^2 + \varepsilon^2)^{1/2}} \frac{\partial \varphi}{\partial q_l}(P_0, Q) \, \mathrm{d}\rho \; ,$$

104   THREE DIMENSIONAL ANALYSIS

as $\tilde{\mathcal{U}}_{,\zeta}(P_0, Q)$ and $\tilde{\mathcal{U}}_{,\varphi}(P_0, Q)$ are independent of ρ. The angle derivatives depend upon the chosen spherical coordinate system. To take a specific example, namely $l = 1$ and Z as the polar axis,

$$
\frac{\partial \zeta}{\partial q_1}(P_0, Q) = -\frac{q_2 - p_2}{(q_1 - p_1)^2 + (q_2 - p_2)^2}\Big|_{\varepsilon=0} = -\frac{1}{\rho}\frac{a_2}{a_1^2 + a_2^2} \tag{4.137}
$$

$$
\frac{\partial \varphi}{\partial q_1}(P_0, Q) = \frac{(q_1 - p_1)(q_3 - p_3)}{r_0^2\left[(q_1 - p_1)^2 + (q_2 - p_2)^2\right]^{1/2}}\Big|_{\varepsilon=0} = \frac{1}{\rho}\frac{a_1 a_3}{a^2\sqrt{a_1^2 + a_2^2}} \, ,
$$

where $r_0 = \|Q - P_0\|$. All angle derivatives, evaluated at (P_0, Q) are in fact of the form γ/ρ and we use the factor of ρ from the polar coordinate Jacobian to allow setting $\varepsilon = 0$ in the above expressions. The ρ integrand is therefore of the form $1/\sqrt{a^2\rho^2 + \varepsilon^2}$, and Eq.(4.136) is found to be

$$
\left[-\frac{\partial \tilde{\mathcal{U}}}{\partial \zeta}(P_0, Q)\frac{a_2}{a_1^2 + a_2^2} + \frac{\partial \tilde{\mathcal{U}}}{\partial \varphi}(P_0, Q)\frac{a_1 a_3}{a^2\sqrt{a_1^2 + a_2^2}}\right]\left(\frac{\log(2a\rho_L)}{a} - \frac{\log(\varepsilon^2)}{2a}\right) . \tag{4.138}
$$

The $\log(\varepsilon^2)$ term self-cancels in the integration over θ, as in Eq.(4.124). This will follow again from the symmetry $\tilde{\mathcal{U}}(-\mathbf{a}) = \tilde{\mathcal{U}}(\mathbf{a})$, or equivalently $\tilde{\mathcal{U}}(\zeta, \varphi) = \tilde{\mathcal{U}}(\zeta + \pi, \pi - \varphi)$, as differentiating yields

$$
\frac{\partial \tilde{\mathcal{U}}}{\partial \zeta}(\zeta + \pi, \pi - \varphi) = \frac{\partial \tilde{\mathcal{U}}}{\partial \zeta}(\zeta, \varphi)
$$

$$
\frac{\partial \tilde{\mathcal{U}}}{\partial \varphi}(\zeta + \pi, \pi - \varphi) = -\frac{\partial \tilde{\mathcal{U}}}{\partial \varphi}(\zeta, \varphi) . \tag{4.139}
$$

Before proceeding to the second term in Eq.(4.134), it is worth noting that the reason multiple spherical coordinate systems will be required is evident from Eq.(4.137). With Z as the polar axis, an *artificial* singularity at $q_1 - p_1 = q_2 - p_2 = 0$, is introduced, and thus this system should not be employed if this situation can arise on the element E. This will be discussed further in the next Section.

The 'non-limit' ρ integral from Eq.(4.134) is of the form

$$
\int_0^{\rho_L} \frac{\rho}{(a^2\rho^2 + \varepsilon^2)^{1/2}}\frac{\partial}{\partial q_l}\left[\tilde{\mathcal{U}}(P_\varepsilon, Q) - \tilde{\mathcal{U}}(P_0, Q)\right]\mathrm{d}\rho . \tag{4.140}
$$

Not surprisingly, this integral is well behaved and can be treated analogously to the limit term for $\tilde{\mathcal{U}}$. However, the precise details are more involved and, moreover, depend upon the value of l and the spherical coordinate system; for illustration we again choose $l = 1$ and the polar axis along Z. The derivative with respect to q_l is

$$
\left(\frac{\partial \tilde{\mathcal{U}}}{\partial \zeta}(P_\varepsilon, Q)\frac{\partial \zeta}{\partial q_l}(P_\varepsilon, Q) - \frac{\partial \tilde{\mathcal{U}}}{\partial \zeta}(P_0, Q)\frac{\partial \zeta}{\partial q_l}(P_0, Q)\right) + \tag{4.141}
$$

$$
\left(\frac{\partial \tilde{\mathcal{U}}}{\partial \varphi}(P_\varepsilon, Q)\frac{\partial \varphi}{\partial q_l}(P_\varepsilon, Q) - \frac{\partial \tilde{\mathcal{U}}}{\partial \varphi}(P_0, Q)\frac{\partial \varphi}{\partial q_l}(P_0, Q)\right)
$$

and only the ζ derivative is discussed in detail. Incorporating the jacobian from the change from ρ to s as in the previous section,

$$\frac{\rho}{(a^2\rho^2 + \varepsilon^2)^{1/2}} \left(\frac{\partial \tilde{\mathcal{U}}}{\partial \zeta}(P_\varepsilon, Q) \frac{\partial \zeta}{\partial q_l}(P_\varepsilon, Q) - \frac{\partial \tilde{\mathcal{U}}}{\partial \zeta}(P_0, Q) \frac{\partial \zeta}{\partial q_l}(P_0, Q) \right) \qquad (4.142)$$

$$= \frac{s}{(1 + a^2 s^2)^{1/2}} \left(\frac{\partial \tilde{\mathcal{U}}}{\partial \zeta}(P_0, Q) \frac{1}{s} \frac{a_2}{a_1^2 + a_2^2} - \frac{\partial \tilde{\mathcal{U}}}{\partial \zeta}(P_\varepsilon, Q) \frac{a_2 s - N_2}{D_s^2} \right) ,$$

where $D_s^2 = (a_1 s - N_1)^2 + (a_2 s - N_2)^2$. Recognizing that it will be permissible to set $\varepsilon = 0$, the Green's function evaluation at (P_ε, Q) can be replaced by (P_0, Q), and doing the necessary algebra, the above expression becomes

$$\frac{\partial \tilde{\mathcal{U}}}{\partial \zeta}(\mathbf{a}) \frac{s}{(1 + a^2 s^2)^{1/2} D_s^2} \frac{a_1^2 N_2 - a_2^2 N_2 - 2a_1 a_2 N_1}{(a_1^2 + a_2^2)} . \qquad (4.143)$$

Thus, the behavior for large s is at worst s^{-2}, and the integration over $0 < s < \infty$ is a finite quantity. Once again the transformation $s = \tan(q)/a$ is convenient, finally reducing Eq.(4.140) to

$$\frac{\partial \tilde{\mathcal{U}}}{\partial \zeta}(\mathbf{a}) \int_0^{\pi/2} f_{1,\zeta}(q)\mathrm{d}q = \qquad (4.144)$$

$$\frac{\partial \tilde{\mathcal{U}}}{\partial \zeta}(\mathbf{a}) \int_0^\infty \sin(q) \frac{a_1^2 N_2 - a_2^2 N_2 - 2a_1 a_2 N_1}{(a_1^2 + a_2^2)D_q^2}\mathrm{d}q$$

where $D_q^2 = (a_1 \sin(q) - aN_1 \cos(q))^2 + (a_2 \sin(q) - aN_2 \cos(q))^2$. The analogous $f_{1,\varphi}$ is a bit more complicated, and as a function of s is

$$f_{1,\varphi}(s) = \frac{sa^2(a_1^2 + a_2^2)^{1/2}(a_1 s - N_1)(a_2 s - N_2) - a_1 a_3(1 + a^2 s^2) D_s}{a^2 D_s (1 + a^2 s^2)^{3/2}(a_1^2 + a_2^2)^{1/2}} . \qquad (4.145)$$

This function appears to behave as s^{-1} (instead of the desired s^{-2}); however, if the numerator is rationalized (to replace $D(s)$ by $D^2(s)$) it is seen that the leading order s^6 term cancels, and the resulting fraction is s^5/s^7. An appropriate form for computation can then be obtained by the tangent subsitution. Finally, this second limit term contribution is also independent of ρ_L, and the complete integral can therefore be expressed as in Eq.(4.130).

4.6.3 Spherical Coordinates

As noted above, one cannot employ one fixed spherical coordinate system, as this introduces artificial singularities, and it is enough to contend with the actual ones. The root cause is that with spherical coordinates, the polar axis of the transformation is a singular point: $\varphi = 0, \pi$ and any value of ζ define the same point. Thus, in the computation of the integrals, a polar angle near zero or π must be avoided, which means that the polar axis should not coincide with $\mathbf{a}$ or any linear combination $\mathbf{a} - \gamma \mathbf{N}$. This shows up clearly in the functions f_ζ and f_φ, they contain denominators which must be kept positive. For the Z coordinate system this denominator is $(a_1^2 + a_2^2)D_s$, and for X and Y systems it will take the same form with the indices of $\mathbf{a}$ changed in an obvious manner.

106 THREE DIMENSIONAL ANALYSIS

A simple solution is to choose the polar axis as the Cartesian vector which lies nearest to $\mathbf{a} \otimes \mathbf{N}$. This assures that $f_0(q)$ can be evaluated without difficulty, and since $\mathbf{a}$ and $\mathbf{N}$ are orthogonal, it also guarantees that the denominators in f_ζ and f_φ stay away from zero. The spherical system will necessarily change as a function of θ, but this does not present a problem.

4.6.4 Second integration

The first analytic integration suffices for the most important task, displaying (and removing) the divergent $\log(\varepsilon^2)$ term, Eq.(4.124) and Eq.(4.123). However, it is beneficial to carry out a second exact integration, for two reasons. First, note that Eq.(4.123) and Eq.(4.138) contain a weak logarithmic singularity at $\xi \approx 0$. Suitable numerical methods could be employed for this integrable singularity, but we think it preferable if it is handled via analytic methods. Second, for the analysis of the hypersingular kernel, a second analytic integration is essential, and thus it is important to first check that the procedures are successful in this simpler setting.

As with the Laplace kernel, the dependence of the integrand on θ in Eq.(4.123) is harmless, it is $\xi = 0$ (and hence $\rho_L = 0$) which must be dealt with analytically. Fortunately, the changes of variables and interchange of integrations is accomplished exactly as for the simpler kernel functions, and there is no need to repeat this here. The only thing worth noting for the anisotropic analysis is that with the two changes of variables, $\theta \to t$ and $\{t, \xi\} \to \{\Lambda, \Psi\}$, $\cos(\theta)$ becomes $\cos(\Psi)$ and $\sin(\theta)$ becomes $-\sin(\Psi)$. Thus, $a(\theta)$ becomes simply $a(\Psi)$ and is a constant as far as the Λ integration is concerned, and $\tilde{\mathcal{U}}(P_0, Q) = \tilde{\mathcal{U}}(\mathbf{a}(\theta))$ becomes $\tilde{\mathcal{U}}(\mathbf{a}(\Psi))$, also independent of Λ. Thus, it is a simple matter to carry out the Λ integration for the singular terms, and the remaining two dimensional integral is well behaved and amenable to numerical quadrature. Finally, there is no singularity present in the limit terms, *e.g.*, Eq.(4.130), and thus there is no pressing need to carry out this second integration.

4.6.5 Edge Adjacent Integration

For the $\mathcal{U}$ and $\mathcal{T}$ kernels, the singular adjacent edge and vertex integrals are finite with P on the boundary, and thus a limit process is not mandatory. However, as we wish to demonstrate that the methods in this chapter apply to anisotropic analysis, and to lay the groundwork for analyzing the hypersingular kernel, the $\mathcal{T}$ kernel edge-adjacent integral will now be considered. The simpler $\mathcal{U}$ kernel and the vertex-adjacent integral are handled in a similar fashion. This discussion will be brief, repeating as little as possible of the discussion for the Laplace Green's function.

Recall that the two elements are oriented so that the shared edge is defined by $\xi = 0$ in E_P, and $\xi^* = 0$ for E_Q, and the singularity occurs when $\eta = -\eta^*$. Corresponding to Eq.(4.117) we have the integral

$$\frac{J_P J_Q}{8\pi^2} \int_{-1}^{1} \int_{0}^{e(\eta)} \psi_k(\eta, \xi) \mathrm{d}\eta \mathrm{d}\xi \tag{4.146}$$

$$\int_{-1}^{1} \int_{0}^{e(\eta^*)} \psi_j(\eta^*, \xi^*) \left(-\frac{R_l}{r^3} \tilde{\mathcal{U}}(P, Q) + \frac{1}{r} \frac{\partial}{\partial q_l} \tilde{\mathcal{U}}(P, Q) \right) \mathrm{d}\eta^* \mathrm{d}\xi^* \ .$$

The two polar coordinate transformations employed are

$$\begin{aligned} \eta^* &= \rho\cos(\theta) - \eta \\ \xi^* &= \rho\sin(\theta) , \end{aligned}$$ (4.147)

followed by

$$\begin{aligned} \rho &= \Lambda\cos(\Psi) \\ \xi &= \Lambda\sin(\Psi) , \end{aligned}$$ (4.148)

with the goal of integrating Λ analytically. The combined jacobian for the two transformations is $\cos(\Psi)\Lambda^2$, and examining just the $\tilde{\mathcal{U}}$ term from Eq.(4.146) (the derivative term is handled similarly) this integral takes the form

$$-\frac{a_l}{b^3}\int_0^{\Lambda_L}\psi_j(\Lambda,\Psi)\tilde{\mathcal{U}}(P,Q)\,\mathrm{d}\Lambda , $$ (4.149)

where $r^2 = b^2\Lambda^2$. There is obviously no longer any trace of the singularity, and thus this integral could be computed numerically. However, this will not be the case for the hypersingular kernel, and an analytic integration with respect to Λ will be essential.

Obtaining an integrand that can be integrated exactly can once again be accomplished by an appropriate splitting of the angle term. For the point $P = (\eta,\xi)$, define the point $P_\eta = (\eta,0)$ to be its projection onto the common edge; it is therefore also a point in E_Q (in fact, $(-\eta^*,0)$). The integral is then split as

$$-\frac{a_l}{b^3}\left[\tilde{\mathcal{U}}(P_\eta,Q)\int_0^{\Lambda_L}\psi_j(\Lambda,\Psi)\,\mathrm{d}\Lambda + \int_0^{\Lambda_L}\psi_j(\Lambda,\Psi)\left(\tilde{\mathcal{U}}(P,Q)-\tilde{\mathcal{U}}(P_\eta,Q)\right)\mathrm{d}\Lambda\right] , $$ (4.150)

noting that $\tilde{\mathcal{U}}(P_\eta,Q)$ is a function of η and θ, independent of Λ. The first integral can therefore be evaluated analytically. Moreover, note that $\tilde{\mathcal{U}}(P,Q) - \tilde{\mathcal{U}}(P_\eta,Q)$ vanishes as $\xi \to 0$, and thus this function is of the form $\Lambda f(\Lambda)$. For the hypersingular kernel, this integral will be one order less singular at $\Lambda = 0$, and will therefore be amenable to numerical quadrature.

The analysis of the vertex adjacent singular integral can be modeled on this procedure, with appropriate modification to the choice of polar coordinate transformations.

To summarize this long Section dealing with anisotropic elasticity, it has been shown that the boundary limit and analytic integration procedures can be executed for this complicated Green's function. While this does not establish that every fundamental solution is amenable to these methods, it does seem likely that this is the case.

CHAPTER 5

SURFACE GRADIENT

Synopsis: In many applications, it is necessary to have complete information about first order boundary derivatives, *e.g.*, the potential gradient for Laplace problems or the complete stress tensor in elasticity. This chapter will demonstrate that these derivatives can be computed very efficiently: a full boundary integration is not required, only *local singular integrals* need to be evaluated. Moreover, as these integrals can be evaluated partially analytically, the results are also highly accurate. The key is to exploit the boundary limit definition of the singular integrals, writing the gradient equation as a difference of interior and exterior limits. The chapter concludes with a discussion of cubic Hermite interpolation in two dimensions, a smooth approximation ($\mathcal{C}^1$) that exploits the ability to compute the surface gradient.

5.1 INTRODUCTION

A boundary integral solution yields complete knowledge of the surface functions: the principal function and a normal derivative, *e.g.*, displacement and traction in the case of elasticity. However, in some applications it is necessary to know more than just the normal derivative, all first order derivatives must be obtained, *e.g.*, the gradient of the potential for the Laplace equation or the surface stress tensor for elasticity. This chapter will present an algorithm for the evaluation of these derivatives.

The Symmetric Galerkin Boundary Element Method. By Sutradhar, Paulino and Gray **109**
ISBN 978-3-540-68770-2 ©2008 Springer-Verlag Berlin Heidelberg

To motivate this discussion, an important general class of simulations that require first order derivatives is *moving boundary* (also called *free boundary*) problems. The task in these applications is to track the evolution of the domain, and thus the primary quantity of interest from the solution of the differential equation is the surface velocity. In many instances, this velocity is a function of the surface gradient.

As a specific example, a problem that has been extensively studied using integral equation methods [125, 198] is a nonlinear potential flow model of a water wave, a simple geometry shown in Fig. 5.1. If the fluid in the tank is modeled as inviscid and irrotational, the general Navier-Stokes equations for fluid flow reduce to the Laplace equation for the *velocity potential* ϕ, where $\nabla \phi$ is velocity of the fluid, **u**. In addition to the no-flow $\partial \phi / \partial \mathbf{n} = 0$ boundary conditions on the tank walls, the potential will be specified on the free surface of the fluid. This boundary condition evolves in time through the solution of the Bernoulli equation

$$\frac{\partial}{\partial t}\phi = -gz - \frac{1}{2}\nabla\phi \cdot \nabla\phi - \frac{p_a}{\rho}, \qquad (5.1)$$

where g is gravity, p_a atmospheric pressure, and ρ the fluid density; thus, the gradient also enters into the free surface boundary condition. Figures 5.2, 5.3 and 5.4 show the wave position and the corresponding potential and velocity solutions, at several times for a two-dimensional wave calculation [98].

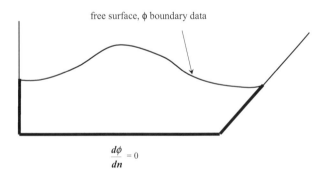

Figure 5.1 Water wave tank.

Given that the boundary solution is of primary concern in a moving boundary simulation, it is not surprising that integral equation methods are an attractive approach. An obvious key advantage is that remeshing the evolving domain boundary is easier than regridding the volume at each time step, and an accurate determination of the surface velocity should be more accurate with a boundary approach. Integral equations, as evidenced below, work directly with the derivatives, as opposed to a numerical differentiation of the initial solution generally required by a volume method. Thus, for equivalent computational effort, a more accurate calculation of the surface gradient should be possible.

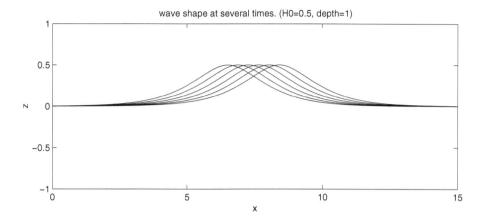

Figure 5.2   Time slices of the water wave position.

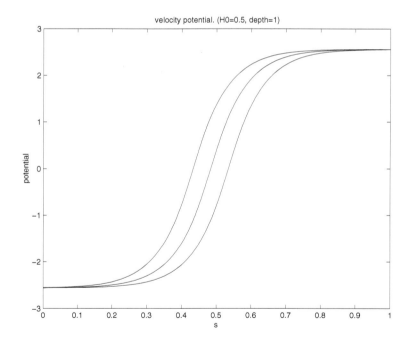

Figure 5.3   Computed potential on the water wave surfaces shown in Fig. 5.2.

An abbreviated list of moving boundary applications that have been investigated using integral equation methods include fluid motion [56, 125], interface motion in solids [219] and void evolution [128]. Two somewhat different but related problems are contact analysis [141] and shape optimization [40]. The goal in these analyses is also to determine a geometry, in these examples either the contact region or the optimal shape, but there is no explicit time variable. The multiple solutions in

112   SURFACE GRADIENT

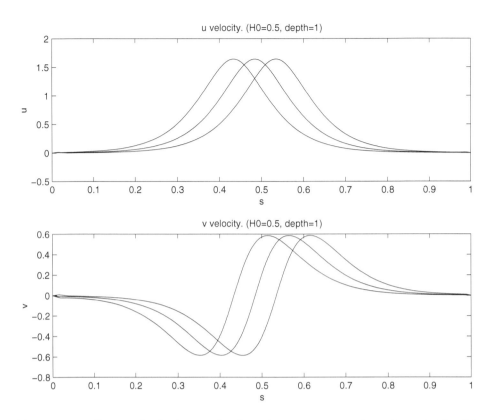

Figure 5.4   Computed velocity $\mathbf{u} = (u,v)$ on the water wave surfaces shown in Fig. 5.2.

this case are the iterations attempting to converge to the correct solution. Thus, similar to a moving boundary problem, the domain will evolve during the course of the iteration.

The boundary integral representation of the gradient will involve hypersingular kernels. As a consequence, many different methods have been developed to evaluate surface derivatives, the papers [106, 300] provide a good entrance to the literature on this subject. It will come as no great shock that the method adopted herein is based upon a boundary limit definition of the integral equations. Although the virtues of the boundary limit have been emphasized in this book, it has also been pointed out that there are other perfectly valid ways for tackling the singular integrals. Gradient evaluation, however, may be the one topic for which limits are far superior: it will be seen that the limit approach is essential for constructing a representation of the gradient that does not require a complete boundary integration. The algorithm based upon this formulation will therefore turn out to be fast and accurate.

5.2   GRADIENT EQUATIONS

To present the equations for the surface derivatives, it suffices, as usual, to employ the Laplace equation (and thus also the terminology 'gradient' to mean first order derivatives). The extension to other equations is straightforward, in particular surface stress evaluation. Moreover, it is important to note that when the Green's

GRADIENT EQUATIONS **113**

function is not a simple algebraic function, the gradient representation presented below will in fact significantly simplify the analysis. This will be brought out in Chapter 6 for the specific case of $3D$ axisymmetric analysis, wherein the Green's function and its derivatives involve complete elliptic integrals. In gradient evaluation for axisymmetric problems, these elliptic integrals will disappear.

We first present the general framework (using the notation for $2D$), and then separately discuss the limit evaluation in two and three dimensions. Differentiating Eq.(2.10) and the corresponding exterior equation Eq.(2.8) with respect to the coordinates of P, we obtain the limit forms of the gradient equations

$$\nabla\phi(P) = \lim_{P_I \to P} \int_\Gamma \left[\nabla G(P_I, Q) \frac{\partial\phi}{\partial\mathbf{n}}(Q) - \phi(Q)\nabla\frac{\partial G}{\partial\mathbf{n}}(P_I, Q) \right] \mathrm{d}Q \quad (5.2)$$

$$0 = \lim_{P_E \to P} \int_\Gamma \left[\phi(Q) \frac{\partial G}{\partial\mathbf{n}}(P_E, Q) - G(P_E, Q)\frac{\partial\phi}{\partial\mathbf{n}}(Q) \right] \mathrm{d}Q . \quad (5.3)$$

Assuming the solution for the unknown boundary data has already been obtained, the integrands only involve known quantities. However, the kernel $\nabla\left(\partial G/\partial\mathbf{n}\right)$ is hypersingular, and thus the C^1 condition (see Chapter 2) makes direct collocation of the *nodal* derivative values difficult.

As a Galerkin approximation successfully bypasses the difficulties with the hypersingular evaluation, consider the Galerkin form of the interior gradient equation

$$\int_\Gamma \hat{\psi}_k(P)\nabla\phi(P) \, \mathrm{d}P = \quad (5.4)$$

$$\lim_{P_I \to P} \int_\Gamma \hat{\psi}_k(P) \int_\Gamma \left[\nabla_P G(P_I, Q) \frac{\partial\phi}{\partial\mathbf{n}}(Q) - \phi(Q)\nabla_P \frac{\partial G}{\partial\mathbf{n}}(P_I, Q) \right] \mathrm{d}Q \, \mathrm{d}P .$$

As stated above, everything on the right hand side is known, and assembling this quantity for every weight function results in a vector b, the above integral being the component b_k. For the the left hand side, interpolation gives

$$\nabla\phi(P) = \sum_j \nabla\phi(P_j)\psi_j(t) \quad (5.5)$$

and thus

$$\int_\Gamma \hat{\psi}_k(P)\nabla\phi(P) \, \mathrm{d}P = \sum_j \nabla\phi(P_j)\psi_k(t)\psi_j(t) = \sum_j a_{k,j}\nabla\phi(P_j) \quad (5.6)$$

This term therefore gives rise to a matrix A, and the system of linear equations for the unknown gradient

$$A\left[\phi_{\mathcal{X}}\right] = b_{\mathcal{X}} , \quad (5.7)$$

where $\mathcal{X}$ denotes a coordinate of P and $\phi_{\mathcal{X}}$ is the gradient component. Although a system of equations has to be solved (for each coordinate), the coefficient matrix A is the same for all components and is symmetric, positive definite and sparse. The matrix elements are comprised simply of integrals of products of shape functions, and thus assembling the matrix is a trivial computation even for large scale problems. The significant drawback, however, is the computational cost of evaluating the right hand side vector b, which requires a complete double integral over the boundary, again for each component.

114   SURFACE GRADIENT

The computational cost of the above algorithm can be dramatically reduced by exploiting the exterior limit equation in Eq.(5.2). It appears to be useless for computing the gradient, as this quantity does not appear anywhere in this expression. However, unlike with potential and normal derivative equations, the exterior gradient equation is in fact different from the interior equation, and it does provide useful information.

Specifically, subtracting this exterior equation from its interior form, we obtain (with shorthand notation)

$$\int_{\Gamma} \hat{\psi}_k(P) \nabla \phi(P) \, \mathrm{d}P = \tag{5.8}$$

$$\left\{ \lim_{P_I \to P} - \lim_{P_E \to P} \right\} \int_{\Gamma} \hat{\psi}_k(P) \int_{\Gamma} \left[\nabla G \frac{\partial \phi}{\partial \mathbf{n}}(Q) - \phi(Q) \nabla \frac{\partial G}{\partial \mathbf{n}} \right] \mathrm{d}Q \, \mathrm{d}P \ .$$

The advantage of this formulation is that now *only the terms that are discontinuous crossing boundary* contribute to the right hand side vector $b_{\mathcal{X}}$ (the coefficient matrix A has not changed). Thus, **all nonsingular integrals**, by far the most time consuming to evaluate, vanish. The integrations that are non-zero are solely the coincident integral and the hypersingular adjacent integral (or adjacent edge in three dimensions). Moreover, for the integrals that do survive, the evaluation for interpolations beyond linear is also greatly simplified. Higher order terms from the shape functions will be continuous crossing the boundary and hence do not contribute. Similarly, the remainder terms that arise from the splitting of the integrands and that would normally have to be evaluated numerically, Section 4.3, are also continuous and can be ignored.

We again emphasize that, without a boundary limit definition, the gradient formulation in Eq.(5.8) would not be possible. Roughly speaking, Eq.(5.8) states that the gradient at a point P only depends on the local potential and flux values at P, and the first order terms in the expansions about this point. Although this is not quite the case, as the local values are coupled through the Galerkin system of linear equations Eq.(5.7), the local nature of this statement is nevertheless mathematically reasonable.

5.2.1   Limit Evaluation in two dimensions

The two dimensional analysis, for a linear element, is discussed in this section. This is naturally somewhat simpler than the corresponding work in three dimensions, and it will turn out that the entire coincident integral can be evaluated analytically. In three dimensions, only a partial analytic evaluation is possible.

The difference of the limits in the gradient equation wipes out all contributions except the coincident and adjacent singular integrals, and in the latter case, only the hypersingular integral contributes. In the following, we examine the equation for the x component of the gradient, there being no difference in the analysis for the y component. The reader should observe that within the limit-difference gradient formulation the hypersingular integral does not 'behave' as hypersingular; the precise meaning of this statement will become clear from the calculations below.

Recall that the linear element shape functions are given by

$$\psi_1(s) = 1 - s \qquad \psi_2(s) = s \tag{5.9}$$

and defined on the parameter space $0 \le s \le 1$. We begin with the coincident integral $E_P = E_Q = E$, where the element E is defined by the two nodes $P_1 = (x_1, y_1)$ and $P_2 = (x_2, y_2)$. The surface and surface potential interpolations are then

$$
\begin{aligned}
P(t) &= P_1\psi_1(t) + P_2\psi_2(t) \\
Q(s) &= P_1\psi_1(s) + P_2\psi_2(s) \\
\phi(s) &= \phi(P_1)\psi_1(s) + \phi(P_2)\psi_2(s) \\
\frac{\partial\phi}{\partial\mathbf{n}}(s) &= \frac{\partial\phi}{\partial\mathbf{n}}(P_1)\psi_1(s) + \frac{\partial\phi}{\partial\mathbf{n}}(P_2)\psi_2(s) \ .
\end{aligned}
\tag{5.10}
$$

5.2.1.1 Coincident The coincident integral therefore results in the integrals

$$
\mathcal{P}_{k,j} = \phi(P_j)J^2 \int_0^1 \psi_k(t) \int_0^1 \psi_j(s)\nabla_{xP}\frac{\partial G}{\partial\mathbf{n}}(P_\varepsilon(t), Q(s)) \, \mathrm{d}s \, \mathrm{d}t
\tag{5.11}
$$

and

$$
\mathcal{F}_{k,j} = \frac{\partial\phi}{\partial\mathbf{n}}\phi(P_j)J^2 \int_0^1 \psi_k(t) \int_0^1 \psi_j(s)\nabla_{xP}G(P_\varepsilon(t), Q(s)) \, \mathrm{d}s \, \mathrm{d}t \ ,
\tag{5.12}
$$

where J is the constant jacobian, $\{k, j\} = 1, 2$, and for simplicity the limit process in Eq.(5.8) is understood. For the interior and exterior limits $R = R(t, s) = Q(s) - P(t)$ is given by

$$
R(s, t) = (a_x(s - t) \pm \varepsilon N_1, a_y(s - t) \pm \varepsilon N_2) \ ,
\tag{5.13}
$$

where the coefficients $\{a_x, a_y\}$ are functions only of the coordinates of P_1 and P_2 and $\mathbf{N}(t) = (N_1, N_2)$ is the unit exterior normal at $P(t)$. Taking the difference of the limits and setting $a^2 = a_x^2 + a_y^2 = \|P_1 - P_2\|^2$, the hypersingular kernel results in

$$
\frac{\partial}{\partial x_P}\frac{\partial G}{\partial\mathbf{n}}(P_\varepsilon, Q) = -\frac{2}{\pi}\varepsilon\frac{a_x(s - t)}{\left(a^2(s - t)^2 + \varepsilon^2\right)^2} \ ,
\tag{5.14}
$$

and for the first derivative of the Green's function we have

$$
\nabla_{xP}G(P_\varepsilon, Q) = \frac{\varepsilon}{\pi}\frac{N_1}{\left(a^2(s - t)^2 + \varepsilon^2\right)} \ .
\tag{5.15}
$$

Note that the most singular terms from the gradient hypersingular kernel, N_1/r^2 and ε^2/r^4, have disappeared: they involve even powers of ε and are therefore continuous crossing the boundary. In effect, the gradient hypersingular kernel is less singular than the corresponding standard (normal derivative) integral.

The above functions are easily integrated completely analytically, yielding simply

$$
\begin{aligned}
\mathcal{P}_{1,1} &= 0 & \mathcal{P}_{1,2} &= -a_x/2a^2 \\
\mathcal{P}_{2,1} &= a_x/2a^2 & \mathcal{P}_{2,2} &= 0
\end{aligned}
\tag{5.16}
$$

and for the flux integral

$$
\begin{aligned}
\mathcal{F}_{1,1} &= -N_1/3a & \mathcal{F}_{1,2} &= -N_1/6a \\
\mathcal{F}_{2,1} &= -N_1/6a & \mathcal{F}_{2,2} &= -N_1/3a \ .
\end{aligned}
\tag{5.17}
$$

116   SURFACE GRADIENT

5.2.1.2 Adjacent   For the adjacent integration, it will turn out that the singularity in the first derivative of the Green's function is sufficiently weak that this integral is continuous, and therefore vanishes in the limit. To evaluate the hypersingular integral, we keep the above element for the outer P integral, and consider the 'right adjacent' element $E_Q = (P_2, P_3)$ (the left adjacent $E_Q = (P_0, P_1)$ follows in a similar fashion). We now have

$$Q(s) = P_2\psi_1(s) + P_3\psi_2(s) \tag{5.18}$$

and the common point is at $t = 1$, $s = 0$. As in Chapter 3, the change of variables

$$\begin{aligned} 1 - t &= \rho\cos(\theta) \\ s &= \rho\sin(\theta) \end{aligned} \tag{5.19}$$

is employed to obtain

$$\begin{aligned} R(s,t) &= Q(s) - P(t) = (b_x\rho \pm \varepsilon N_1, b_y\rho \pm \varepsilon N_2) \\ r^2 &= a_2\rho^2 \pm a_1\varepsilon\rho + \varepsilon^2 \end{aligned} \tag{5.20}$$

the sign depending upon the limit direction. In this case, due to the sign change in r^2, the expressions for the hypersingular kernel for the two limit directions have separate forms, and must be treated separately. Moreover, note that of the four shape function products, only $\psi_2(t)\psi_1(s)$ does not vanish at the common point $\rho = 0$. The other products will therefore contribute a factor of ρ to the integrand, and this is sufficient to kill off the singularity and the difference of the limits necessarily vanishes. Thus, only the $k = 2$, $j = 1$ term contributes, and only through its value ($= 1$) at $\rho = 0$. Carrying out the integration for this term we find the somewhat lengthy expression

$$\begin{aligned} \frac{\dot{\jmath}_P \, \dot{\jmath}_Q}{2\,(4\,a_2 - a_1^2)^{3/2}\,a_2^2} &\big[-8\,a_2^2 n_2\,(b_y\,N_1 + N_2\,b_x) - 16\,a_2^2 n_1\,b_x\,N_1 \\ &-4\,a_2^2 a_1\,n_1 + 12\,a_2\,a_1\,b_x\,(n_2\,b_y + n_1\,b_x) + 4\,a_2^2 a_1\,N_1\,(n_2\,N_2 + n_1\,N_1) \\ &+ n_1\,a1^3 a2 - 2\,a1^3 b_x\,(n_2\,b_y + n_1\,b_x) \big] \ . \end{aligned} \tag{5.21}$$

The variables in this expression are functions of θ, and this integration can be carried out numerically.

An important observation from this calculation is that, in the limit difference, the individual coincident and adjacent hypersingular integrals have *finite* limits: unlike for the normal derivative equation, there are no $\log(\varepsilon^2)$ divergent terms present. This can be readily understood: the expected divergences depend upon ε^2, independent of the sign of ε, and hence cancel in the subtraction of the limits. Thus, the difference of the limits erases the most troublesome singular terms, and does not exhibit the 'usual' hypersingular divergent behavior.

Observe that the automatic disappearance of the divergent terms means that the continuity requirements for the gradient hypersingular evaluation are weaker than C^0 – the gradient limits exist even if ϕ is discontinuous across element boundaries. This would indicate that, with this approach, the evaluation of second order derivatives might be possible. Although we will not pursue this here, this is indeed the case for smooth geometries in two dimensions [185].

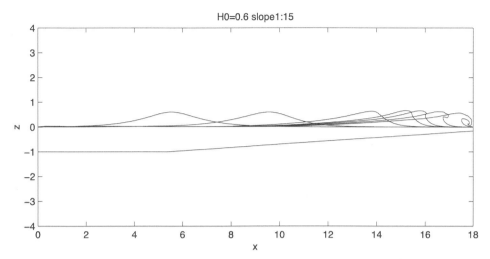

Figure 5.5 Several time slices for the potential flow model of a wave breaking over a sloping beach. The slope in this example is 1 : 15.

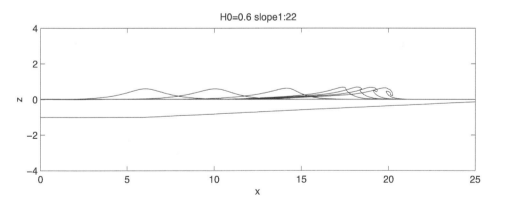

Figure 5.6 Several time slices for the potential flow model of a wave breaking over a sloping beach. The slope in this example is 1 : 22.

Before examining the application of this algorithm to surface stress evaluation, results for the wave modeling problem introduced at the beginning of this chapter are displayed in Figures 5.5 and 5.6. These figures show several time slices of the wave position for a wave approaching a sloping beach (H_0 is the wave height at

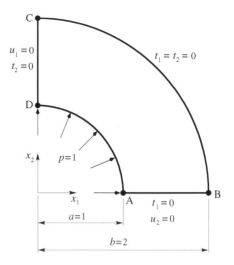

Figure 5.7 A model of the Lamé problem.

crest of the initial solitary wave). The difference in the two calculations is the slope of the sea floor, and note the overturning of the wave. A key point is that a large number of time steps are required for these simulations, and thus the reduction of the gradient calculation to just the evaluation of singular integrals is an important advantage of the algorithm.

5.2.2 Example: Surface Stress

This section will demonstrate the application of the gradient algorithm to a standard problem in elasticity. The Lamé problem is defined as follows: consider a thick-walled cylinder of inner radius a and outer radius b, subjected to a uniform pressure p_i on the inner surface and zero traction on the outer wall. Due to symmetry, only a quarter of the structure is modeled, as shown in Fig. 5.7. The introduced radial surfaces have symmetry boundary conditions, $u_x = 0$ and $\tau_y = 0$ on the y–axis and $u_y = 0$ and $\tau_x = 0$ on the x–axis.

For this example, take $a = 1$, $b = 2$, and $p_i = 1$ in consistent units. The stresses are obviously symmetrical about the cylinder's axis, and thus the stress distribution along a radial line, e.g., line AB in Fig. 5.7, is of interest.

For the initial boundary integral solution, and for the post-processing of the surface stress, a quadratic element has been employed. As noted above, implementing a quadratic interpolation for the gradient requires very little additional effort beyond what is needed for a linear approximation. The higher order terms from the shape functions, as well as the nonsingular terms from a splitting of the kernel functions, all give rise to contributions that are continuous crossing the boundary. Thus, these integrals vanish in the difference of the limits and can be ignored.

The interior and exterior stress equations developed in Chapter 2 are

$$\sigma_{lk}(P_I) = \int_\Gamma [D_{lkm}(P_I, Q)\tau_m(Q) - S_{lkm}(P_I, Q)u_m(Q)] \, dQ$$

$$0 = \int_\Gamma [D_{lkm}(P_E, Q)\tau_m(Q) - S_{lkm}(P_E, Q)u_m(Q)] \, dQ , \quad (5.22)$$

where $P_I = P - \varepsilon\mathbf{N}$, $P_E = P + \varepsilon\mathbf{N}$ are interior and exterior points respectively, and $\varepsilon > 0$. As with the Laplace equation, we take the difference of these two equations and then apply a Galerkin approximation. Computed boundary stresses on the boundary segments AB, BC, CD are compared with the analytical solutions available from the literature [277] in Figs. 5.8-5.11.

To study the convergence of the algorithm, the same number M of uniform quadratic elements used to mesh each of the four segments (resulting in a coarser mesh on segment BC). The $\mathcal{L}_2$ norm and maximum errors in the stress are plotted as functions of M in Figs. 5.12 and 5.13. These graphs show that the numerical results for boundary stresses converge very well with the mesh refinement.

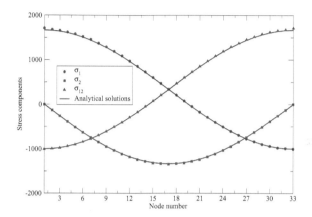

Figure 5.8 Comparison of stress components along AB

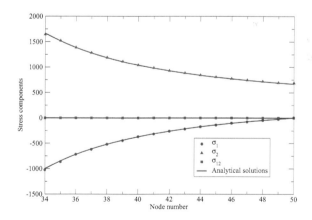

Figure 5.9 Comparison of stress components along BC

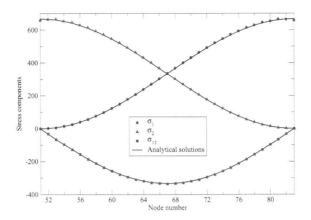

Figure 5.10 Comparison of stress components along CD

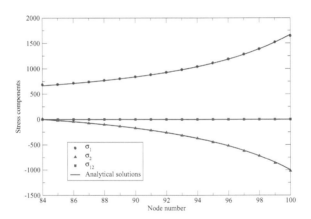

Figure 5.11 Comparison of stress components along DA

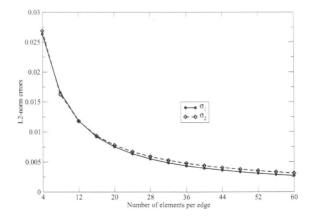

Figure 5.12 L2-norm errors.

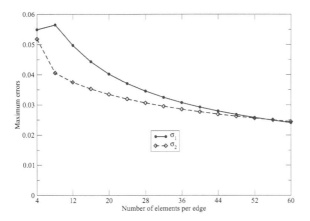

Figure 5.13 Maximum errors.

5.2.3 Limit Evaluation in three dimensions

At the risk of being overly repetitious, we wish to emphasize again that the most notable difference between the limit-difference gradient equation

$$\int_\Sigma \psi_k(P)\nabla\phi(P)\,dP = \tag{5.23}$$
$$\left\{\lim_{P_I \to P} - \lim_{P_E \to P}\right\} \int_\Sigma \hat{\psi}_k(P) \int_\Sigma \left[\nabla G \frac{\partial \phi}{\partial \mathbf{n}}(Q) - \phi(Q)\nabla\frac{\partial G}{\partial \mathbf{n}}\right] dQ\,dP$$

and the normal derivative equation

$$\int_\Sigma \psi_k(P)\frac{\partial \phi}{\partial \mathbf{N}}(P)\,dP = \tag{5.24}$$
$$\left\{\lim_{\varepsilon \to 0}\right\} \int_\Sigma \hat{\psi}_k(P) \int_\Sigma \left[\frac{\partial G}{\partial \mathbf{N}}(P_\varepsilon,Q)\frac{\partial \phi}{\partial \mathbf{n}}(Q) - \phi(Q)\frac{\partial^2 G}{\partial \mathbf{n}\partial \mathbf{N}}(P_\varepsilon,Q)\right] dQ\,dP,$$

relates to the divergences in the hypersingular integrals. For the flux equation, the coincident and edge-adjacent hypersingular integrals are not separately finite. However, the divergent quantities are of the form $\log(\varepsilon^2)$, independent of the sign of ε (i.e., the limit direction), and they therefore cancel in the limit-difference, Eq.(5.23). Thus, for gradient evaluation, the coincident and edge-adjacent hypersingular integrals are independently finite quantities, and in this sense Eq.(5.23) is much simpler to deal with than the normal derivative equation.

As in two dimensions, only the coincident and adjacent-edge integrals in Eq.(5.23) will survive, and in the latter case, only the hypersingular kernel is sufficiently singular to produce a nonvanishing integral. The singular integration algorithms for evaluating these terms are almost entirely the same as presented in Chapter 4, and for the gradient discussion here it once again suffices to employ a linear interpolation. The one key difference is, not surprisingly, with the coincident hypersingular integration, and thus the discussion will focus solely on this calculation.

122 SURFACE GRADIENT

Denoting a particular coordinate direction by $\mathbf{E}_k$, the gradient hypersingular kernel function is given by

$$\frac{\partial^2 G}{\partial \mathbf{E}_k \partial \mathbf{n}}(P, Q) = \frac{1}{4\pi}\left(\frac{\mathbf{n}\cdot\mathbf{E}_k}{r^3} - 3\frac{(\mathbf{n}\cdot\mathbf{R})(\mathbf{E}_k\cdot\mathbf{R})}{r^5}\right). \tag{5.25}$$

As before, $\mathbf{R} = Q - (P \pm \varepsilon\mathbf{N})$, $r = \|\mathbf{R}\|$, and $\mathbf{N}$ is the unit outward normal on the P element. Transferring to parameter spaces, Eq.(5.23) becomes the four dimensional integral

$$\sum_{j=1}^{3}\phi(Q_j)\int_{-1}^{1}\int_{0}^{e(\eta)}\psi_k(\eta,\xi)\int_{-1}^{1}\int_{0}^{e(\eta^*)}\psi_j(\eta^*,\xi^*)J_P^2\frac{\partial^2 G}{\partial \mathbf{E}_k\partial\mathbf{n}}\,\mathrm{d}\xi^*\mathrm{d}\eta^*\,\mathrm{d}\xi\mathrm{d}\eta\ , \tag{5.26}$$

where $e(\eta) = \sqrt{3}(1 - |\eta|)$. Rapidly summarizing the algorithm in Chapter 4, evaluation of this integral involves two polar coordinate transformations and analytic integration of the radial variables. The first step is to replace $\{\eta^*, \xi^*\}$ with a polar coordinate system centered at (η, ξ),

$$\begin{aligned}\eta^* - \eta &= \rho\cos(\theta) \\ \xi^* - \xi &= \rho\sin(\theta)\ ,\end{aligned} \tag{5.27}$$

and integrate ρ analytically. With this transformation,

$$\mathbf{R} = (\,a_1\rho \pm N_1\varepsilon, a_2\rho \pm N_2\varepsilon, a_3\rho \pm N_3\varepsilon\,) \tag{5.28}$$

and thus, independent of limit direction, $r^2 = (a^2\rho^2 + \varepsilon^2)$. For what follows it is again important to note that the coefficients in $\mathbf{R}$ are of the form

$$a_k = a_k(\theta) = a_{k,c}\cos(\theta) + a_{k,s}\sin(\theta) \tag{5.29}$$

where $a_{k,c}$ and $a_{k,s}$ are functions only of the nodal coordinates of the element E.

For purposes herein, the shape function $\psi_j(Q)$ can be ignored, as the only term of interest (the most singular) comes from $\psi_j(\eta^*, \xi^*)$ evaluated at $\rho = 0$. Similarly, it is convenient to drop the factor of J_P^2. The integrand, the difference of the interior and exterior kernel functions, is then

$$-6\varepsilon\frac{a_k\rho}{\left(\varepsilon^2 + a^2\,\rho^2\right)^{5/2}}\ , \tag{5.30}$$

and integrating $0 < \rho < Q_R$ yields

$$-2\frac{1}{\varepsilon}\frac{Q_R^3\,a_k}{(\,\varepsilon^2 + a^2\,Q_R^2\,)^{3/2}}\ . \tag{5.31}$$

This quantity behaves as ε^{-1} as $\varepsilon \to 0$, and is clearly a problem. This type of expression does not show up in the integration of the normal derivative kernel, and the treatment of this apparently divergent quantity is therefore the new aspect of the analysis. Fortunately, as will now be shown, this divergence will go away, once again by suitably integrating completely around the singular point.

To see this, note that the limiting form of Eq.(5.31) for ε small, obtained by replacing $(\varepsilon^2 + a^2Q_R^2)^{-3/2}$ with $a^{-3}Q_R^{-3}$, is $-2a_k/(a^3\varepsilon)$ and satisfies

$$-\int_{0}^{2\pi}\frac{2a_k}{a^3\varepsilon}\,\mathrm{d}\theta = -\frac{2}{\varepsilon}\int_{0}^{2\pi}\frac{a_k}{a^3}\,\mathrm{d}\theta = 0\ . \tag{5.32}$$

This follows from Eq.(5.29), noting that $a_k(\theta+\pi) = -a_k(\theta)$ and $a(\theta+\pi) = a(\theta)$; as the integrand is independent of Q_R, this integration around the singular point can be carried out. It is therefore permissible to subtract this limiting quantity from Eq.(5.31), resulting in

$$-2\frac{1}{\varepsilon}\frac{a_k\left(Q_R^3 a^3 - (\varepsilon^2 + a^2 Q_R^2)^{3/2}\right)}{a^3(\varepsilon^2 + a^2 Q_R^2)^{3/2}}. \tag{5.33}$$

It is tempting at this point to invoke a Taylor series expansion in ε of the numerator. However, during the course of the outer integration P will come close to the element edges, in which case Q_R also becomes small, and thus a Taylor expansion is not applicable. The expression in Eq.(5.33) is not yet a viable form for examining the limit $\varepsilon \to 0$, this must wait until after the second integration. Nevertheless, progress has been made, it is clear that this quantity is less singular at $\varepsilon = 0$ than its predecessor Eq.(5.31).

After the appropriate changes of variables, first replacing θ with t,

$$\theta = -\frac{\pi}{2} + \tan^{-1}\left(\frac{t-\eta}{\xi}\right), \tag{5.34}$$

and then the polar coordinate transformation $\{\Lambda, \Psi\}$,

$$\begin{aligned} t &= \Lambda\cos(\Psi) + \eta \\ \xi &= \Lambda\sin(\Psi), \end{aligned} \tag{5.35}$$

the integration of Eq.(5.33) with respect to Λ can be carried out. Letting $\varepsilon \to 0$, finally yields the finite quantity

$$\frac{4a_k\sin(\Psi)}{a^4}. \tag{5.36}$$

Thus, for the gradient evaluation, the only additional step is that required to produce Eq.(5.33).

In conclusion, a effective post-processing gradient algorithm has been developed, the key being to utilize both interior and exterior limit forms of the boundary integral representation. The ability to rapidly compute surface derivatives can be utilized in the development of a cubic Hermite interpolation, and this will be discussed in the next section.

5.3 HERMITE INTERPOLATION IN TWO DIMENSIONS

5.3.1 Introduction

One benefit of having a relatively inexpensive means of computing the surface gradient is the ability to construct an effective boundary integral approximation based upon *Hermite* interpolation. The gradient is used to set (in two dimensions) a unique nodal tangential derivative, resulting in an approximation that is differentiable ($\mathcal{C}^1$) crossing element boundaries. The shape functions in this case are cubic but, unlike the standard quadratic or cubic, the Hermite element employs onlys two nodes, the same as a linear element. The four conditions that determine the cubic coefficients are the interpolation of the potential and the tangential derivative at the two nodes.

124 SURFACE GRADIENT

Compared to the post-processing of the gradient discussed above, a new feature here is that the gradient equations will now be solved *in conjunction with* the potential and flux integral equations. This leads to a much larger system of linear equations, and one part of the discussion below will concern an iterative algorithm that mitigates the computational cost of the linear algebra.

Cubic Hermite interpolation for boundary integral analysis was first introduced by Watson [284], and further development and application has been carried out [85, 122, 279], most notably by Rudolphi, Muci-Küchler and co-workers [188, 189, 238] (see [187] for additional references). Almost all of this work has employed collocation; a Galerkin analysis, to be followed herein, was presented in [111].

Hermite interpolation can be attractive for the solution moving boundary problems mentioned earlier in this chapter. The gradient will be needed anyway for computing the surface velocity, and the extra computational cost of the Hermite approach can therefore be partially balanced against the elimination of the post-processing step. The cubic will also be better for accuracy: in general, the boundary integral analysis must do the best it can with the (discrete) boundary specified by the evolution algorithm, thus this can be important.

5.3.2 Hermite Interpolation

The basic approach in two dimensions is presented in this section. The Hermite approximation will be applied to both the potential and boundary interpolations, and thus the approximate boundary contour will also be $\mathcal{C}^1$ continuous at the nodes. To achieve this smooth boundary approximation, it is assumed that the normal vector is known at the nodes, this information available, for example, from a CAD file, or from the Level Set method [246, 247] in a moving boundary problem.

The boundary interpolation is accomplished by setting

$$Q(t) = (x(t), y(t)) = \sum_{j=1}^{2}(x_j, y_j)\psi_j(t) + \sum_{j=3}^{4}(a_j, b_j)\psi_j(t) , \qquad (5.37)$$

with the shape functions $\psi_j(t)$, $0 < t < 1$,

$$
\begin{aligned}
\psi_1(t) &= (1 + 2t)(1 - t) \\
\psi_2(t) &= t^2(3 - 2t) \\
\psi_3(t) &= t(1 - t)^2 \\
\psi_4(t) &= -t^2(1 - t) .
\end{aligned}
\qquad (5.38)
$$

All values of ψ_j and ψ_j' at $t = 0$ and $t = 1$ are zero (the prime indicating differentiation with respect to t) except $\psi_1(0) = \psi_2(1) = 1$ and $\psi_3'(0) = \psi_4'(1) = 1$. The coefficients (a_3, b_3) and (a_4, b_4) chosen to obtain the correct normals at the nodes,

$$
\begin{aligned}
(b_3, -a_3) &= j_h \mathbf{N}(Q_1) \\
(b_4, -a_4) &= j_h \mathbf{N}(Q_2)
\end{aligned}
\qquad (5.39)
$$

where $\mathbf{N}(Q_k)$ is the specified outward unit normal at Q_k. Note that these values are not uniquely defined, as specifying the unit normal leaves open the value of the jacobian at the nodes. Herein we have made what appears to be a reasonable choice, choosing the jacobian to be the value that it would be for linear interpolation, $j_h = \|Q_2 - Q_1\|$.

HERMITE INTERPOLATION IN TWO DIMENSIONS **125**

Note that if Q_1 is a node on a smooth part of the boundary, then

$$Q'(Q_1) = (a_3, b_3) \tag{5.40}$$

for either of the two elements containing Q_1. Thus the interpolation provides a unique unit tangent at each node.

To approximate the potential, note that

$$\frac{d}{dt}\phi(t) = \frac{d}{dt}\phi(x(t), y(t)) = \frac{\partial\phi}{\partial x}x'(t) + \frac{\partial\phi}{\partial y}y'(t) \,, \tag{5.41}$$

and thus from Eq.(5.37) and Eq.(5.39)

$$\begin{aligned}
\frac{d\phi}{dt}(Q_1) &= a_3 \frac{\partial\phi}{\partial x}(Q_1) + b_3 \frac{\partial\phi}{\partial y}(Q_1) \\
\frac{d\phi}{dt}(Q_2) &= a_4 \frac{\partial\phi}{\partial x}(Q_2) + b_4 \frac{\partial\phi}{\partial y}(Q_2) \,.
\end{aligned} \tag{5.42}$$

The gradient values are of course initially unknown, they will be determined, using the gradient equations, simultaneously with the unknown potential and flux values. The approximation for the surface potential is then given by

$$\phi(Q(t)) = \sum_{j=1}^{2}\phi(Q_j)\psi_j(Q) + \sum_{j=3}^{4}\frac{d\phi}{dt}(Q_{j-2})\psi_j(Q) \,. \tag{5.43}$$

As noted above, $\{a_j, b_j\}$ depend only on the node (through the normal vector), not the particular element being considered, and the same will be true for the values of the gradient. Thus, the Hermite interpolation produces a unique tangential derivative of ϕ at the nodes.

5.3.3 Iterative Solution

A distinguishing feature of a cubic Hermite boundary integral approximation is that the standard boundary integral equations are solved simultaneously with the gradient equations. When the limit-difference gradient formulation developed in Section 5.2 is implemented for unknown boundary values (as opposed to a post-processing calculation), the equations are *sparse*. This is a consequence of the fact that only the singular integrals survive, together with the sparsity of the Galerkin matrix A, Eq.(5.7). Thus, the construction of the extra equations is a relatively minor expense.

However, the solution of the larger system of linear equations is a serious drawback. For a two-dimensional scalar problem discretized with N boundary nodes, the resulting coefficient matrix is of order $3N$, compared to N for a linear approximation, N equations being required for each gradient component. For a vector problem these numbers are $6N$ versus $2N$, and in three dimensions the matrix order would increase by a factor of 4 over a linear analysis.

The sparsity of the gradient equations suggests an iterative solution, and as will be demonstrated below, a 'two-level' iterative scheme can significantly reduce the computational costs associated with Hermite. The basic idea is to separate the solution of the gradient equations from the boundary integral equation, taking

126　　SURFACE GRADIENT

advantage of two features of the gradient equations: the principal sub-matrix for the gradient unknowns, namely A, is well-conditioned, and the equations for the separate components of the gradient are only weakly coupled.

If the boundary is discretized with N nodes, the Hermite algorithm results in a $3N$ by $3N$ system of linear equations $Ax = b$. It is convenient to write these equations in the form

$$
\begin{pmatrix}
A_{11} & A_{12} & A_{13} \\
A_{21} & A_{22} & A_{23} \\
A_{31} & A_{32} & A_{33}
\end{pmatrix}
\begin{pmatrix}
x_1 \\
x_2 \\
x_3
\end{pmatrix}
=
\begin{pmatrix}
b_1 \\
b_2 \\
b_3
\end{pmatrix},
\tag{5.44}
$$

where each A_{ij} submatrix is $N \times N$. The vector x_1 represents the unknown boundary values of potential or flux, and the first block row of equations is the discretized form of the potential and/or flux boundary integral equation. Similarly, x_2 and x_3 are vectors containing the $x-$ and $y-$ components of the gradient, and the second and third rows are obtained from Eq.(5.8).

As noted above, the gradient equations only involve local singular integrals, and thus the second and third block rows, A_{2j} and A_{3j}, are sparse matrices. Moreover, if a linear interpolation had been employed to approximate the potential in the gradient equations, then the only matrix elements multiplying the gradient values would be from the free term in Eq.(5.8). Consequently, the diagonal blocks A_{22} and A_{33} would be symmetric positive definite and moreover $A_{23} = A_{32} = 0$. As it is expected that the Hermite sub-matrices are 'small' perturbations of the corresponding linear ones, the $x-$ and $y-$component gradient equations should be only weakly coupled, and A_{22}, A_{33} should be well-conditioned.

The proposed algorithm therefore assumes an initial guess for x_2 and x_3 and then solves for x_1 from

$$
A_{11}x_1 = b_1 - A_{12}x_2 - A_{13}x_3 .
\tag{5.45}
$$

The values of x_2 and x_3 can then be updated by solving

$$
\begin{aligned}
A_{22}x_2 &= b_2 - A_{21}x_1 - A_{23}x_3 \\
A_{33}x_3 &= b_3 - A_{31}x_1 - A_{32}x_2
\end{aligned}
\tag{5.46}
$$

and the whole process (termed an *outer iteration*) repeated until convergence. As just noted, the two systems in Eq.(5.46) should be well-conditioned, and the coupling between them, provided by A_{23} and A_{32}, very weak. Thus, there is reason to expect that this 'two-level' iteration will first of all converge, and second, be efficient.

If all sparse matrix-vector multiplications are ignored, $3N^2$ operations are required for each matrix-vector multiplication in applying an iterative solver to the full Hermite matrix. On the other hand, Eq.(5.45) requires only N^2 operations per iteration, plus $2N^2$ operations to compute the right hand side. Thus, depending upon the convergence, the two-level scheme could provide significant savings.

As an illustration of the possible savings using this iterative algorithm, results for Dirichlet problems posed on the unit disk $x^2 + y^2 = 1$ and the ellipse $x^2 + 4y^2 = 1$ are reported. The applied boundary potential in each case is $\phi = x^2 - y^2$. Uniform meshes with N nodes are employed for the disk, whereas the discretizations for the ellipse employ a constant central angle, and are therefore non-uniform. The two-level algorithm is compared with a direct application of the iterative solver to the full Hermite matrix, and in both cases the tolerance for the residual was $\epsilon = 10^{-10}$.

HERMITE INTERPOLATION IN TWO DIMENSIONS

Table 5.1 Computational cost (in units of N^2) and residuals for iterative solution methods, $\epsilon = 10^{-10}$. The domain is the unit disk.

N	Full Matrix		Two-Level	
	C_I	Residual	C_I	Residual
100	219	6.96 (-11)	60	6.31 (-12)
200	483	4.67 (-11)	75	6.52 (-12)
300	387	3.62 (-11)	84	2.37 (-11)
400	555	3.98 (-11)	100	5.58 (-12)
500	459	3.93 (-11)	100	1.22 (-12)
600	555	3.34 (-11)	107	7.37 (-12)
800	771	3.16 (-11)	93	1.60 (-11)

Tables 5.1 and 5.2 list the computational work – again, ignoring all sparse matrix operations – and the residuals for solving the Dirichlet problem. The costs are given in terms of N^2, and thus C_I for the full matrix method is three times the number of iterations required. For the two-level scheme, C_I consists of the total number of matrix-vector multiplies involving A_{11}, plus five times the number of outer iterations. This is due to the $2N^2$ needed per outer iteration to re-compute the right hand side in Eq.(5.45), plus $3N^2$ to compute the global residual.

Table 5.2 Computational cost (in units of N^2) and residuals for iterative solution methods, $\epsilon = 10^{-10}$. The domain is an ellipse.

N	Full Matrix		Two-Level	
	C_I	Residual	C_I	Residual
100	1100	3.33 (-11)	163	6.31 (-11)
150	990	4.45 (-11)	118	2.19 (-11)
200	894	2.38 (-11)	134	4.15 (-12)
300	1100	2.64 (-11)	174	1.97 (-12)
600	1518	4.48 (-11)	82	5.38 (-11)

With a few exceptions, either iterative scheme involves much less work than the $(3N)^3/3$ required by direct factorization. However, the 'divide and conquer' approach of the two-level algorithm is clearly significantly faster than working with the entire matrix. The success of this approach stems from the fact that, at least in these tests, very few outer iterations are required to reach convergence. Table 5.3 lists the number of outer iterations required for the problem on the unit disk, along with the total number of matrix-vector multiplications required to solve the two systems in Eq.(5.46).

The very small number of outer iterations is clearly key to the efficiency of the algorithm, and this is likely due to weak coupling with the gradient equations: initially solving A_{11} by itself must get very close to the correct x_1. What is also noteworthy about these numbers is that they are, compared to the iterations required to solve Eq.(5.45), largely independent of the mesh. In fact, for the ellipse

128 SURFACE GRADIENT

Table 5.3 Outer iterations and iterations required to solve the gradient equations for the problem posed on the unit disk, $\epsilon = 10^{-10}$.

N	Outer	A_{22}	A_{33}
100	3	27	51
200	3	27	43
300	3	27	43
400	3	27	43
500	3	27	35
600	3	27	35
800	3	27	51

tests these numbers are constants, the values corresponding to a row in Table 5.3 being 3, 27 and 43. The good conditioning of A_{22} and A_{33} certainly plays a role here, and again justifies the splitting of the A matrix. Moreover, it also indicates that the challenge of further expediting the linear algebra by developing a good pre-conditioner (note that no pre-conditioning has been employed) can focus solely on the A_{11} matrix.

The two iterative procedures have also been applied to a Neumann problem posed on the unit disk. The differences compared to the Dirichlet problem are in the first block row A_{1k} and column A_{k1} matrices, the integrals that comprise these matrices now originating from the integrals of the Green's function in Eq.(2.8), rather than its normal derivative. The results are shown in Table 5.4, and again indicate that the two-level algorithm is quite effective.

Table 5.4 Computational cost (in units of N^2) and residuals for iterative solution methods, $\epsilon = 10^{-10}$. The problem is a circle with Neumann boundary data.

N	Full Matrix		Two-Level	
	C_I	Residual	C_I	Residual
100	75	3.74 (-13)	42	3.73 (-11)
200	51	6.88 (-11)	42	2.94 (-12)
300	75	1.09 (-11)	42	6.51 (-12)
400	75	4.59 (-14)	28	6.09 (-11)
500	75	3.22 (-12)	28	2.42 (-11)
600	51	2.16 (-11)	28	5.77 (-11)
800	75	3.87 (-12)	28	1.43 (-11)

CHAPTER 6

AXISYMMETRY

Synopsis: There are several distinguishing aspects of a boundary integral analysis for three-dimensional axisymmetric geometries. The axisymmetric Green's function $G(Q, P)$ is, first of all, not symmetric with respect to Q and P, and thus a symmetric-Galerkin formulation is not immediately obvious. Second, $G(Q, P)$, defined in terms of complete elliptic integrals, is not a simple algebraic function of r and, moreover, it has a different singular behavior when P is on the axis of symmetry. These difficulties can be surmounted by employing suitable Galerkin weight functions, these alternative weights smoothing out the axis singularity and restoring symmetry to the formulation. Finally, all integrands will contain a logarithmic singularity, and these integrals are converted to a nonsingular form and evaluated partially analytically and partially numerically. The modified weight functions, together with a boundary limit definition, also result in an effective algorithm for the post-processing of the surface gradient.

6.1 INTRODUCTION

A Galerkin boundary integral implementation for three-dimensional axisymmetric problems is the subject of this chapter. Beginning in the mid-1970's [65, 145, 178], collocation solutions of integral equations for axisymmetric problems have been extensively considered in the literature [11]. Recent work has focused fracture and contact analysis for axisymmetric elasticity [69, 70, 105].

The Symmetric Galerkin Boundary Element Method. By Sutradhar, Paulino and Gray
ISBN 978-3-540-68770-2 ©2008 Springer-Verlag Berlin Heidelberg

130 AXISYMMETRY

Two key aspects of an axisymmetric formulation are that the Green's function $G(Q, P)$ is no longer a symmetric function of Q and P, and moreover it has a different singular behavior when the source point P is on the symmetry axis. (Note that in this chapter symmetry will have several different meanings). For P off the axis, the Green's function represents, by rotating this point around the symmetry axis, the effect of ring source in three dimensions. However, P on the axis, the ring source degenerates to a single point, and thus some difference in behavior is to be expected.

The consequences for a collocation approximation are significant. In a standard collocation analysis, an equation is written with the source point on the symmetry axis, and thus the singular integration for this equation is necessarily different from that in other equations. This adds an additional complication to the analysis, and a variety of techniques have been employed, including the use of a different fundamental solution on the axis (see [105] and references therein). While having to work with a second Green's function clearly involves additional work when implementing the method, this special axis fundamental solution is, on the other hand, a simpler function than the general expression.

It will be shown below that for Galerkin, with a modified definition of the weight functions, a separate treatment at the axis of symmetry is not required. This comes about partly due to the weak formulation, as no equation is written 'directly' at the axis, and partly due to the ability to choose appropriate weight functions. The chosen weight functions are in fact zero at the symmetry axis, and this effectively counterbalances the different behavior of the Green's functions. This on-axis singular integration analysis for both first and second order derivatives of the Green's function is also greatly simplified. Moreover, the modified weight functions also restore symmetry to the formulation, permitting a symmetric-Galerkin formulation.

An axisymmetric analysis is also somewhat unique in that all Green's function kernels, as opposed to their standard two-dimensional counterparts, contain a logarithmic singularity. This is a consequence of the presence of complete elliptic integrals in the definition of the kernels. The integrable singularity obviously requires an appropriate numerical treatment, and this is impeded by the complicated form of these weakly singular terms. To handle this a hybrid analytic/numeric method is employed for the evaluation of these integrals. The method is based upon reformulating the integral as a nonsingular double integral, as in [67], and then evaluating one part of the integral analytically.

As noted in the previous chapter, the complete elliptic integrals will not pose any problem when it comes to the evaluation of the surface gradient. For the boundary limit approach, all of the complexity of the axisymmetric kernel functions disappears: the kernels for gradient evaluation are no more difficult than for the simple two-dimensional Laplace equation.

We focus on the analysis of the integral equation for surface potential ϕ, $\nabla^2\phi = 0$. However, these techniques will apply directly to the hypersingular equation for surface flux, and moreover, to the much more complicated axisymmetric formulation for elasticity [11, 70, 105].

6.2 AXISYMMETRIC FORMULATION

The basic procedure for deriving the boundary integral formulation for the three-dimensional axisymmetric Laplace equation is to start with the standard boundary integral equation for surface potential [13, 167, 292]. If the Cartesian coordinates (x, y, z) are replaced with cylindrical coordinates (r, θ, z), the integration with respect to θ can then be carried out. As the boundary potential and flux are independent of θ, the interior and exterior integral equations for the potential take the form

$$\phi(\hat{r}, \hat{z}) = \int_\Gamma r \left(\frac{\partial \phi}{\partial \mathbf{n}}(r, z) G(\hat{r}, \hat{z}; r, z) - \phi(r, z) \frac{\partial G}{\partial \mathbf{n}}(\hat{r}, \hat{z}; r, z) \right) d\Gamma_{rz} \quad (6.1)$$

$$0 = \int_\Gamma r \left(\frac{\partial \phi}{\partial \mathbf{n}}(r, z) G(\hat{r}, \hat{z}; r, z) - \phi(r, z) \frac{\partial G}{\partial \mathbf{n}}(\hat{r}, \hat{z}; r, z) \right) d\Gamma_{rz} ,$$

the kernel functions to be defined below. (Employing r for a coordinate, whereas previously it was always the distance between P and Q, is admittedly a possible source of confusion. Nevertheless, this is the standard notation.) In the first equation, $(\hat{r}, \hat{z})$ is a point *interior* to the domain, and in the second $(\hat{r}, \hat{z})$ lies outside. As for the standard two and three dimensional formulations, these equations are valid and identical for $(\hat{r}, \hat{z}) \in \Gamma_{rz}$, with an appropriate definition of the singular integrals. The line integral is with respect to the field point (r, z) (subsequently the subscript on Γ will be omitted), and the boundary contour Γ is the $x > 0$ section of the intersection of the three dimensional boundary surface with the $y = 0$ plane (see Figure 6.1). For what is to follow, it is worth noting that the θ integral is over a circle of radius r, and thus the r factor in Eq.(6.1) comes from the jacobian of this integration.

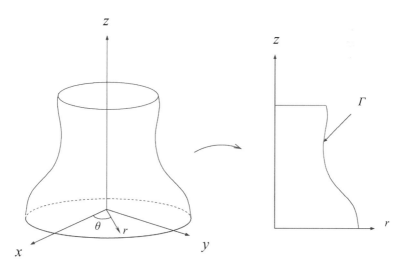

Figure 6.1 Axisymmetric Solution domain.

The axisymmetric kernel functions $G(\hat{r}, \hat{z}; r, z)$ and its normal derivative (note that the full 'Green's function' includes the factor of r in Eq.(6.1)) are the θ integrals of the three-dimensional functions and are defined in terms of the complete elliptic integrals of the first and second kind, $K(m)$ and $E(m)$,

$$G(\hat{r}, \hat{z}; r, z) = \frac{1}{\pi} \frac{1}{(a+b)^{1/2}} K(m) \tag{6.2}$$

$$\frac{\partial G}{\partial \mathbf{n}}(\hat{r}, \hat{z}; r, z) = \frac{1}{\pi} \left[\frac{n_r}{2r(a+b)^{1/2}} \left\{ E(m) - K(m) \right\} - \frac{\mathbf{n} \cdot \mathbf{R}}{(a-b)(a+b)^{1/2}} E(m) \right] .$$

Here $a = r^2 + \hat{r}^2 + \Delta z^2$, $b = 2r\hat{r}$, $\Delta r = r - \hat{r}$, $\Delta z = z - \hat{z}$, $\mathbf{R} = (\Delta r, \Delta z)$ and $\mathbf{n} = \mathbf{n}(r, z)$ is the unit outward normal at the field point. Adopting the notation in [183],

$$K(m) = \int_0^{\pi/2} \frac{\mathrm{d}\theta}{\left(1 - m \sin^2(\theta)\right)^{1/2}} \tag{6.3}$$

$$E(m) = \int_0^{\pi/2} \left(1 - m \sin^2(\theta)\right)^{1/2} \mathrm{d}\theta ,$$

where the parameter m and its complementary parameter $m_1 = 1 - m$ are defined by

$$m = \frac{2b}{a+b} = \frac{4r\hat{r}}{(r+\hat{r})^2 + \Delta z^2}$$

$$m_1 = \frac{a-b}{a+b} = \frac{\Delta r^2 + \Delta z^2}{(r+\hat{r})^2 + \Delta z^2} . \tag{6.4}$$

The formula for the normal derivative of G can be derived by using the relations [287]

$$\frac{\mathrm{d}}{\mathrm{d}k} \tilde{K}(k) = \frac{\tilde{E}(k)}{k(1-k^2)} - \frac{\tilde{K}}{k} \tag{6.5}$$

$$\frac{\mathrm{d}}{\mathrm{d}k} \tilde{E}(k) = \frac{\tilde{E}(k) - \tilde{K}(k)}{k} ,$$

where $\tilde{K}(k) = K(k^2)$ and $\tilde{E}(k) = E(k^2)$.

To evaluate $E(m)$ and $K(m)$, we will use the polynomial approximations developed by Hastings [133],

$$K(m) \approx \sum_{\nu=0}^{4} a_\nu m_1^\nu - \log(m_1) \sum_{\nu=0}^{4} b_\nu m_1^\nu \tag{6.6}$$

$$E(m) \approx 1 + \sum_{\nu=1}^{4} c_\nu m_1^\nu - \log(m_1) \sum_{\nu=1}^{4} d_\nu m_1^\nu ,$$

the error in these expansions being less than 2×10^{-8}; the coefficients $\{a_\nu, b_\nu, c_\nu, d_\nu\}$ can be found in [183]. Thus, as expected, G has a logarithmic singularity for $(\hat{r}, \hat{z}) \to (r, z)$ $(m_1 = 1 - m = 0)$, and its normal derivative behaves as a 'Cauchy Principal Value' (CPV) integral, $\|(\Delta r, \Delta z)\|^{-1}$. The logarithmic singularity is, however, also present in the normal derivative of G, and thus the numerical treatment of this integral will also have to take into account the presence of this integrable singularity.

AXISYMMETRIC FORMULATION **133**

Moreover, as noted and emphasized above, the singular behavior of the kernel functions is different at the symmetry axis. Note that $a + b$ appears in the denominator in Eq.(6.2) and Eq.(6.3), and $a + b = r^2 + \hat{r}^2 + \Delta z^2 = 0$ when $r = \hat{r} = \Delta z = 0$. In this regard, a Galerkin approximation has an immediate advantage, in that unlike collocation, an equation is not written precisely at the axis. Moreover, the standard Galerkin weight functions will be modified, and this will eliminate any difficulties in handling this axis singularity. This will be the case not only for the equation for surface potential, but also for the (hypersingular) derivative equation for surface flux.

Galerkin Approximation. Denoting the field point (r, z) by Q and the source point $(\hat{r}, \hat{z})$ by P, the *exterior limit* form of Eq.(6.1) can be written as

$$\mathcal{P}(P) \equiv \lim_{\varepsilon \to 0^+} \int_\Gamma r \left(\frac{\partial \phi}{\partial \mathbf{n}}(Q) G(P_\varepsilon, Q) - \phi(Q) \frac{\partial G}{\partial \mathbf{n}}(P_\varepsilon, Q) \right) d\Gamma_Q = 0 , \qquad (6.7)$$

where $P_\varepsilon = (\hat{r}_\varepsilon, \hat{z}_\varepsilon) = (\hat{r}, \hat{z}) + \varepsilon \mathbf{N}$, $\mathbf{N} = \mathbf{N}(P)$ being the unit outward normal at $P = (\hat{r}, \hat{z})$.

The two irritating aspects of this axisymmetric boundary integral equation were discussed above. First, although $G(P, Q) = G(Q, P)$ is a symmetric function of field and source points, $r G$ is clearly not, and the matrix resulting from this integral will not be symmetric if a standard Galerkin procedure is applied. Second, the kernel functions contain an additional singularity at the axis $r = \hat{r} = 0$, due to the presence of the $a + b$ term in the denominators. This singularity causes some difficulty for collocation approximations, and the same would be true here if, once again, standard Galerkin weight functions were employed. Fortunately there is leeway in the choice of the weight functions, and this flexibility will be exploited herein.

The weight function $\hat{\psi}_k(P)$ in the Galerkin formulation

$$0 = \int_\Gamma \hat{\psi}_k(P) \, \mathcal{P}(P) \, d\Gamma_P . \qquad (6.8)$$

usually consists of all shape functions that are nonzero at a particular node P_k. However, to regain symmetry (and thereby allow a symmetric-Galerkin formulation), and at the same time to ameliorate the axis singularity, the obvious course of action is to take the standard weight functions $\hat{\psi}_k(P)$ and multiply by $\hat{r}$. Thus, the equations to be solved take the form

$$0 = \lim_{\varepsilon \to 0^+} \int_\Gamma \hat{r} \, \hat{\psi}_k(P) \int_\Gamma r \left(\frac{\partial \phi}{\partial \mathbf{n}}(Q) G(P_\varepsilon, Q) - \phi(Q) \frac{\partial G}{\partial \mathbf{n}}(P_\varepsilon, Q) \right) d\Gamma_Q d\Gamma_P \qquad (6.9)$$

Moreover, the equation for surface flux is

$$0 = \lim_{\varepsilon \to 0^+} \int_\Gamma \hat{r} \, \hat{\psi}_k(P) \int_\Gamma r \left(\frac{\partial \phi}{\partial \mathbf{n}}(Q) \frac{\partial G}{\partial \mathbf{N}}(P_\varepsilon, Q) - \phi(Q) \frac{\partial^2 G}{\partial \mathbf{n} \partial \mathbf{N}}(P_\varepsilon, Q) \right) d\Gamma_Q d\Gamma_P ,$$

$$\qquad (6.10)$$

where derivatives with respect to $\mathbf{N}(P)$ are with respect to $P = (\hat{r}, \hat{z})$. With the additional factor of $\hat{r}$, these equations contain the same symmetry properties as those for the standard Laplace equation, and a symmetric-Galerkin formulation is possible.

134   AXISYMMETRY

The concern in modifying the weight functions in this manner is that now *all* weight functions are zero at the axis, possibly wiping out information needed to solve for the axis unknowns (*i.e.*, it is possible that an ill-conditioned systems of equations might result). A plausible explanation for why this doesn't happen is that non-axis points represent a ring source, whereas an axis point is a degenerate single point; thus, assigning a weight of 0 to the equation precisely at this point is physically reasonable, and turns out to be computationally sound. In fact this approach corresponds to starting with a 3-D Galerkin boundary integral formulation, and then, as in the derivation of Eq.(6.1), integrating out the angular variable in the cylindrical representation of P. The Jacobian of the cylindrical coordinates would produce an $\hat{r}$ factor in the boundary integral, as in the above equations.

Moreover, as there is no extra difficulty on the axis in the fully three dimensional formulation, it is not surprising that the $\hat{r}$ factor helps to mollify the kernel function behavior on the axis. As will be seen below in the gradient discussion, this is especially useful for treating the derivative of Eq.(6.7) (*e.g.*, hypersingular equation for surface flux).

6.3   SINGULAR INTEGRATION

For the most part, the boundary limit evaluation of the singular integrals follows the procedures described in Chapter 3. In particular, the important CPV integral, the most singular part of $\partial G/\partial \mathbf{n}$, comes solely from the last term in Eq.(6.3),

$$\frac{\mathbf{n} \cdot \mathbf{R}}{(a-b)(a+b)^{1/2}} E(m) \tag{6.11}$$

and moreover only from, expanding at the singular point, $E(1) = 1$. For the coincident Galerkin integral, $(a+b)^{1/2}$ can be replaced by $2\hat{r}$, and with these substitutions the above function is then essentially the same as the CPV kernel for two dimensional Laplace equation. There is therefore no difficulty in the analytic evaluation of the integral and the boundary limit. Herein we therefore focus on the one major new aspect present in the axisymmetric analysis: there is now, in both kernel functions, a logarithmic singularity having a fairly complicated coefficient. While integrable, these logarithm terms cannot be accurately evaluated with standard Gauss quadrature, and a suitable algorithm must be developed.

The troublesome $\log(m_1)$ expression is

$$\log\left(\frac{a-b}{a+b}\right) = \log\left(\Delta r^2 + \Delta z^2\right) - \log\left((r+\hat{r})^2 + \Delta z^2\right) \tag{6.12}$$

and it is the first term on the right that is the primary concern. We first show that for both coincident and adjacent singular integrals, the log integrals take the form

$$\int_0^\beta \mathcal{F}(x) \log(Ax) \, dx , \tag{6.13}$$

where A is a constant with respect to x, but may be a function of other variables (Eq.(6.13) being just one part of a multidimensional integral). A possible approach would be to write

$$\int_0^\beta \mathcal{F}(x) \log(Ax) \, dx = \log(A) \int_0^\beta \mathcal{F}(x) \, dx + \int_0^\beta \mathcal{F}(x) \log(x) \, dx , \tag{6.14}$$

SINGULAR INTEGRATION **135**

thereby removing the constant A; however, this would require a separate numerical integration of the first term on the right, and thus it is preferable to treat the more general situation in Eq.(6.13).

The function $\mathcal{F}(x)$ arising from the axisymmetric kernels is sufficiently complicated that analytic integration is not possible. Again a possible algorithm would be to use a Taylor series (at $x = 0$) for $\mathcal{F}$, integrating the polynomial part analytically and the remainder numerically. However, given the complexity of the logarithm functions for axisymmetric analysis (in particular, considering future applications in elasticity), this would be fairly tedious to implement for the potential equation, and much worse in implementing the equation for surface flux. Moreover, for higher order interpolation, the jacobians are not constants, and this would complicate matters even further. We have therefore implemented a numerical treatment based upon extending the procedure in [67] designed to handle the simple case $A = \beta = 1$.

We first briefly demonstrate that, the CPV integral aside, the remaining singular integrations reduce to integrals of the form Eq.(6.13), further details can be found in [109]. The two types of Galerkin singular integrals, coincident and adjacent, are discussed separately.

6.3.1 Adjacent Integration

Assume that the adjacent elements are $E_Q = (P_1, P_2)$ and $E_P = (P_2, P_3)$, the reverse situation (E_P precedes E_Q) can be handled similarly. If s and t denote the parameters for the Q and P integrations, respectively, the singularity $Q = P = P_2$ is at $1 - s = 0$, $t = 0$. By introducing polar coordinates $\{\rho, \vartheta\}$

$$t = \rho \cos(\vartheta) \qquad 1 - s = \rho \sin(\vartheta) \tag{6.15}$$

the singularity is then identified by $\rho = 0$, and $\Delta r^2 + \Delta z^2 = \alpha^2 \rho^2$, α being a function of ϑ. Thus, the integration of the logarithm term $\log(\Delta r^2 + \Delta z^2) = 2\log(\alpha\rho)$ takes the form given in Eq.(6.13),

$$2 \int_0^{\pi/4} d\vartheta \int_0^{1/\cos(\vartheta)} \rho\, f(\rho, \vartheta)\, \log(\alpha\rho)\, d\rho + \tag{6.16}$$
$$2 \int_{\pi/4}^{\pi/2} d\vartheta \int_0^{1/\sin(\vartheta)} \rho\, f(\rho, \vartheta)\, \log(\alpha\rho)\, d\rho$$

6.3.2 Coincident Integration

For the coincident integration, $E_P = E_Q = E$, and the integral is singular when $s = t$. Replacing s by the variable w, $s = w + t$, the coincident integral is

$$2 \int_0^1 dt \int_{-t}^{1-t} f(w) \log(\alpha|w|)\, dw =$$
$$2 \int_0^1 dt \int_0^{1-t} f(w) \log(\alpha w)\, dw + 2 \int_0^1 dt \int_0^t f(-w) \log(\alpha w)\, dw\ . \tag{6.17}$$

These integrals are also of the form in Eq.(6.13).

6.3.3 Axis singularity

In addition to the algebraic singularities due to $(a+b)^{-1}$ appearing in Eq.(6.2) and Eq.(6.3), an additional logarithmic singularity is present on the axis, the last term in Eq.(6.12). Thus the coincident integration of an axis element contains this additional weak singularity (this is the only part of the calculation for which an axis element is treated differently from other elements). However, the procedure for handling this $\log(a+b)$ term when $E_P = E_Q$ is the same as for the adjacent singular integration. For this term, arrange the parameters so that the node on the axis corresponds to $s = t = 0$, and proceed as in the adjacent case: introduce polar coordinates $t = \rho\cos(\vartheta)$, $s = \rho\sin(\vartheta)$. This leads to $a+b = \alpha_+^2 \rho^2$ and $a-b = \alpha_- \rho^2$ and thus $m_1 = \alpha_-/\alpha_+$, independent of ρ. Incorporating the shape functions, the ρ integrals in this case take the simple form $\rho^j \log(\alpha_+\rho)$, and can be computed analytically. This leaves just the ϑ integration to be evaluated numerically.

6.3.4 Log Integral Transformation

As shown above, the singular integration of the complete elliptic integrals requires evaluating integrals of the form Eq.(6.13). The basic idea is to transform these singular integrals into a double integral involving just the nonsingular function $\mathcal{F}$; the price that is paid for the overall simplicity of this approach is the computational effort required to evaluate the double integrals.

After the transformation to the double integral, the expectation is that one could use low order Gauss quadrature for this evaluation. However, numerical tests have shown that the function $\mathcal{F}$ in the coincident integral varies sufficiently rapidly that a large number of Gauss points are required to obtain a converged value of the integral. We will therefore, in the coincident case, integrate one part of the double integral analytically. This makes the implementation somewhat more involved, but on the other hand reduces the computational work.

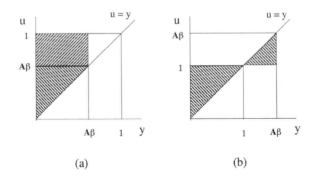

Figure 6.2 The domains for changing the order of integration in the logarithm integrals.

To accomplish the transformation of Eq.(6.13), we first assume $A\beta < 1$. The case $A\beta > 1$ (which does not occur in the coincident integration, but may occur in the adjacent) is handled similarly. We have

$$\int_0^\beta \mathcal{F}(x) \log(Ax)\,\mathrm{d}x = \frac{1}{A}\int_0^{A\beta} \mathcal{F}(y/A)\log(y)\,\mathrm{d}y \tag{6.18}$$

$$= \frac{1}{A}\int_0^{A\beta} \mathcal{F}(y/A)\int_1^y \frac{1}{u}\,\mathrm{d}u\,\mathrm{d}y$$

and interchanging the order of integration

$$\int_0^\beta \mathcal{F}(x) \log(Ax)\,\mathrm{d}x = -\frac{1}{A}\int_0^{A\beta}\int_0^u \mathcal{F}(y/A)\frac{1}{u}\,\mathrm{d}y\,\mathrm{d}u \tag{6.19}$$

$$-\frac{1}{A}\int_{A\beta}^1\int_0^{A\beta} \mathcal{F}(y/A)\frac{1}{u}\,\mathrm{d}y\,\mathrm{d}u$$

$$= -\int_0^{A\beta}\int_0^{1/A} \mathcal{F}(uz)\,\mathrm{d}z\,\mathrm{d}u - \int_{A\beta}^1\int_0^{\beta/u} \mathcal{F}(uz)\,\mathrm{d}z\,\mathrm{d}u$$

the last line coming from the change of variables $z = y/(Au)$. The interchange of the order of integrals is illustrated in Fig. 6.2. The corresponding result for $A\beta > 1$ is

$$\int_0^\beta \mathcal{F}(x) \log(Ax)\,\mathrm{d}x = -\int_0^1\int_0^{1/A} \mathcal{F}(uz)\,\mathrm{d}z\,\mathrm{d}u + \int_1^{A\beta}\int_{1/A}^{\beta/u} \mathcal{F}(uz)\,\mathrm{d}z\,\mathrm{d}u \,. \tag{6.20}$$

Note that for the simple case $A = \beta = 1$ there is only one double integral, so the minor additional complications are the two integrals, and the two cases depending upon the value of $A\beta$.

6.3.4.1 Analytic integration: Coincident

For the logarithm integrals stemming from the coincident integration, it is necessary to proceed further: numerical evaluation of the double integral using Gauss quadrature turns out to be ineffective. The function $\mathcal{F}$, the coefficient of the logarithm, is a polynomial in $m_1 = (a-b)/(a+b)$, and this becomes a rational function of the integration variable. For the coincident integral, $a - b = c^2 w^2$ and $a + b = c^2 w^2 + bw + a$, where $w = s - t$. Thus, the function $\mathcal{F}$ for the flux integral is

$$(\psi_{j,0} + w\psi_{j,1})\frac{\alpha + \beta w}{(w^2 + bw + a)^{1/2}}\sum_{\nu=0}^4 b_\nu\left(\frac{w^2}{w^2 + bw + a}\right)^\nu, \tag{6.21}$$

where $r = \alpha + \beta w$ and $\psi_{j,0} + w\psi_{j,1}$ is the linear shape function $\psi_j(Q)$ expressed as a polynomial in w. The potential integral lacks this r coefficient, but on the other hand involves both $E(m)$ and $K(m)$, and is therefore

$$\frac{\psi_{j,0} + w\psi_{j,1}}{(w^2 + bw + a)^{1/2}}\sum_{\nu=0}^4 (d_\nu - b_\nu)\left(\frac{w^2}{w^2 + bw + a}\right)^\nu. \tag{6.22}$$

Using the procedures of the previous section, w is replaced by uz for the double integral over u and z; this changes the coefficients, but leaves the form of the rational function unaltered, e.g.,

$$\left(\frac{u^2 z^2}{u^2 z^2 + buz + a}\right)^\nu = \left(\frac{z^2}{z^2 + b'z + a'}\right)^\nu, \tag{6.23}$$

138 AXISYMMETRY

which has the same form as the w expression, only now $b' = b/u$ and $a' = a/u^2$. These rational functions, together with the appropriate function outside the summations in Eq.(6.21) and Eq.(6.22), can be integrated analytically with respect to z.

Similar expressions can be obtained, in terms of the variable ρ, for the adjacent integration. However, in these cases, analytic integration is less imperative, as the singularity is only at one point in the double integral. Moreover, note that the polar coordinate transformation combines the P and Q integration, and thus both $\hat{r}(P)$ and $\psi_k(P)$ must be incorporated when integrating ρ; the required analytic expressions are therefore much longer, and the computational advantage of eliminating a numerical integration may in fact disappear. Finally, numerical experiments indicate that the double integral can be accurately evaluated with a low number of Gauss points. As the analytic approach for these integrals appears to have no advantages, and actually some increased complexity in implementation, these integrals have been computed entirely numerically using Eq.(6.18) and Eq.(6.20).

6.3.5 Analytic integration formulas

A minor difficulty in implementing the anlytic integration of the rational functions is that the simplest formulas (provided, say, by integration tables or Maple) breakdown at $\alpha^2 = 4a - b^2 = 0$. This in fact occurs when the element is horizontal, $z(1) = z(2)$. However, it is a simple matter to derive formulas that are valid for all α.

First note that by applying an appropriate change of variables, the required integrals are linear combinations of the simpler integrals of the form

$$\int \frac{Z^j}{(Z^2 + \alpha^2)^{k+1/2}} \, dZ \,, \tag{6.24}$$

where $0 < k < 4$ and $0 < j < 2k + 2$. The values of $j > 2k$ are necessary to incorporate the additional w factors in r and $\psi_j(Q)$. Here $\alpha^2 = a - b^2/4 \geq 0$, and for various values of $\{k, j\}$, the simplest analytic expressions contain α in their denominators. For example, when $k = 2$ and $j = 0$,

$$\int \frac{1}{(Z^2 + \alpha^2)^{5/2}} \, dZ = \frac{Z \left(3\alpha^2 + 2Z^2 \right)}{3\alpha^4 \left(Z^2 + \alpha^2 \right)^{3/2}} \tag{6.25}$$

which diverges at $\alpha = 0$. However, it is possible to add a constant to this indefinite integral, and subtracting off the coefficient of the $Z^3/(Z^2 + \alpha^2)^{3/2}$ term (the limiting value as $Z \to \infty$), namely $2/(3\alpha^4)$, results in

$$\frac{3Z\alpha^2 + 2Z^3 - 2\left(Z^2 + \alpha^2 \right)^{3/2}}{3\left(Z^2 + \alpha^2 \right)^{3/2}\alpha^4} \,. \tag{6.26}$$

The final result, clearly well behaved at $\alpha = 0$,

$$\int \frac{1}{(Z^2 + \alpha^2)^{5/2}} \, dZ = -\frac{3Z^2 + 4\alpha^2}{3\left(Z^2 + \alpha^2 \right)^{3/2}\left(3Z\alpha^2 + 2Z^3 + 2\left(Z^2 + \alpha^2 \right)^{3/2} \right)} \tag{6.27}$$

is now just a matter of rationalizing the numerator.

6.4 GRADIENT EVALUATION

The algorithm presented in the previous chapter reduced the boundary integration for gradient evaluation down to local singular integrals. Another advantage is that this procedure, as will be demonstrated for axisymmetric analysis, also simplifies the handling of complicated kernel functions. By taking the difference of the interior and exterior limit integral equations for the gradient, the calculation can be reduced to terms that are 'discontinuous' crossing the boundary, and the implications of this for axisymmetric analysis, are substantial. The elliptic integrals, $K(m)$ and $E(m)$, and most especially the log singular terms that required so much attention in solving the potential equation, do not appear at all in the gradient evaluation. The elliptic integrals appear solely through $E(1) = 1$, and as a consequence, the integrations can be carried out entirely analytically.

6.4.1 Gradient Equations

The surface gradient equations are obtained by differentiating, with respect to $\hat{r}$ and $\hat{z}$, the interior and exterior limit potential equations, Eq.(6.1), resulting in

$$\frac{\partial}{\partial \mathcal{X}} \phi(P) = \lim_{\varepsilon \to 0^-} \int_{\Gamma} r \left(\frac{\partial \phi}{\partial \mathbf{n}}(Q) \frac{\partial G}{\partial \mathcal{X}}(P_\varepsilon, Q) - \phi(Q) \frac{\partial^2 G}{\partial \mathcal{X} \partial \mathbf{n}}(P_\varepsilon, Q) \right) d\Gamma_Q$$

(6.28)

$$0 = \lim_{\varepsilon \to 0^+} \int_{\Gamma} r \left(\frac{\partial \phi}{\partial \mathbf{n}}(Q) \frac{\partial G}{\partial \mathcal{X}}(P_\varepsilon, Q) - \phi(Q) \frac{\partial^2 G}{\partial \mathcal{X} \partial \mathbf{n}}(P_\varepsilon, Q) \right) d\Gamma_Q$$

(6.29)

where $\mathcal{X}$ is either $\hat{r}$ or $\hat{z}$. Expressions for the kernel functions can be obtained by using Eq.(6.5), and are given by

$$\frac{\partial G}{\partial \hat{r}} = \frac{1}{\pi} \frac{1}{2\hat{r}(a+b)^{1/2}} \left[\frac{r^2 - \hat{r}^2 + \Delta z^2}{a - b} E(m) - K(m) \right]$$

$$\frac{\partial G}{\partial \hat{z}} = \frac{1}{\pi} \frac{\Delta z}{(a-b)(a+b)^{1/2}} E(m) ,$$

(6.30)

$$\frac{\partial^2 G}{\partial \mathcal{X} \partial \mathbf{n}} = \frac{1}{\pi} \left[\left(\frac{n_r}{2r} - \frac{\mathbf{n} \cdot \mathbf{R}}{a - b} \right) \frac{\partial}{\partial \mathcal{X}} \frac{E(m)}{(a+b)^{1/2}} - \frac{n_r}{2r} \frac{\partial}{\partial \mathcal{X}} \frac{K(m)}{(a+b)^{1/2}} \right.$$

$$\left. + \left(\frac{n_{\mathcal{X}}}{a - b} - \frac{\mathbf{n} \cdot \mathbf{R} \Delta \mathcal{X}}{(a-b)^2} \right) \frac{E(m)}{(a+b)^{1/2}} \right]$$

(6.31)

and

$$\frac{\partial}{\partial \mathcal{X}} \frac{K(m)}{(a+b)^{1/2}} = \pi \frac{\partial G}{\partial \mathcal{X}}$$

$$\frac{\partial}{\partial \hat{r}} \frac{E(m)}{(a+b)^{1/2}} = \frac{1}{2\hat{r}(a+b)^{3/2}} \left[(\Delta r^2 - 4\hat{r}^2 + \Delta z^2) E(m) - (r^2 - \hat{r}^2 + \Delta z^2) K(m) \right]$$

(6.32)

$$\frac{\partial}{\partial \hat{z}} \frac{E(m)}{(a+b)^{1/2}} = \frac{\Delta z}{(a+b)^{3/2}} [2E(m) - K(m)] .$$

140   AXISYMMETRY

In Galerkin form, once again employing the modified weight functions, the 'limit-difference' gradient equation takes the form

$$\int_\Gamma \hat{r}\hat{\psi}_k(P)\frac{\partial}{\partial \mathcal{X}}\phi(P)\,\mathrm{d}\,\Gamma \;=\; \left\{\lim_{\varepsilon\to 0^-} - \lim_{\varepsilon\to 0^+}\right\}\int_\Gamma \hat{r}\hat{\psi}_k \int_\Gamma r\left(\frac{\partial\phi}{\partial\mathbf{n}}(Q)\frac{\partial G}{\partial \mathcal{X}}(P_\varepsilon,Q)\right.$$
$$\left. -\, \phi(Q)\frac{\partial^2 G}{\partial \mathcal{X}\partial\mathbf{n}}(P_\varepsilon,Q)\right)\,\mathrm{d}\,\Gamma_Q\,\mathrm{d}\,\Gamma_P\;. \qquad (6.33)$$

Clearly any nonsingular integral is continuous crossing the boundary and vanishes in the difference of the limits, leaving just the consideration of coincident and adjacent integrals. A further simplification, available for equations such as Laplace or elasticity wherein a constant function is a solution, is to avoid the adjacent integral as well. To see that the adjacent integrals can also be bypassed, first note that in this case the only contribution is from the hypersingular $\phi(Q)$ integral. The polar coordinate transformation establishes that the adjacent integral of the CPV kernel is integrable for $\varepsilon = 0$, hence continuous at the boundary, and thus must vanish. Similarly, for the hypersingular kernel, the only nonzero contributions must come at the common (singular) node: integrals involving shape functions that are zero at the common node are likewise well defined for $\varepsilon = 0$ and must vanish. Thus, if $\phi(P_k)$ happens to be zero, the adjacent integral does not contribute to the gradient integral for this node. However, $\phi(P_k) = 0$ can always be arranged: if the potential is shifted by a constant it remains a valid solution of the Laplace equation, and moreover, the gradient is unaltered. Therefore, in writing the gradient equation at P_k, all one has to do is to shift the potential by $-\phi(P_k)$, and the adjacent integral can be ignored.

6.4.2 Coincident Integration

Let the element E for the coincident integral be defined by the two nodes $P_1 = (r_1, z_1)$ and $P_2 = (r_2, z_2)$, and define $c_r = r_2 - r_1$, $c_z = z_2 - z_1$ and $c^2 = c_r^2 + c_z^2$.

Although the expressions for the kernel functions are complicated, for gradient evaluation they become very simple. We begin by examining the integral of the flux, and from Eq.(6.30) we obtain

$$r\hat{r}\frac{\partial G}{\partial \hat{r}} = \frac{1}{\pi}\frac{r\hat{r}}{2\hat{r}(a+b)^{1/2}}\left[\left(1+\frac{2r\Delta r}{a-b}\right)E(m) - K(m)\right] \rightarrow \frac{1}{\pi}\frac{r\Delta r}{2(a-b)} \qquad (6.34)$$

the arrow signifying the result of ignoring any term that is not sufficiently singular to survive the difference of the limits. Note that at the singular point $(\Delta r, \Delta z) \to 0$, and thus $\sqrt{a+b} \to 2\hat{r}$. In a similar fashion, for the derivative with respect to $\hat{z}$,

$$r\hat{r}\frac{\partial G}{\partial \hat{z}} \rightarrow \frac{1}{\pi}\frac{r\Delta z}{2(a-b)} \qquad (6.35)$$

For the gradient equation at P_k, the contribution from the flux integral is therefore $\partial\phi/\partial\mathbf{n}(P_j)\mathcal{I}_{kj}^{\mathcal{X}}$, $k,j = 1,2$, where once again $\mathcal{X}$ is either $\hat{r}$ or $\hat{z}$ and

$$\mathcal{I}_{kj}^{\mathcal{X}} = \left\{\lim_{\varepsilon\to 0^-} - \lim_{\varepsilon\to 0^+}\right\}\frac{1}{\pi}\int_E\int_E \psi_k(P)\,\psi_j(Q)\frac{r\Delta\mathcal{X}}{2(\Delta r^2 + \Delta z^2)}\,\mathrm{d}\,\Gamma_Q\,\mathrm{d}\,\Gamma_P\;. \qquad (6.36)$$

These integrals can be computed analytically, and for the derivative with respect to $\hat{r}$,

$$
\begin{array}{ll}
\mathcal{I}_{11}^{\hat{r}} = c_z \left(3\,r_1 + r_2\right)/12 & \mathcal{I}_{12}^{\hat{r}} = c_z \left(r_1 + r_2\right)/12 \\
\mathcal{I}_{21}^{\hat{r}} = c_z \left(r_1 + r_2\right)/12 & \mathcal{I}_{22}^{\hat{r}} = c_z \left(r_1 + 3\,r_2\right)/12 \ ,
\end{array}
\tag{6.37}
$$

and for $\hat{z}$,

$$
\begin{array}{ll}
\mathcal{I}_{11}^{\hat{z}} = -c_r \left(3\,r_1 + r_2\right)/12 & \mathcal{I}_{12}^{\hat{z}} = -c_r \left(r_1 + r_2\right)/12 \\
\mathcal{I}_{21}^{\hat{z}} = -c_r \left(r_1 + r_2\right)/12 & \mathcal{I}_{22}^{\hat{z}} = -c_r \left(r_1 + 3\,r_2\right)/12 \ .
\end{array}
\tag{6.38}
$$

The 'simplification' of the hypersingular integral, *i.e.*, ignoring all nonsingular terms, proceeds as above, with one change. In this case the order of the singularity is -2, and the substitution $\sqrt{a+b} \to 2\hat{r}$ is no longer appropriate. The appropriate expansion in this case is

$$
\begin{aligned}
\frac{1}{(a+b)^{1/2}} &\approx \frac{1}{r+\hat{r}} = \frac{1}{2\hat{r}} + \left(\frac{1}{r+\hat{r}} - \frac{\Delta r}{2\hat{r}(r+\hat{r})}\right) \\
&= \frac{1}{2\hat{r}} - \frac{\Delta r}{2\hat{r}}\left(\frac{1}{2\hat{r}} + \left(\frac{1}{r+\hat{r}} - \frac{\Delta r}{2\hat{r}(r+\hat{r})}\right)\right) \approx \frac{1}{2\hat{r}} - \frac{\Delta r}{4\hat{r}^2} \ .
\end{aligned}
\tag{6.39}
$$

For the hypersingular kernels we therefore obtain

$$
\begin{aligned}
r\hat{r}\frac{\partial^2 G}{\partial \hat{r} \partial \mathbf{n}} &= \frac{1}{4\pi}\left[\frac{n_z \Delta z}{(\Delta r^2 + \Delta z^2)} + (r+\hat{r})\left(\frac{n_r}{\Delta r^2 + \Delta z^2} - 2\frac{\mathbf{n}\bm{\cdot}\mathbf{R}\Delta r}{(\Delta r^2 + \Delta z^2)^2}\right)\right] \\
r\hat{r}\frac{\partial^2 G}{\partial \hat{z} \partial \mathbf{n}} &= \frac{1}{4\pi}\left[-\frac{n_r \Delta z}{(\Delta r^2 + \Delta z^2)}\right. \\
&\quad + \left. (r+\hat{r})\left(\frac{n_z}{\Delta r^2 + \Delta z^2} - 2\frac{\mathbf{n}\bm{\cdot}\mathbf{R}\Delta z}{(\Delta r^2 + \Delta z^2)^2}\right)\right]
\end{aligned}
\tag{6.40}
$$

$$
\tag{6.41}
$$

Note that with the additional $\hat{r}$ factor from the weight function, these kernel functions present no problem at the axis.

The integrals corresponding to Eq.(6.36) are

$$
\mathcal{J}_{kj}^{\mathcal{X}} = \left\{\lim_{\varepsilon \to 0^-} - \lim_{\varepsilon \to 0^+}\right\}\int_E \int_E \psi_k(P)\,\psi_j(Q)r\hat{r}\frac{\partial^2 G}{\partial \mathcal{X} \partial \mathbf{n}}\mathrm{d}\,\Gamma_Q\,\mathrm{d}\,\Gamma_P
\tag{6.42}
$$

and analytic evaluation of the hypersingular integrals yields, for $\hat{r}$ derivatives,

$$
\begin{array}{ll}
\mathcal{J}_{11}^{\hat{r}} = \left(c_r^2 + 3\,r_1\,c_r\right)/(6\,c) & \mathcal{J}_{12}^{\hat{r}} = \left(-3\,r_1\,c_r - 3\,r_2\,c_r + c_r^2\right)/(12\,c) \\
\mathcal{J}_{21}^{\hat{r}} = \left(3\,r_2\,c_r + 3\,r_1\,c_r + c_r^2\right)/(12\,c) & \mathcal{J}_{22}^{\hat{r}} = \left(-3\,r_2\,c_r + c_r^2\right)/(6\,c) \ ,
\end{array}
\tag{6.43}
$$

and for $\hat{z}$,

$$
\begin{array}{ll}
\mathcal{J}_{11}^{\hat{z}} = c_z \left(c_r + 3\,r_1\right)/(6\,c) & \mathcal{J}_{12}^{\hat{z}} = c_z \left(c_r - 3\,r_1 - 3\,r_2\right)/(12\,c) \\
\mathcal{J}_{21}^{\hat{z}} = c_z \left(c_r + 3\,r_2 + 3\,r_1\right)/(12\,c) & \mathcal{J}_{22}^{\hat{z}} = c_z \left(c_r - 3\,r_2\right)/(6\,c) \ .
\end{array}
\tag{6.44}
$$

The simplified kernel expressions for gradient evaluation should be especially useful for axisymmetric elasticity, where the kernel functions are significantly more complicated [12, 69, 105].

142   AXISYMMETRY

6.5   NUMERICAL RESULTS

As a check on the above methods, the results of some simple numerical tests are presented below. In the first problem, the two-dimensional geometry is a circle of radius one centered at $(2,0)$, thus a torus in three dimensions. This problem is therefore simple in that any possible difficulties near the axis are avoided. The Dirichlet boundary condition is the harmonic function $\phi = x^2 + y^2 - 2z^2 = r^2 - 2z^2$, and the computed flux and post-processed surface gradient are compared with the easily obtained exact solutions. Listed in Table 6.1 are the discretized $\mathcal{L}^2$ errors,

$$\left[\frac{1}{N} \sum_{j=1}^{N} (f_c(n_j) - f_x(n_j))^2 \right]^{1/2} \tag{6.45}$$

where f_c and f_x are the computed and exact values at the nodes n_j.

Table 6.1 $\mathcal{L}^2$ errors for the Dirichlet problem on the torus.

Elements	Flux	$\hat{r}$ derivative	$\hat{z}$ derivative
50	0.337E-02	0.649E-02	0.737E-02
100	0.847E-03	0.161E-02	0.183E-02
150	0.377E-03	0.715E-03	0.812E-03
200	0.213E-03	0.402E-03	0.457E-03
250	0.137E-03	0.257E-03	0.292E-03
300	0.961E-04	0.178E-03	0.202E-03
350	0.714E-04	0.131E-03	0.148E-03
400	0.556E-04	0.100E-03	0.113E-03
450	0.450E-04	0.790E-04	0.893E-04

Table 6.2 $\mathcal{L}^2$ errors for the Dirichlet problem on the sphere, with a point source located at $(0, 1.2)$.

Elements	Flux	$\hat{r}$ derivative	$\hat{z}$ derivative
50	0.120E-00	0.122E-00	0.151E-00
100	0.435E-01	0.108E-01	0.497E-01
150	0.235E-01	0.285E-02	0.258E-01
200	0.154E-01	0.121E-02	0.165E-01
250	0.112E-01	0.655E-03	0.118E-01
300	0.878E-02	0.414E-03	0.919E-02
350	0.728E-02	0.288E-03	0.756E-02
400	0.632E-02	0.215E-03	0.652E-02

Two things should be noted. First, the convergence is approximately quadratic, which is the expected behavior when using linear elements. Second, there is no point in proceeding too much further in the refinement of the boundary. As the interpolation and integration errors decay, eventually the approximation of the Green's functions via Eq.(6.6), must begin to be a significant component of the error in the solution (as noted previously, the pointwise error in the four term expansions in Eq.(6.6) is $< 2 \times 10^{-8}$). It can therefore be expected that the quadratic convergence with mesh size eventually disappears.

CHAPTER 7

INTERFACE AND MULTIZONE

Synopsis: Domains containing an 'internal boundary', such as a bi-material interface, arise in many applications, notably in composite materials and geophysical simulations. In this case, the physical quantities are known to satisfy continuity conditions across the interface, but no boundary conditions are specified there. This chapter will demonstrate that the SG approach has a natural extension to interface geometries. Moreover, this formulation is computationally very efficient: the coefficient matrix is of reduced size, solving for values only on one side of the interface, and the symmetry can be invoked to lessen the computational work required to set up the equations. This chapter describes how to achieve the formulation. A prototype numerical example is employed to validate the formulation and corresponding numerical procedure.

7.1 INTRODUCTION

The interface problem refers to any domain containing an 'internal boundary', one for which no boundary data is specified. Typical geometries are illustrated by Figure 7.1. These problems arise in many important applications, with composite materials [230] being an area of much current interest. A particular example is provided by the analysis of composite beams, as shown in Figure 7.1(b). In this figure, the composite consists, for example, of fibre reinforced layers joined at the interfaces. As the fibre orientation in the layers is not the same, the material properties are different. Assuming perfect bonding, the constraints on the interface are continuity

The Symmetric Galerkin Boundary Element Method. By Sutradhar, Paulino and Gray **145**
ISBN 978-3-540-68770-2 ©2008 Springer-Verlag Berlin Heidelberg

of displacements and equilibrium conditions (equal and opposite tractions), but there are no prescribed values.

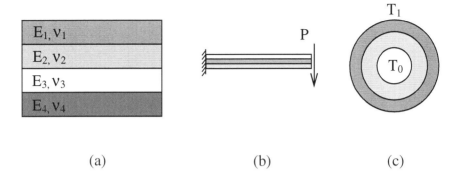

Figure 7.1 Examples of interface problems: (a) Multilayered elastic rock mass; E_i and ν_i, $i = 1...4$ are Young's modulus and Poisson's ratio of each layer. (b) Cantilever 'sandwich' beam with a concentrated load at one endpoint. (c) Heat transfer in a cross-section of a hollow shaft consisting of two concentric materials; T_0 and T_1 denote temperatures in the interior and exterior of the shaft, respectively

The physical conditions at the interface are usually continuity of the principal function and its derivative, e.g. displacement and traction in elasticity, or potential and flux in potential theory. For the displacement-based FEM, enforcing the continuity of traction (or flux) is a very difficult task. For boundary integral equations however, the derivative quantity appears directly in the formulation, and thus the interface conditions can be incorporated simply and accurately. This issue is quite important in practical applications. For instance, for a bi-material interface crack problem, the interface lies in the critical region ahead of the crack tip where an accurate solution is imperative.

For the BEM, there is another important class of problems for which an efficient interface algorithm is required, namely for long thin domains [143]. These problems can be solved by means of a multi-zone boundary element analysis. Although adding surface area is generally a bad idea, in this situation it is computationally efficient to decompose the domain into small subdomains by means of artificial interfaces (Figure 7.2). The reason is that constructing the boundary integral equations for each subdomain only requires integrating over a small piece of the boundary, and this more than compensates for the presence of the additional boundary surface.

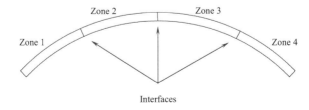

Figure 7.2 Division of a thin geometry into zones for multi-zone boundary element analysis.

7.2 SYMMETRIC GALERKIN FORMULATION

We start with the basic Laplace problem. The discussion below will consider the Laplace equation for the potential, $\nabla^2 \phi = 0$, but it will be clear that the method is generally valid. This choice is not only for simplicity of notation, but also because an issue (to be discussed below) involving the material parameters arises in the potential formulation that does not appear in elasticity. This will require a slight reformulation of the BIEs for potential problems. The interior limit singular BIE and the hypersingular BIE are,

$$\phi(P) + \int_\Gamma \frac{\partial G}{\partial \mathbf{n}}(P,Q)\phi(Q)dQ = \int_\Gamma G(P,Q)\frac{\partial \phi}{\partial \mathbf{n}}(Q)dQ \qquad (7.1)$$

and

$$\frac{\partial \phi(P)}{\partial \mathbf{N}} + \int_\Gamma \frac{\partial^2 G}{\partial \mathbf{N}\partial \mathbf{n}}(P,Q)\phi(Q)dQ = \int_\Gamma \frac{\partial G}{\partial \mathbf{N}}(P,Q)\frac{\partial \phi}{\partial \mathbf{N}}(Q)dQ \qquad (7.2)$$

Previously we have employed the exterior limit equation; here we have chosen the interior form simply to emphasize once again that either can be used. For the following discussion, recall that this function, and its derivatives, satisfy certain symmetry properties, i.e.

$$\begin{aligned} G(P,Q) &= G(Q,P) \\ \nabla_Q G(P,Q) &= -\nabla_P G(P,Q) = \nabla_P G(Q,P) \\ \frac{\partial^2 G}{\partial \mathbf{N}\partial \mathbf{n}}(P,Q) &= \frac{\partial^2 G}{\partial \mathbf{N}\partial \mathbf{n}}(Q,P) \,. \end{aligned} \qquad (7.3)$$

The above HBIE Eq.(7.2) is an equation for the normal derivative, but it is the flux

$$\mathcal{F}(P) \equiv \kappa \frac{\partial \phi(P)}{\partial \mathbf{N}} \qquad (7.4)$$

which satisfies the continuity conditions across a bi-material interface. The constant κ, which goes by various names (e.g. thermal conductivity) depending upon the application (e.g. heat transfer, electrostatics, potential flow), different in each material.

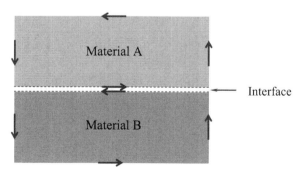

Figure 7.3 Bi-material interface problem.

The continuity equations on the interface are then

$$\phi_A = \phi_B \qquad (7.5)$$

148 INTERFACE AND MULTIZONE

and

$$\kappa_A \frac{\partial \phi_A}{\partial \mathbf{n}} = -\kappa_B \frac{\partial \phi_B}{\partial \mathbf{n}} \tag{7.6}$$

where the subscripts A and B refer to the top and bottom layer in Figure 7.3. By contrast, the material properties in an elasticity formulation are already embedded in the fundamental solution, and the hypersingular equation is most often written directly for the traction [116,118]. Thus, rather than Eq.(7.6), the interface equation would simply be $\tau_A = -\tau_B$, where τ denotes the traction vector.

For the SGBEM, it will be convenient to rewrite the BIE, Eq.(7.1), and the HBIE, Eq.(7.2), in terms of the flux, i.e.

$$\phi(P) + \int_\Gamma \phi(Q) \frac{\partial G}{\partial \mathbf{n}}(P,Q) dQ = \int_\Gamma \left[\frac{1}{\kappa} G(P,Q) \right] \mathcal{F}(Q) dQ \tag{7.7}$$

and

$$\mathcal{F}(P) + \int_\Gamma \phi(Q) \left[\kappa \frac{\partial^2 G}{\partial \mathbf{N} \partial \mathbf{n}}(P,Q) \right] dQ = \int_\Gamma \frac{\partial G}{\partial \mathbf{N}}(P,Q) \mathcal{F}(Q) dQ \tag{7.8}$$

respectively. Note that including the constant κ with G and $\partial^2 G/\partial \mathbf{N} \partial \mathbf{n}$ does not alter the basic properties required for the symmetric Galerkin method, Eq.(7.3). These kernel functions are still symmetric with respect to P and Q, and the integrals containing a single derivative of G (which also play a role in achieving symmetry) remain unchanged. While this is a trivial rewriting of the boundary integral equations, it is key for directly embedding the interface continuity equations into the formulation and for obtaining a symmetric matrix.

Following the Galerkin approximation, using Eq.(7.7) and Eq.(7.8) with the weighting functions as the basis shape functions ψ_l (e.g. linear, quadratic) employed in the approximation of ϕ and $\mathcal{F}$ and on the boundary,

$$\int_\Gamma \psi_l(P) \phi(P) dP + \int_\Gamma \psi_l(P) \int_\Gamma \phi(Q) \frac{\partial G}{\partial \mathbf{n}}(P,Q) dQ dP$$
$$= \int_\Gamma \psi_l(P) \int_\Gamma \left[\frac{1}{\kappa} G(P,Q) \right] \mathcal{F}(Q) dQ dP \tag{7.9}$$

and

$$\int_\Gamma \psi_l(P) \mathcal{F}(P) dP + \int_\Gamma \psi_l(P) \int_\Gamma \phi(Q) \left[\kappa \frac{\partial^2 G}{\partial \mathbf{N} \partial \mathbf{n}}(P,Q) \right] dQ dP$$
$$= \int_\Gamma \psi_l(P) \int_\Gamma \frac{\partial G}{\partial \mathbf{N}} \mathcal{F}(Q) dQ dP \tag{7.10}$$

The additional boundary integration, with respect to P, is the last ingredient required to obtain a symmetric matrix. After discretization, the set of equations in matrix form can be written as $[H]\{\phi\} = [G]\{\mathcal{F}\}$ and in block-matrix form these equations become

$$\begin{bmatrix} H_{11} & H_{12} \\ H_{21} & H_{22} \end{bmatrix} \begin{Bmatrix} \phi_{bv} \\ \phi_u \end{Bmatrix} = \begin{bmatrix} G_{11} & G_{12} \\ G_{21} & G_{22} \end{bmatrix} \begin{Bmatrix} \mathcal{F}_u \\ \mathcal{F}_{bv} \end{Bmatrix}. \tag{7.11}$$

The first row represents the BIE written on the Dirichlet surface, and the second represents the HBIE on the Neumann surface. Similarly, the first and second columns arise from integrating over Dirichlet and Neumann surfaces. The subscripts in the vectors therefore denote known boundary values (bv) and unknown (u) quantities. Rearranging Eq.(7.1) into the form $[\mathbf{A}]\{x\} = \{b\}$, and multiplying the hypersingular equations by -1, one obtains

$$\begin{bmatrix} -G_{11} & H_{12} \\ G_{21} & -H_{22} \end{bmatrix} \left\{ \begin{array}{c} \mathcal{F}_u \\ \phi_u \end{array} \right\} = \left\{ \begin{array}{c} -H_{11}\phi_{bv} + G_{12}\mathcal{F}_{bv} \\ H_{21}\phi_{bv} - G_{22}\mathcal{F}_{bv} \end{array} \right\}. \tag{7.12}$$

The symmetry of the coefficient matrix, $G_{11} = G_{11}^T$, $H_{22} = H_{22}^T$, and $H_{12} = G_{21}^T$, now follows from the properties of the kernel functions (Eq.(7.3)).

7.3 INTERFACE AND SYMMETRY

To describe the interface algorithm, it suffices to deal with a geometry having a single common boundary, as in Figure 7.3. The extension to more complicated interface geometries, if not immediately obvious, is nevertheless seen to be relatively straightforward after a little thought. This topic will be further discussed in the next section. For convenience, we will refer to the two subdomains A and B as the top and bottom regions, respectively, as shown in Figure 7.3. The basic idea is to write the usual symmetric Galerkin equations on the non-interface boundaries in each subdomain, together with an appropriate combination of equations on the common boundary, as indicated by Table 7.1. In addition, only one set of variables is employed on the interface, i.e. potential and flux on the top side. Whenever the interface flux on the bottom side appears in an equation, it is replaced by the negative of the top flux. It will again be convenient to use block-matrix notation, and only the coefficient matrix (the analogue of the left-hand side in Eq.(7.12)) will be considered. For the interface problem, the matrix is partitioned into a 4×4 block structure and takes the form

$$\begin{bmatrix} S_{AA} & 0 & S_{A\mathcal{F}_I} & S_{A\phi_I} \\ 0 & S_{BB} & -S_{B\mathcal{F}_I} & S_{B\phi_I} \\ \hline S_{\mathcal{F}_I A} & S_{\mathcal{F}_I B} & S_{\mathcal{F}_I \mathcal{F}_I} & S_{\mathcal{F}_I \phi_I} \\ S_{\phi_I A} & S_{\phi_I B} & S_{\phi_I \mathcal{F}_I} & S_{\phi_I \phi_I} \end{bmatrix} \tag{7.13}$$

The first row corresponds to the symmetric Galerkin equations written for the non-interface boundary in the top (A) domain, Eq.(7.9) or Eq.(7.10), depending upon the boundary conditions. The first column therefore indicates integrations involving unknown quantities on the top (non-interface) boundary. The second row and column (B) are the analogous entries for the bottom material. Thus, S_{AA} and S_{BB} are (square) symmetric matrices, a consequence of the SG procedure. The off-diagonal $(1,2)$ and $(2,1)$ blocks are, as shown, equal to zero, as the top equations do not involve the bottom geometry, and vice versa.

The third and fourth columns refer to the unknown interface flux $(\mathcal{F}_I)$ and interface potential (ϕ_I), respectively. The top and bottom equations (i.e. the first and second rows) integrate over the interface and thus the $(1, 3)$, $(1, 4)$, $(2, 3)$ and $(2,4)$, entries are in general non-zero. At the risk of being repetitive, these rows do not include equations for the interface, i.e. Eq.(7.9) and Eq.(7.10) with the weighting functions ψ_l centered on an interface node. The minus sign multiplying $S_{B\mathcal{F}_I}$ is due to the change in sign for the interface flux, as mentioned above. The key

150 INTERFACE AND MULTIZONE

to achieving a symmetric matrix is to fill in rows three and four with appropriate interface equations (Galerkin weighting functions ψ_l centered on interface nodes) for determining ϕ_I and $\mathcal{F}_I$. The successful procedure is as follows (see Table 7.1).

Table 7.1 Symmetric Galerkin integral equations for an interface or multi-zone problem

Surface	Integral Equations		
Dirichlet		BIE	
Neumann		HBIE	
Interface $(\mathcal{F}_I)$	BIE(top)	$-$	BIE(bottom)
(ϕ_I)	HBIE(top)	$+$	HBIE(bottom)

For the interface flux, row three, write the BIE (Eq.(7.9)) for the top domain, minus this equation constructed for the bottom domain. This is reasonable, as equations for flux in the usual SG procedure indicate that potential is specified, and therefore the BIE is employed. Thus, it follows that $S_{\mathcal{F}_I A} = S_{A\mathcal{F}_I}^T$, the T superscript indicating transpose, as this is the symmetric Galerkin procedure for the top domain. One can convince oneself of this simply by assuming that the top boundary is either all Dirichlet or all Neumann, and noting which integration kernel comes into play in forming this matrix. Symmetry also holds for the $S_{\mathcal{F}_I B}$ contribution, as taking the negative BIE over the bottom domain compensates for the negative sign in the (2; 3) position. It is required that the diagonal block $S_{\mathcal{F}_I \mathcal{F}_I}$ be symmetric on its own, and this easily follows from the observation that this entry comes from integrating the symmetric kernel $G(P,Q)/\kappa$ over the interface.

For the interface potential, row four, write the HBIE (Eq.(7.10)) for the top domain, plus this equation constructed for the bottom domain. Again, as these equations are thought of as the equations to determine the interface potential, the use of the hypersingular equation is consistent with the usual SG procedure. That $S_{\phi_I A} = S_{A\phi_I}^T$ and $S_{\phi_I B} = S_{B\phi_I}^T$ follows again from the basic symmetric Galerkin algorithm. The symmetry of the (4,4) block follows immediately by noting that this matrix arises from integrating the symmetric kernel $\kappa\partial^2 G(P,Q)/\partial\mathbf{N}\partial\mathbf{n}$ over the interface.

It therefore remains to show that $S_{\mathcal{F}_I \phi_I} = S_{\phi_I \mathcal{F}_I}^T$. Note first that $S_{\mathcal{F}_I \phi_I}$ originates from the left hand side of Eq.(7.9). However, as the single integral term is the same for both top and bottom equations, this term drops out, leaving just the $\partial G/\partial\mathbf{n}$ integration. On the other hand, $S_{\phi_I \mathcal{F}_I}$ comes from the flux terms in Eq.(7.10). The single integral term once again drops out, this time because of the change in sign of the flux across the interface, and thus both matrices come from the first derivative kernel. Moreover, the difference of the two equations used for the third row is once again matched by the change in sign in the flux, and so the symmetry follows.

It is interesting to note that, for the Laplace equation at a flat interface, $S_{\phi_I \mathcal{F}_I} = 0$. The normal derivative of the Green's function is given by

$$\frac{\partial G}{\partial\mathbf{n}} = -\frac{1}{2\pi}\frac{\mathbf{n}\cdot\mathbf{R}}{r^2} \tag{7.14}$$

where $r = \|\mathbf{R}\| = \|\mathbf{Q} - \mathbf{P}\|$. Due to the $\mathbf{n} \cdot \mathbf{R}$ factor, the only surviving term in integrating over a flat surface is the singular contribution. However, as in the single integral term, the singular part is the same on both sides of the interface (note that the material constant κ is not present), and therefore cancels. In some applications, this observation could be invoked to reduce the computational effort.

7.3.1 Multiple Interfaces

The extension of the interface algorithm to more complicated geometries is not difficult, but nevertheless warrants some further discussion. Consider the problem with three zones, labeled A, B and C, illustrated in Figure 7.4.

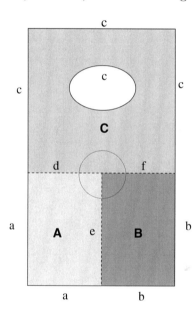

Figure 7.4 A more complicated interface problem.

The noninterface boundary segments for each subdomain are denoted by a, b, c, respectively, and d, e, f denote the interfaces. The usual symmetric Galerkin procedure is followed on the non-interface segments, and combinations of BIE and HBIE (see Table 7.1) are used on the interfaces. Thus, for example, the equations written for the interface segment d are potential and flux equations for regions A and C, while the equations for e involve regions A and B. The coefficient matrix for this problem takes the form

$$\begin{bmatrix} S_{aa} & 0 & 0 & S_{ad} & S_{ae} & 0 \\ 0 & S_{bb} & 0 & 0 & S_{be} & S_{bf} \\ 0 & 0 & S_{cc} & S_{cd} & 0 & S_{cf} \\ \hline S_{da} & 0 & S_{dc} & S_{dd} & S_{de} & S_{df} \\ S_{ea} & S_{eb} & 0 & S_{ed} & S_{ee} & S_{ef} \\ 0 & S_{fb} & S_{fc} & S_{fd} & S_{fe} & S_{ff} \end{bmatrix} \quad (7.15)$$

Here, to simplify notation, the unknown potential and flux on the interface segments have been combined. Thus, the fourth block column corresponds to potential and flux on the *d* interface. The symmetry of the upper-left 3 × 3 block again follows from the symmetric Galerkin procedure. Verification of the symmetry for the rest of the matrix follows along the same lines.

7.3.2 Corners

Multiple interface configurations can produce difficult corner problems. For three-dimensional problems, even more complicated corner/edge configurations can occur. With respect to the A and B subregions of Figure 7.4, there are two unknown fluxes to be determined at the corner point where all three subdomains meet. This situation is analogous to a boundary value problem in which Dirichlet data is supplied at a boundary corner. For a collocation approximation, this situation requires special treatment [115], as additional equations must be written to determine the unknown values. Another distinct advantage of the Galerkin formulation is that the extra equations are easily handled by proper definition of the Galerkin weighting functions at the corner point. All that is required is appropriate use of 'double nodes' in the discretization process. Thus, symmetric Galerkin provides an effective treatment of corners with conforming elements, which includes the type of corners produced by interface configurations.

7.3.3 Free interface

Another special situation worth mentioning is illustrated in Figure 7.5. This configuration arises in applications such as the transmission problem in wave scattering [236], and in this terminology the scattering object is completely embedded in the host material. In this case the interface forms the entire boundary surface for the inclusion subregion. The SG algorithm in this case is straightforward, the boundary integral equations for the object consisting solely of interface equations. The coefficient matrix will take the form of Eq.(7.13), with the first row and column removed, and hence, is clearly symmetric.

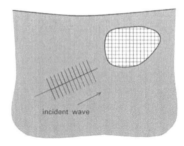

Figure 7.5 The interface boundary can comprise the entire boundary of a subdomain, as in this wave scattering/transmission problem.

7.3.4 Computational Aspects

As in the symmetric Galerkin formulation for a non-interface problem, the primary computational advantage of the symmetric formulation is in the reduced time required to solve the system of equations. For a direct solution, this is of the order of

$N^3/6$ versus $N^3/3$ for a non-symmetric matrix. Thus, for sufficiently large problems, this factor of two can translate into significantly less computational time. Some savings can be realized in constructing the matrix, but in general this is unfortunately limited: although the coefficient matrix is symmetric, the contributions to the right-hand side vector must still be computed for each equation. Thus, the number of calculations that can be bypassed due to symmetry is, speaking somewhat loosely (as this can depend upon the equation being solved), not large relative to the total operations required.

For interface problems, however, the symmetry of the coefficient matrix can be used to good advantage. Note that in Eq.(7.13) the upper right 2×2 block need not be calculated, due to symmetry. These matrix elements come from the boundary integral equations written for the non-interface boundaries, integrating over the interface. As indicated above, the only reason for now integrating over the interface in these equations is for constructing the right-hand side vector. However, as there are no boundary conditions on the interface, there is *no contribution* to the right-hand side. Thus, in forming the equations for the non-interface boundary, *integration over the interface is not required.* Depending upon the geometry of the problem, this can result in considerable computational savings. For instance, for the prototype problem considered in the next section, invoking this symmetry, i.e. bypassing the interface integrations for the non-interface equations, produced roughly a 13 per cent reduction in total computation time. For this relatively small problem (102 nodes), the matrix solution time is minimal, and thus the percent reduction (measured with total time) will be less when larger matrices are involved. Nevertheless, it is clear that a significant amount of computation has been avoided. The 13 per cent number is consistent with related symmetric Galerkin fracture calculations (Chapter 9). In this case, integrations over the crack surface can be skipped when forming the non-crack equations.

7.4 NUMERICAL EXAMPLES

A prototype numerical example modelling heat transfer in an eccentric annulus geometry is considered. To validate the interface algorithm, several variations in the boundary conditions and material properties are examined. This problem has characteristics which make it very suitable for testing purposes, e.g. curved boundaries, a non-convex region, and an interesting (i.e. non-trivial) corner situation at the interface junctions. The basic interpolation approximation employed in these tests consists of standard isoparametric, conforming, linear elements.

Figure 7.6 shows the geometry, boundary conditions, and material properties for the first calculation. The thermal conductivities for the top and bottom parts are denoted by k_A and k_B, respectively. If only one material is present, i.e. $k_A = k_B$, then there is a closed-form solution for this problem. This solution can be obtained by means of conformal mapping and complex variable techniques, and is given by (see, for example, the book by Greenberg [123])

$$\phi = 100 \left[1 - \frac{\ln(u^2 + v^2)}{2 \ln a} \right], \quad a = 2 + \sqrt{3} \tag{7.16}$$

where
$$u(x,y) = \frac{(ax-1)(x-a)+ay^2}{(ax-1)^2+a^2y^2}, \quad v(x,y) = \frac{(a^2-1)y}{(ax-1)^2+a^2y^2} \quad (7.17)$$

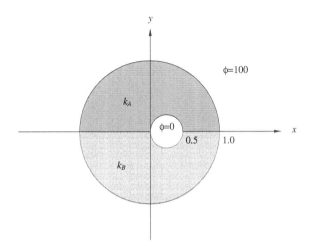

Figure 7.6 The geometry and the boundary condition of eccentric annulus

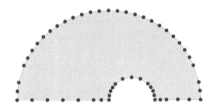

Figure 7.7 Boundary element mesh (58 nodes and 54 elements) considering the symmetric part of the eccentric annulus. The corresponding mesh for the interface calculations, obtained by symmetry with respect to the horizontal line ($y = 0$) has 116 nodes and 108 elements.

To verify the computer code, the material constants k_A and k_B were set equal ($k_A = k_B = 1.0$), and the problem solved using interfaces as shown in Figure 7.6. The boundary element mesh employed for the top-half of the domain is shown in Figure 7.7, the discretization for the bottom-half is identical. A comparison of the theoretical solution (Eq.(7.16) and Eq.(7.17)) with the numerical solution is provided in Tables 7.2-7.5 for some representative points along the boundary. The actual matching interface fluxes are zero, and the SGBEM values were of the order of 10^{-12} or less. These results indicate that the SG interface algorithm is correct.

The more interesting situations, e.g., $k_A \neq k_B$, with different boundary conditions have also been investigated [117]. For example, if $k_A = 1/2$ and $k_B = 1.0$, then the numerical solution on the bottom semicircles is essentially the same as before, but now the fluxes on the top semicircles are half of what they were before. This happens because the actual matching flux across the interface is zero. The boundary element solution in this case was consistent with this observation, and thus provides additional validation of the present numerical solution procedure for interface problems.

Table 7.2 Flux at top semi-circle with radius = 1.0

Angle(rad)	Theory	$\partial\phi/\partial\mathbf{n}$ SGBEM	Error(%)
0	131.52	133.77	-1.71
$\pi/4$	101.72	101.27	0.44
$\pi/2$	65.76	65.42	0.51
$3\pi/4$	48.58	48.45	0.27
π	43.84	44.77	-2.14

Table 7.3 Flux at top semi-circle with radius = 0.25

Angle(rad)	Theory	$\partial\phi/\partial\mathbf{n}$ SGBEM	Error(%)
0	-350.71	-346.17	1.30
$\pi/4$	-334.39	-329.72	1.40
$\pi/2$	-300.61	-300.25	0.12
$3\pi/4$	-273.03	-269.24	1.40
π	-263.04	-259.48	1.35

Table 7.4 Potential at the LHS interface ($y = 0$)

x	Theory	ϕ SGBEM	Error(%)
-1.0000	100.0000	100.0000	0.00
-0.8333	92.0218	92.2483	-0.25
-0.6667	82.3856	82.6837	-0.37
-0.5000	70.4045	70.8167	-0.59
-0.3333	54.8778	55.5480	-1.22
-0.1667	33.4085	34.0673	-1.97
-0.0833	18.8852	19.4347	-2.91
0.0000	0.0000	0.0000	0.00

7.5 REMARKS

This chapter presented a fully symmetric Galerkin boundary integral equation formulation for interface and multi-zone problems. This approach on the non-interface boundaries, together with a strategic use of both equations on the interface, yields a symmetric system, while automatically incorporating the continuity conditions on the interface. The present interface formulation is therefore a natural extension of the symmetric Galerkin procedure. The two most important benefits provided by

Table 7.5 Potential at the RHS interface ($y = 0$)

x	Theory	SGBEM	Error(%)
		ϕ	
0.5000	0.0000	0.0000	0.00
0.5625	19.5925	20.1178	-2.68
0.6250	35.7162	36.3434	-1.76
0.7500	61.6269	62.2651	-1.04
0.8750	82.3856	82.8032	-0.51
1.0000	100.0000	100.0000	0.00

this boundary integral approach to interface problems are (a) a symmetric coupling of boundary and finite elements, and (b) a computationally efficient algorithm. Although a two-dimensional scalar field formulation was discussed herein, there is no difficulty in extending this method to three-dimensional, vector-field problems and fracture problems. This algorithm provides an effective framework for studying interfacial cracks. Coupling this approach with special elements for capturing crack tip singularities, or interface corner singularities (as occur in multi-material thermoelastic problems), is straightforward. The interface formulation should also be directly applicable to non-linear analysis, in particular the symmetric Galerkin formulation for plasticity developed by Maier and co-workers [174–176].

CHAPTER 8

ERROR ESTIMATION AND ADAPTIVITY

Synopsis: This chapter presents a simple a posteriori error estimator and an effective adaptive mesh refinement procedure for the symmetric Galerkin boundary element method. *Galerkin residuals*, which are intrinsic to the symmetric Galerkin boundary integral approach form the basis of the present error estimation scheme. This estimator is then coupled with an *h*-adaptive strategy. This technique can be used for various problem configurations including mixed boundary conditions, corners, and nonconvex domains. The SGBEM code BEAN (Boundary Element ANalysis) for potential (Laplace Equation) problems is introduced, and example calculations employing *h*-adaptivity are discussed.

8.1 INTRODUCTION

Whenever a numerical method is utilized to solve the governing differential equations of a problem, error is introduced by the discretization process which reduces the continuous mathematical model to one having a finite number of degrees of freedom. The discretization errors are defined as the difference between the actual solution of the differential equation and its numerical approximation. Reliable estimation of these errors is essential to guarantee a certain level of accuracy of the numerical solution, and is the key component of adaptive procedures. Estimation of the discretization error in the SGBEM is the focus of this chapter. Other types of errors, such as roundoff errors and uncertainties in material, geometry and boundary conditions are not considered.

The Symmetric Galerkin Boundary Element Method. By Sutradhar, Paulino and Gray   **157**
ISBN 978-3-540-68770-2   ©2008 Springer-Verlag Berlin Heidelberg

158   ERROR ESTIMATION AND ADAPTIVITY

An important feature of the theory of singular integral equations is that the problem for the boundary unknowns may be formulated in different ways (e.g., Banerjee [13]). The error estimation method proposed herein relies on this feature. Thus, one may formulate two distinct boundary integral equations, e.g., the singular BIE and the hypersingular BIE (HBIE), to represent the same boundary value problem. A natural measure of the error, presented by Paulino *et al.* [210], rests on the use of both the BIE and the HBIE. For instance, suppose that an approximate solution, using one of the BIEs, has been obtained. Then, one expects that the residual obtained when this approximate solution is substituted in the other BIE is related to the error. Numerical experiments have suggested that this is indeed the case [180, 208]. The above error estimation method, developed originally for the collocation BEM (e.g., Paulino *et al.* [210]), comes as a natural extension for the case of the symmetric Galerkin BEM.

This method is the natural setting for this type of residual error estimates for two basic reasons. First, symmetric Galerkin formulation by definition employs both equations (i.e., the BIE and the HBIE) in the problem solution, so it is natural to think of using the alternate equation to compute a residual. Second, as mentioned above, hypersingular equations are most easily dealt with by means of a Galerkin approximation, and thus both the problem solution and the residuals can be computed in the same fashion. Based on the error estimation method, a SGBEM code for 2D Laplace equation with self-adaptive mesh capability is developed in the code BEAN.

8.2 BOUNDARY INTEGRAL EQUATIONS

First we setup the formulation for the Laplace problem. Consider the solution of Laplace equation in a region with mixed boundary conditions, formally stated as

$$\nabla^2 \phi = 0 \text{ in } \Omega \in \mathbf{R}^2$$

$$\phi = g_1 \quad \text{on } \Gamma_1 \quad (\textit{Dirichlet type})$$

$$\mathcal{F} = \frac{\partial \phi}{\partial \mathbf{n}} = g_2 \text{ on } \Gamma_2 \quad (\textit{Neumann type}) \tag{8.1}$$

where $\Gamma = \Gamma_1 \cup \Gamma_2$ for a well-posed problem. The solution of the boundary value problem consists of finding $\mathcal{F}(\frac{\partial \phi}{\partial \mathbf{n}} = \nabla\phi.\mathbf{n})$ on Γ_1 (Dirichlet surface) and ϕ on Γ_2 (Neumann surface).

Within the framework the interior limit singular BIE can be written in the form

$$\phi(P) + \int_\Gamma \frac{\partial G}{\partial \mathbf{n}}(P,Q)\phi(Q)dQ = \int_\Gamma G(P,Q)\mathcal{F}(Q)dQ \tag{8.2}$$

where $\mathbf{n} \equiv \mathbf{n}(Q)$ is the unit normal at a point Q on the domain boundary G, and $\mathbf{n}$ denotes the normal derivative with respect to Q. The free-space Green's function or fundamental solution is taken as the point source potential

$$G(P,Q) = -\frac{1}{2\pi}\log r \tag{8.3}$$

where $r = \|r\| = \|Q - P\|$. The corresponding HBIE is obtained by differentiating Eq.(8.2) with respect to P in the direction $\mathbf{N} \equiv n(P)$, the normal to the boundary at P (according to the limit from the interior representation). This results in

$$\frac{\partial\phi(P)}{\partial\mathbf{N}} + \int_\Gamma \frac{\partial^2 G}{\partial\mathbf{N}\partial\mathbf{n}}(P,Q)\phi(Q)dQ = \int_\Gamma \frac{\partial G}{\partial\mathbf{N}}(P,Q)\mathcal{F}(Q)dQ \qquad (8.4)$$

where $\partial(\bullet)/\partial\mathbf{N}$ indicates the normal derivative with respect to P. The collocation method is inherently nonsymmetric, as the computation involves an integration for every point (P) and element (Q) pair.

When using the Galerkin method in a BEM context, each matrix element is composed of double surface integrals on the boundary — an outer integration with respect to P and an inner integration with respect to Q. Thus, the integral equations Eq.(8.2) and Eq.(8.4) are enforced in a weighted sense in the form

$$\int_{\gamma_P} \psi_k(P) \int_{\Gamma_1} G(P,Q)\mathcal{F}_1(Q)dQ\,dP - \int_{\gamma_P} \psi_k(P) \int_{\Gamma_2} \frac{\partial G}{\partial\mathbf{n}}(P,Q)\phi_2(Q)dQ\,dP$$

$$= \int_{\gamma_P} \psi_k(P)g_1(P)dP + \int_{\gamma_P} \psi_k(P) \int_{\Gamma_1} \frac{\partial G}{\partial\mathbf{n}}(P,Q)g_1(Q)dQ\,dP$$

$$- \int_{\gamma_P} \psi_k(P) \int_{\Gamma_2} G(P,Q)g_2(Q)dQ\,dP, \quad \gamma_P \in \Gamma_1 \qquad (8.5)$$

and

$$\int_{\gamma_P} \psi_k(P) \int_{\Gamma_1} \frac{\partial G}{\partial\mathbf{N}}(P,Q)\mathcal{F}_1(Q)dQ\,dP - \int_{\gamma_P} \psi_k(P) \int_{\Gamma_2} \frac{\partial^2 G}{\partial\mathbf{N}\partial\mathbf{n}}(P,Q)\phi_2(Q)dQ\,dP$$

$$= \int_{\gamma_P} \psi_k(P)g_2(P)dP + \int_{\gamma_P} \psi_k(P) \int_{\Gamma_1} \frac{\partial^2 G}{\partial\mathbf{N}\partial\mathbf{n}}(P,Q)g_1(Q)dQ\,dP$$

$$- \int_{\gamma_P} \psi_k(P) \int_{\Gamma_2} \frac{\partial G}{\partial\mathbf{N}}(P,Q)g_2(Q)dQ\,dP, \quad \gamma_P \in \Gamma_2 \qquad (8.6)$$

respectively, where γ_P denotes the support of the weighting function $\psi_k(P)$. The weighting functions are chosen to be the basis shape functions $\psi_k(P)$ (e.g., linear: $k = 1, 2$; quadratic: $k = 1, 2, 3$ employed in the approximation of ϕ and $\mathcal{F}$ on the boundary. This procedure does not by itself guarantee symmetry of the coefficient matrix, but combined with the appropriate use of the BIE and HBIE, a symmetric set of algebraic equations will be generated. This choice of equation is dictated by the symmetry properties of the kernel functions.

$$\begin{aligned}
G(P,Q) &= G(Q,P) \\
\nabla_Q G(P,Q) &= -\nabla_P G(P,Q) = \nabla_P G(Q,P) \\
\frac{\partial^2 G}{\partial\mathbf{N}\partial\mathbf{n}}(P,Q) &= \frac{\partial^2 G}{\partial\mathbf{N}\partial\mathbf{n}}(Q,P)\ .
\end{aligned} \qquad (8.7)$$

Thus, for a Dirichlet problem, a Galerkin formulation of the BIE will produce a symmetric coefficient matrix, while the HBIE will be appropriate for a Neumann problem. The above relationships for the first derivatives of G also guarantee symmetry for a mixed boundary value problem, provided the BIE is employed on Γ_1 and the HBIE on Γ_2, as illustrated by Fig. 8.1. After discretization, the set of Eq.(8.5) and Eq.(8.6) can be written in matrix form as

$$[A]\{w\} = \{f\} \tag{8.8}$$

The matrix elements are composed of double integrals, and P and Q are treated on an equal footing. Note that $[A]$ is the symmetric system matrix obtained from the terms on the lefthand side, and $\{f\}$ is the known right-hand side obtained from the terms on the right-hand side of both Eq.(8.5) and Eq.(8.6). Moreover, the approximate numerical solution $\{w\}$ obtained from Eq.(8.8) can be decomposed as

$$w = [w_1, w_2]^T$$

where $w_1 \equiv \mathcal{F}_1, w_2 \equiv \phi_2$, and T denotes the transpose of the matrix.

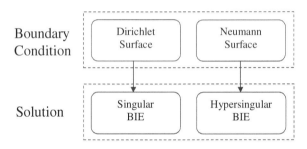

Figure 8.1 Symmetric Galerkin BEM solution phase

8.3 GALERKIN RESIDUALS AND ERROR ESTIMATES

'Galerkin residuals' gives a good estimate for discretization errors in numerical solutions obtained by the SGBEM. The key concept for obtaining the Galerkin residuals is the duality of the pair of integral integrations (i.e., standard and hypersingular BIEs), which has been described by Paulino et al. [210]. In the SGBEM, both the standard and the hypersingular BIEs are employed, the choice being dictated by the prescribed boundary condition. The interchange in the role of the two equations is the basis for the error estimation, as illustrated by Fig. 8.2 (cf., Fig. 8.1). Thus, on the Dirichlet parts of the boundary, the error estimate $\epsilon_1(P)$ is defined as the residual that arises when the approximate solution is substituted in the HBIE,

$$\begin{aligned}
\epsilon_1(P) &= -\int_{\gamma_P} \psi_k(P) w_1(P) dP \\
&+ \int_{\gamma_P} \psi_k(P) \int_{\Gamma_1} \frac{\partial G}{\partial \mathbf{N}}(P,Q) w_1(Q) dQ \, dP \\
&+ \int_{\gamma_P} \psi_k(P) \int_{\Gamma_2} \frac{\partial G}{\partial \mathbf{N}}(P,Q) g_2(Q) dQ \, dP \\
&- \int_{\gamma_P} \psi_k(P) \int_{\Gamma_1} \frac{\partial^2 G}{\partial \mathbf{N} \partial \mathbf{n}}(P,Q) g_1(Q) dQ \, dP \\
&- \int_{\gamma_P} \psi_k(P) \int_{\Gamma_2} \frac{\partial^2 G}{\partial \mathbf{N} \partial \mathbf{n}}(P,Q) w_2(Q) dQ \, dP.
\end{aligned} \tag{8.9}$$

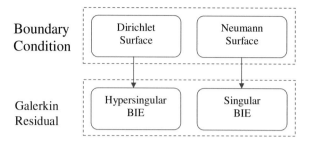

Figure 8.2 Combined Singular-Hypersingular Residual Calculation Phase.

As the weight function is centered on the node P_k and is nonzero only on the neighborhood of this node, this is taken to be an estimate of the local error in the computed flux at this point. Similarly, on the Neumann parts of the boundary, the error estimate $\epsilon_2(P)$ is defined as the residual that arises when the approximate solution is substituted in the BIE,

$$\begin{aligned}
\epsilon_2(P) &= -\int_{\gamma_P} \psi_k(P) w_2(P) dP \\
&\quad - \int_{\gamma_P} \psi_k(P) \int_{\Gamma_1} \frac{\partial G}{\partial \mathbf{n}}(P,Q) g_1(Q) dQ\, dP \\
&\quad - \int_{\gamma_P} \psi_k(P) \int_{\Gamma_2} \frac{\partial G}{\partial \mathbf{n}}(P,Q) w_2(Q) dQ\, dP \\
&\quad + \int_{\gamma_P} \psi_k(P) \int_{\Gamma_1} G(P,Q) w_1(Q) dQ\, dP \\
&\quad + \int_{\gamma_P} \psi_k(P) \int_{\Gamma_2} G(P,Q) g_2(Q) dQ\, dP.
\end{aligned}$$

(8.10)

This is likewise interpreted as an estimate of the error in the computed value of $\phi(P_k)$. Moreover

$$\epsilon = \epsilon_1 \cup \epsilon_2$$

Only the magnitude (and not the sign) of the Galerkin residuals (ϵ) is employed in the error estimation and adaptive procedure developed below. In general, error estimates are defined in terms of appropriate norms of the residuals (e.g., Paulino et al. [210]) in this case, only the magnitude of the residuals is needed.

8.4 SELF ADAPTIVE STRATEGY

The self-adaptive mesh refinement strategy employed in this work is the h-version, which generates a sequence of meshes of increasing refinement. The self-adaptive procedure is performed according to the flowchart of Fig. 8.3. The goal is to efficiently develop a well-graded final mesh, leading to a reliable numerical solution, in as simple a manner as possible. To avoid loss of numerical accuracy (Crouch and Starfield [61]; Rencis and Jong [232]; Guiggiani [126]), elements should not be graded such that large elements appear close to small elements. To solve this

162 ERROR ESTIMATION AND ADAPTIVITY

problem, Guiggiani [126] has adopted an additional rule, called a compatibility condition, so that whenever the ratio between the length of two adjacent elements was out of the range 0.25 to 4.0, the longer element was bisected. Because of its arbitrariness, this type of rule has not been used in the present work. Moreover, bad mesh gradation (in the sense described above) has not occurred for the examples presented in this chapter. This is a result of the quality of the error estimators using Galerkin residuals.

8.4.1 Local Error Estimation

Once the "Galerkin residuals" have been obtained (Eq.(8.9) and Eq.(8.10)) at each nodal point, they are normalized as

$$\bar{\epsilon} = \left| \frac{\epsilon_i}{\epsilon_{max}} \right|, \quad i = 1, \ldots n_n \tag{8.11}$$

where,

$$\epsilon_{max} = max(|\epsilon_1|, |\epsilon_2|, \ldots, |\epsilon_{n_n}|) \tag{8.12}$$

and n_n denotes the total number of nodes. In this work, linear boundary elements with shape functions

$$N_1(\xi) = 1 - \xi; \quad N_2(\xi) = \xi; \tag{8.13}$$

have been used, and $\psi_k = N_k(\xi), k = 1, 2$. However, the error estimation method is general and is not limited to linear elements. The error indicator for the boundary element (i) is denoted as $\epsilon^{(i)}$ and is obtained as

$$\epsilon^{(i)} = L^{(i)} \int_0^1 (N_1 \bar{\epsilon}^{(i)}_{node\ 1} + N_2 \bar{\epsilon}^{(i)}_{node\ 2}) d\xi$$

$$\epsilon^{(i)} = L^{(i)} \frac{\bar{\epsilon}^{(i)}_{node\ 1} + \bar{\epsilon}^{(i)}_{node\ 2}}{2}, \quad i = 1, \ldots, n_e \tag{8.14}$$

where n_e denotes the number of elements, $L^{(i)}$ is the element, length, and $\bar{\epsilon}^{(i)}_{node\ 1}$ and $\bar{\epsilon}^{(i)}_{node\ 2}$ are the values of the normalized error indicators at the beginning and end nodes of the boundary element (i).

8.4.2 Element Refinement Criterion

A simple criterion for mesh refinement consists of bisecting the element for which its error indicator is larger than a reference value. Here, this reference quantity is taken as the average error indicator given by

$$\epsilon_{ref} = \frac{1}{n_e} \sum_{i=1}^{n_e} \epsilon^{(i)} \tag{8.15}$$

If the inequality

$$\epsilon^{(i)} > \gamma \epsilon_{ref}, \quad i = 1, \ldots, n_e \tag{8.16}$$

is satisfied, then the element is divided into two elements (bisection). The parameter γ in Eq.(8.16) is a weighting coefficient that allows one to control the "refining velocity." The standard procedure consists of using $\gamma = 1$. Cases where $\gamma \neq 1$ are discussed next.

If $\gamma > 1$, then the number of elements to be refined is less than with $\gamma = 1$. By selecting $\gamma > 1$, one can control the total number of elements at each step and avoid too many refinements. The numerical solution from the next step (Fig. 8.3) is expected to be more accurate than that of the current step; however, the increase on the total number of elements is comparatively smaller. This approach is useful when the total number of elements is expected to be less than a certain number. The disadvantage is that the convergence rate is slower than that with $\gamma = 1$.

If $\gamma < 1$, the number of elements to be refined is larger than with $\gamma = 1$. The advantage is that the refinement rate should increase. However, the computational efficiency would decrease owing to the likely generation of an excessive number of elements.

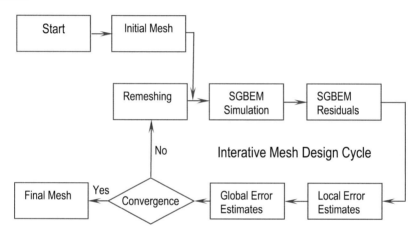

Figure 8.3 Self-Adaptive Analysis Algorithm (h-version).

8.4.3 Global Error Estimation

The adaptive mesh refinement process is carried out iteratively (Fig. 8.3). Although the refinement may be terminated by restrictions on storage and computing time, the stopping criterion is generally a specified level of accuracy. In this case, one may be interested either in the global error of the approximation or in a pointwise error bound. An indication of the overall convergence may be obtained by

$$\epsilon_{global} = \sum_{i=1}^{n_e} \epsilon^{(i)} \qquad (8.17)$$

or

$$\epsilon_{global} = \epsilon_{ref} \qquad (8.18)$$

Both Eq.(8.17) and Eq.(8.18) can be very easily obtained. The goal of the adaptive procedure is to obtain well-distributed meshes (i.e., near optimal). Ideally, as the iterative meshing progresses, the error estimates should decrease both locally and globally.

8.4.4 Solution Algorithm for Adaptive Meshing

The solution algorithm for adaptive meshing is summarized below and depicted in Figure 8.3.

1. Solve Eq.(8.5) and Eq.(8.6) simultaneously (in discretized form) to obtain the unknown values of potential and flux on the boundary.

2. In a postprocessing stage, calculate the "Galerkin residuals" at the nodal points by means of Eq.(8.9) and Eq.(8.10).

3. Compute element nodal errors using Eq.(8.14).

4. Compute average error indicator using Eq.(8.15).

5. Perform element refinement according to the criterion of Eq.(8.16).

6. Check global stopping criterion with reference to ϵ_{global} given by, for example, Eq.(8.18). If the global stopping criterion is satisfied, then stop. Otherwise, repeat steps 1 to 5.

8.5 NUMERICAL EXAMPLE

Eccentric Annulus. The first example models heat conduction in an eccentric annulus geometry. This problem has characteristics that make it very suitable for testing purposes, e.g., curved boundaries, nonconvex region, and corners. Figure 8.4 shows the geometry and boundary conditions. There is a closed-form solution for this problem, obtained by means of conformal mapping and complex variable techniques, given by (see, for example, the book by Greenberg [123])

$$\phi = 100 \left[1 - \frac{\ln(u^2 + v^2)}{2 \ln a} \right], \quad a = 2 + \sqrt{3} \tag{8.19}$$

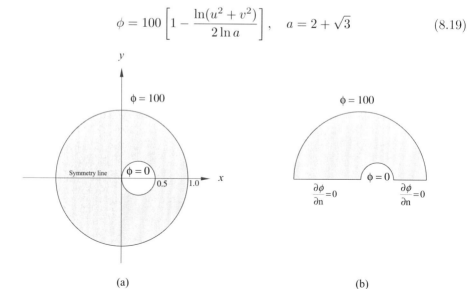

Figure 8.4 Eccentric Annulus: (a) Geometry; (b) Boundary Conditions.

where

$$u(x,y) = \frac{(ax-1)(x-a) + ay^2}{(ax-1)^2 + a^2y^2}, \quad v(x,y) = \frac{(a^2-1)y}{(ax-1)^2 + a^2y^2} \qquad (8.20)$$

For this problem, one can take advantage of symmetry and model only the top (or bottom) part of the problem. This approach is followed here, and only the top part of the annulus [Figure 8.4(b)] is considered in the adaptive analysis. Note that, when symmetry is used, the discretization error is greater than when symmetry is not employed since elements are placed on the symmetry axis. Also, by using the symmetry, the problem becomes a mixed boundary value problem.

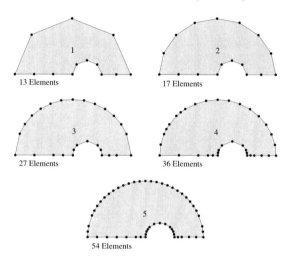

Figure 8.5 Adaptive Meshes for Eccentric Annulus Problem (First Four Refinement Steps)

The sequence of meshes obtained during the self-adaptive mesh refinement procedure is given in Fig. 8.5. A well-graded mesh (i.e., mesh 5) is generated in four iterative steps. Note that this mesh possesses well-distributed elements with a strong gradation at the corners. Comparison of the theoretical solution provided by Eq.(8.19) and Eq.(8.20) with the SGBEM solution reveals that for most nodal points the error is much less than 1%. The numerical results obtained with this mesh are practically the same as those reported in the numerical example in Chapter 7. However, in Chapter 7, the problem is modeled using a double region in order to validate a symmetric Galerkin multizone formulation, and here it is modeled using a single region [Figure 8.4(b)]. Table 8.1 shows the values of the two global error estimators given by Eq.(8.17) and Eq.(8.18) for each of the meshes in Fig. 8.5. Again, both estimates monotonically decrease with the number of elements.

Rectangular Region with Discontinuous Boundary Conditions. This problem has a rectangular region with two discontinuities in the boundary condition. Fig. 8.6 shows the rectangular geometry and the applied boundary conditions, respectively, for the test case. This problem has been studied by Shi *et al.* [255] in an *h*-adaptive BEM procedure (collocation based) for potential problems using linear elements and mesh sensitivities as error indicators.

Table 8.1 Global Error Estimates for Eccentric Annulus Problem

Mesh	n_e	Eq.(8.17)	Eq.(8.18)
		Global Estimates	
1	13	3.19378	0.24568
2	17	1.81541	0.10679
3	27	1.59126	0.05894
4	36	1.18794	0.03300
5	54	0.75659	0.01401

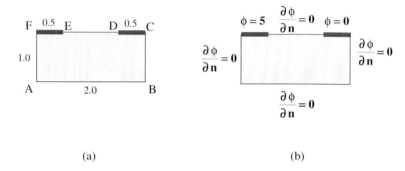

Figure 8.6 Rectangular region: (a) Geometry; (b) Boundary Conditions.

The initial mesh discretization, consisting of 14 elements, is the same as the one adopted by Shi et al. [255]. Both global error measures are monotonically decreasing with mesh refinement. The corresponding meshes, shown in Fig. 8.7, display a smooth mesh gradation at each iteration. The difficult areas for the calculation are the geometrical corners and the neighborhood of points D and E, where there is a discontinuity in the flux. Note that the meshes in these regions are progressively refined. The number of elements at the first four refinement steps are 14, 18, 22, 30, and 40. The final mesh (i.e., mesh 5), shown in Fig. 8.7, has the same number of elements (i.e., 40) and a similar element distribution to the final mesh obtained by Shi et al. [255] in their second solution for this problem.

As there is no analytical solution available for this problem (Fig. 8.6), the reference mesh, shown in Fig. 8.8, is used to assess the quality of the numerical solution obtained with the adaptive SGBEM. The comparison of the solution obtained at each step (Fig. 8.7) with an approximate reference solution (Fig. 8.8) is given in Figs. 8.5. These graphs show that the numerical solution improves consistently as the mesh is refined; i.e., the solution for mesh i is better than the solution for mesh i−1 (i = 2, . . . , 5). In the present problem, points D and E (Fig. 8.6) are of special interest because there is a discontinuity in the value of the flux from a nonzero value on the Dirichlet sides (CD and EF) to a zero value on the Neumann side (DE). The oscillations in the flux distribution on CD (near point D) and on EF (near point E) are expected because there is no special treatment of the discontinuity condition at points D and E. Thus, the reference mesh data in Figure 8.5 should be considered with caution because the discontinuity condition has not been

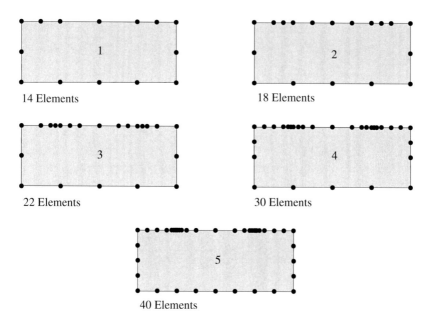

Figure 8.7 Adapted meshes for Rectangular region problem (first four refinement steps).

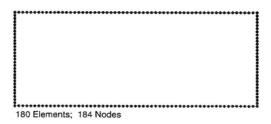

Figure 8.8 Reference mesh for rectangular geometry with discontinuous boundary conditions.

treated in the numerical scheme. Note that the oscillations are localized and the flux decreases rapidly away from points D (toward C) and E (toward F).

8.6 BEAN CODE

This section introduces the educational Matlab code BEAN for the solution of 2D Laplace problems. The use and capabilities of this code will be discussed more fully in Chapter 11. BEAN is capable of analyzing single and multi-region problems of arbitrary geometry. Interior results at selected points can be obtained as well. In the present version only linear elements are used. It has a graphical user interface to generate the model, view the results which includes contour plots, line plots as well as the output file.

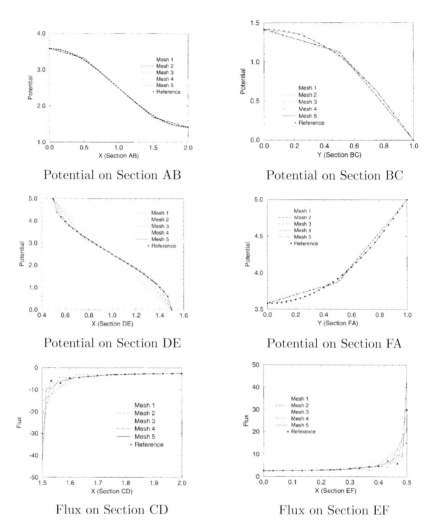

Figure 8.9 Plots of Potential on Section AB, BC, DE, FA and Flux on Section CD and EF. No special treatment of discontinuity condition at Point D (X=1.5) and Point E (X=0.5) for either the 'Reference' mesh or meshes 1 to 5.

In order to demonstrate BEAN, the classical 'Motz Problem', represented in Figure 8.10 in potential theory is solved. The Motz problem is a benchmark Laplace equation problem that is very often used for testing various numerical methods proposed in the literature for the solution of elliptical boundary value problem with boundary singularities (see, for example, the book by París and J. Cañas [205]) The dimensions and the boundary conditions of the problem (same as [205]) are represented in the Figure 8.10.

Figure 8.11 shows the initial mesh of the problem. The BEAN mesh adaptation is illustrated in Figure 8.12. The adaptive strategy picks up the singularity at the center of the lower side and thus nodes are added there as the mesh is refined. Figure 8.13 shows the potential distribution on the lower edge where flux is zero.

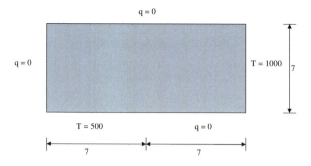

Figure 8.10 Geometry and Boundary Conditions of the Motz problem

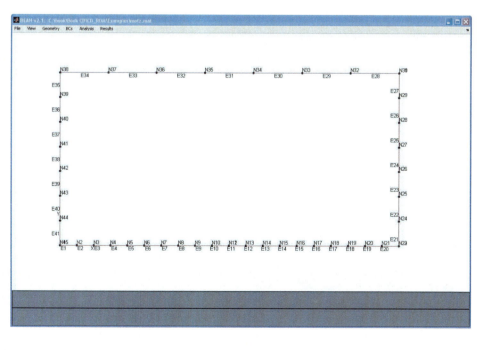

Figure 8.11 Mesh of the Motz problem

170 ERROR ESTIMATION AND ADAPTIVITY

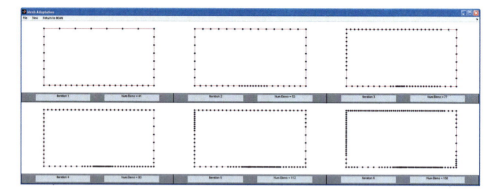

Figure 8.12 Mesh adaptation of the Motz problem

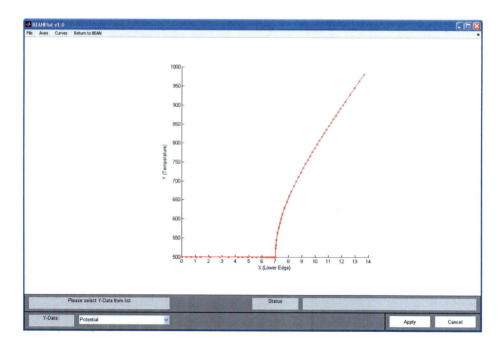

Figure 8.13 Variation of temperature along the lower edge of the rectangle.

CHAPTER 9

FRACTURE MECHANICS

Synopsis: One of the most successful application areas for boundary integral methods is fracture analysis. Fracture problems arise in many important engineering areas and boundary integral methods have inherent advantages for these calculations. It is therefore essential that an efficient symmetric Galerkin formulation be developed for this class of problems. This chapter will demonstrate the excellence and efficiency of the technique. The basic framework for the 2D elasticity boundary value problem is described first followed by the fracture algorithm. A path-independent integral technique known as interaction integral method is used for evaluating mixed-mode stress intensity factors (SIFs) and T-stress for 2D crack problems. Both standard quarter-point crack-tip elements and modified quarter-point crack-tip elements are used in the implementation. Some numerical examples are examined to demonstrate the accuracy, efficiency and the robustness of the method.

9.1 INTRODUCTION

Fracture behaviour is generally characterized by a single parameter such as the stress intensity factors (SIFs) or path independent J-integral [233]. These quantities provide a measure of the dominant behaviour of the stress field in the vicinity of a crack tip. In order to understand the effect of the structural and loading configuration on the "constraint" [5] conditions at the crack tip, another parameter is required. A second fracture parameter often used is the elastic T-stress. In

The Symmetric Galerkin Boundary Element Method. By Sutradhar, Paulino and Gray **171**
ISBN 978-3-540-68770-2 ©2008 Springer-Verlag Berlin Heidelberg

172 SYMMETRIC GALERKIN

two dimensions, the T-stress is defined as a constant stress acting parallel to the crack and its magnitude is proportional to the nominal stress in the vicinity of the crack. Various studies have shown that the T-stress has significant influence on crack growth direction, crack growth stability, crack-tip constraint and fracture toughness [59, 79, 203, 261, 280, 288].

For a few idealized cases, analytical solutions for T-stress and SIFs are available. However, for practical problems involving finite geometries with complex loading, numerical methods need to be employed. Compared to domain-type methods (e.g. finite element method (FEM) and meshless methods [169]), boundary method significantly reduces the problem size and the problem setup. Additionally, remeshing of propagating cracks, which only involves local operations on the propagating tip/front region, is simple and easy. The Galerkin BEM has several advantages over the collocation BEM in dealing with fracture problems. The biggest advantage is that the hypersingular integrals can be evaluated using standard C^0 elements without any compromise or ambiguity [25]. Moreover, the weighted averaging formula in the Galerkin BEM provides a smooth and reliable solution in the neighborhood of geometric discontinuities such as corners and junctions. By exploiting symmetry [23] and using faster algorithms [113] in setting up the coefficient matrix, computational effort can be reduced more. A unified scheme is developed by using the interaction integral method for calculating both the T-stress and the SIFs for mixed-mode cracks by means of the SGBEM.

9.2 FRACTURE PARAMETERS: STRESS INTENSITY FACTORS (SIFs) AND T-STRESS

Williams' asymptotic solution [289] for crack-tip stress fields in any linear elastic body is given by a series of the form:

$$\sigma_{ij}(r, \theta) = A_1 r^{-1/2} f_{ij}^{(1)}(\theta) + A_2 f_{ij}^{(2)}(\theta) + A_3 r^{1/2} f_{ij}^{(3)}(\theta) + \text{higher order terms}, \quad (9.1)$$

where σ_{ij} is the stress tensor, r and θ are polar coordinates with the origin at the crack tip as shown in Figure 9.2, $f_{ij}^{(1)}$, $f_{ij}^{(2)}$, $f_{ij}^{(3)}$ are universal functions of θ, and A_1, A_2, A_3 are parameters proportional to the remotely applied loads. In the vicinity of the crack ($r \to 0$), the leading term which exhibits a square-root singularity dominates. The amplitude of the singular stress fields is characterized by the SIFs, i.e.

$$\sigma_{ij} = \frac{K_I}{\sqrt{2\pi r}} f_{ij}^I(\theta) + \frac{K_{II}}{\sqrt{2\pi r}} f_{ij}^{II}(\theta), \quad (9.2)$$

where K_I and K_{II} are the mode I and mode II SIFs, respectively.

The second term in the Williams' series solution (Eq.(9.1)) is a non-singular term, which is defined as the elastic T-stress. Thus the above expression (9.2) can be expanded to include this term as follows

$$\sigma_{ij} = \frac{K_I}{\sqrt{2\pi r}} f_{ij}^I(\theta) + \frac{K_{II}}{\sqrt{2\pi r}} f_{ij}^{II}(\theta) + T\delta_{1i}\delta_{1j}. \quad (9.3)$$

The angular functions $f_{ij}(\theta)$ in Eq.(9.2) are given by

$$
\begin{aligned}
f_{11}^{I}(\theta) &= \cos\frac{\theta}{2}\left(1 - \sin\frac{\theta}{2}\sin\frac{3\theta}{2}\right), \\
f_{11}^{II}(\theta) &= -\sin\frac{\theta}{2}\left(2 + \cos\frac{\theta}{2}\cos\frac{3\theta}{2}\right), \\
f_{22}^{I}(\theta) &= \cos\frac{\theta}{2}\left(1 + \sin\frac{\theta}{2}\sin\frac{3\theta}{2}\right), \\
f_{22}^{II}(\theta) &= \sin\frac{\theta}{2}\cos\frac{\theta}{2}\cos\frac{3\theta}{2}, \\
f_{12}^{I}(\theta) &= \sin\frac{\theta}{2}\cos\frac{\theta}{2}\cos\frac{3\theta}{2}, \\
f_{12}^{II}(\theta) &= \cos\frac{\theta}{2}\left(1 - \sin\frac{\theta}{2}\sin\frac{3\theta}{2}\right)
\end{aligned}
\tag{9.4}
$$

The T-stress varies with different crack geometries and loadings. It plays a dominant role on the shape and size of the plastic zone, the degree of local crack-tip yielding, and also in quantifying fracture toughness. For mixed-mode problems, the T-stress contributes to the tangential stress and, as a result, it affects the crack growth criteria. By normalizing the T-stress with the applied load $\sigma_0(=K_I/\sqrt{\pi a})$, a non-dimensional parameter B can be defined by [5, 160]

$$
B = T\sqrt{\pi a}/K_I,
\tag{9.5}
$$

where a is the crack length. The dependence on geometrical configurations can be best indicated by the biaxiality parameter B.

9.3 SGBEM FORMULATION

The basic SGBEM framework for two dimensional (2D) elastic boundary value problems is introduced in this section. First, the algorithm for bodies without cracks is provided. Then, the fracture algorithm is developed followed by a brief section on crack-tip elements.

9.3.1 Basic SGBEM formulation for 2D elasticity

The boundary integral equation (BIE) for a source point P interior to the domain for linear elasticity without body forces [235] is given by

$$
u_k(P) - \int_{\Gamma_b} \left[U_{kj}(P,Q)\,\tau_j(Q) - T_{kj}(P,Q)\,u_j(Q) \right]\,dQ = 0 ,
\tag{9.6}
$$

where Q is a field point, τ_j and u_j are traction and displacement vectors respectively, U_{kj} and T_{kj} are the Kelvin kernel tensors and Γ_b denotes the boundary of the domain.

174 SYMMETRIC GALERKIN

For plane strain problems (see, e.g., [235]), the Kelvin kernels are

$$U_{kj} = \frac{1}{8\pi\mu(1-\nu)}\left[r_{,k}r_{,j} - (3-4\nu)\delta_{kj}\ln(r)\right], \tag{9.7}$$

$$T_{kj} = -\frac{1}{4\pi(1-\nu)r}\left[\{(1-2\nu)\delta_{kj} + 2r_{,k}r_{,j}\}\frac{\partial r}{\partial n} - (1-2\nu)(n_j r_{,k} - n_k r_{,j})\right], \tag{9.8}$$

where ν is Poisson's ratio, μ is shear modulus, δ_{ij} is the Kronecker delta and,

$$r_k = x_k(Q) - x_k(P), \qquad r^2 = r_i r_i, \qquad r_{,k} = r_k/r \quad \text{and} \quad \partial r/\partial n = r_{,i}n_i. \tag{9.9}$$

For a point P interior to the domain, the displacement gradient can be obtained by differentiating Eq.(9.6) with respect to the source point P. As P approaches the boundary, the limit of right-hand-side of Eq.(9.6) exists [118]. For $P \in \Gamma_b$, the BIE is defined in the limiting sense. Using the Somigliana identity, by substituting the displacement gradient into the Hooke's law, we get the HBIE for the boundary stresses:

$$\sigma_{k\ell}(P) - \int_{\Gamma_b} \left[D_{kj\ell}(P,Q)\,\tau_j(Q) - S_{kj\ell}(P,Q)\,u_j(Q)\right]\,dQ = 0, \tag{9.10}$$

where the kernels are given by [235]

$$D_{kj\ell} = \frac{1}{4\pi(1-\nu)r}\left[(1-2\nu)(\delta_{kj}r_{,\ell} + \delta_{j\ell}r_{,k} - \delta_{\ell k}r_{,j}) + 2r_{,k}r_{,j}r_{,\ell}\right], \tag{9.11}$$

$$\begin{aligned} S_{kj\ell} = \frac{\mu}{2\pi(1-\nu)r^2}\Big[& 2\frac{\partial r}{\partial n}\{(1-2\nu)\,\delta_{\ell k}r_{,j} + \nu\,(\delta_{kj}r_{,\ell} + \delta_{j\ell}r_{,k}) - 4r_{,k}r_{,j}r_{,\ell}\} \\ & +2\nu\,(n_k r_{,j}r_{,\ell} + n_\ell r_{,k}r_{,j}) + (1-2\nu)\,(2n_j r_{,\ell}r_{,k} + \delta_{kj}n_\ell + \delta_{j\ell}n_k) \\ & - (1-4\nu)\,\delta_{\ell k}n_j\Big]. \end{aligned} \tag{9.12}$$

In the collocation approach the BIE (9.6) and HBIE (9.10) are enforced at discrete source points. In a Galerkin approximation, the error in the approximate solution is orthogonalized against the shape functions. The shape functions are the weighting functions and the integral equations (9.6) and (9.10) are enforced in the 'weak sense', i.e.

$$\int_{\Gamma_b} \psi_m(P)\left[u_k(P) - \int_{\Gamma_b}[U_{kj}(P,Q)\,\tau_j(Q) - T_{kj}(P,Q)\,u_j(Q)]\,dQ\right]dP = 0, \tag{9.13}$$

$$\int_{\Gamma_b} \psi_m(P)\left[\sigma_{k\ell}(P) - \int_{\Gamma_b}[D_{kj\ell}(P,Q)\,\tau_j(Q) - S_{kj\ell}(P,Q)\,u_j(Q)]\,dQ\right]dP = 0, \tag{9.14}$$

respectively. As a result, the Galerkin technique possesses the important property of the *local support*. A symmetric coefficient matrix in the symmetric Galerkin approximation can be obtained by using Eq.(9.13) on the boundary $\Gamma_{b(u)}$ where displacements u_{bv} are prescribed, and using Eq.(9.14) on the boundary $\Gamma_{b(\tau)}$ with prescribed tractions τ_{bv} (See Figure 9.1). Note that, $\Gamma_b = \Gamma_{b(u)} + \Gamma_{b(\tau)}$ for a well-posed boundary value problem.

The additional boundary integration is the key to obtaining a symmetric coefficient matrix, as this ensures that the source point P and field point Q are treated in the same manner in evaluating the kernel tensors U_{kj}, T_{kj}, D_{kjl} and S_{kjl}. After discretization, the resulting equation system can be written in block-matrix form [118] as

$$\begin{bmatrix} H_{11} & H_{12} \\ H_{21} & H_{22} \end{bmatrix} \begin{Bmatrix} u_{bv} \\ u_* \end{Bmatrix} = \begin{bmatrix} G_{11} & G_{12} \\ G_{21} & G_{22} \end{bmatrix} \begin{Bmatrix} T_* \\ T_{bv} \end{Bmatrix}. \tag{9.15}$$

Here, the first and second rows represent, respectively, the BIE written on $\Gamma_{b(u)}$ and the HBIE written on $\Gamma_{b(T)}$. Further, u_* and τ_* denote unknown displacement and traction vectors. Rearranging Eq.(9.15) into the form $[A]\{x\} = \{b\}$, and multiplying the HBIE by -1, one obtains

$$\begin{bmatrix} -G_{11} & H_{12} \\ G_{21} & -H_{22} \end{bmatrix} \begin{Bmatrix} T_* \\ u_* \end{Bmatrix} = \begin{Bmatrix} -H_{11}u_{bv} + G_{12}T_{bv} \\ H_{21}u_{bv} - G_{22}T_{bv} \end{Bmatrix} \tag{9.16}$$

The symmetry of the coefficient matrix, $G_{11} = G_{11}^T$, $H_{22} = H_{22}^T$, and $H_{12} = G_{21}^T$ now follows from the symmetry properties of the kernel tensors [118].

9.3.2 Fracture analysis with the SGBEM

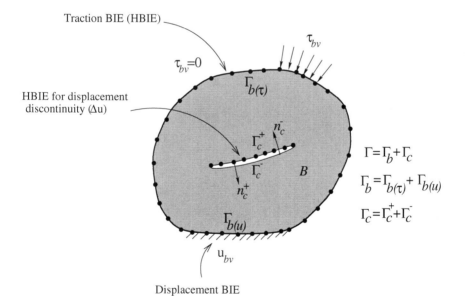

Figure 9.1 Configuration of the fracture scheme using the SGBEM. Notice that on the crack surface, only the upper surface is discretized and the unknown is the displacement discontinuity. The bullets denote the nodal points of the BEM mesh.

Consider a body of arbitrary shape B which contains a crack, as shown in Figure 9.1. The boundary Γ of the body B is composed of non-crack boundary Γ_b and the crack surface Γ_c. The portion of the boundary Γ_b with prescribed displacements is denoted by $\Gamma_{b(u)}$, and the portion with prescribed traction boundary is denoted by $\Gamma_{b(\tau)}$. Fracture is typically idealized as having zero thickness, it is essential to remember that it is composed of two distinct surfaces, and that there

176 SYMMETRIC GALERKIN

are two unknowns values at each point of the crack. The crack surface Γ_c consists of two coincident surfaces Γ_c^+ and Γ_c^-, where Γ_c^+ and Γ_c^- represent the upper and lower crack surfaces respectively. The outward normals to the crack surfaces, designated by n_c^+ and n_c^- are oriented in opposite directions and at any point on the crack $n_c^- = -n_c^+$ (see Figure 9.1). If we write the BIEs and HBIEs in terms of a 3x3 block matrix for the fracture geometry then, the first block row will represent the outer or non-crack boundary equations. The equation for a particular node is chosen according to the prescribed boundary data as per the usual Galerkin procedure. If we write dual equations (both HBIE and BIE) on the crack surface then second and third rows will denote, respectively, the HBIE and the BIE on the crack surface. With these definitions, the equations take the abbreviated form

$$\begin{pmatrix} h_{11} & h_{12} & h_{13} \\ h_{21} & h_{22} & h_{23} \\ h_{31} & h_{32} & h_{33} \end{pmatrix} \begin{pmatrix} \Omega_1 \\ u_c^+ \\ u_c^- \end{pmatrix} = \begin{pmatrix} g_{11} & g_{12} & g_{13} \\ g_{21} & g_{22} & g_{23} \\ g_{31} & g_{32} & g_{33} \end{pmatrix} \begin{pmatrix} \hat{\Omega}_1 \\ \tau_c^+ \\ \tau_c^- \end{pmatrix} \tag{9.17}$$

The vector of unknowns on the non-crack boundary can be a mixture of displacement and traction, and is therefore denoted by Ω_1. The corresponding vector of prescribed boundary values is indicated by $\hat{\Omega}_1$. On the fracture geometry, u represents the vector of unknown displacement values, τ the specified traction, and the superscripts $\{+, -\}$ label the two sides of the crack. The matrix $\mathcal{H}$ on the left therefore multiplies the vector of unknowns, and the right hand side consists of known quantities.

Notice that, the only difference between the two coincident crack surfaces is the orientation of the normals $(n_c^- = -n_c^+)$ and, as a result, $h_{13} = -h_{12}$, $h_{23} = -h_{22}$ and also $g_{13} = g_{12}$, $g_{23} = g_{22}$. These relationships between the second and third columns are a consequence of the integration over the two sides of the fracture differing by a sign.

$$\left(\begin{array}{cc|c} h_{11} & h_{12} & -h_{12} \\ h_{21} & h_{22} & -h_{22} \\ \hline h_{31} & h_{32} & -h_{32} \end{array}\right) \left(\begin{array}{c} \Omega_1 \\ u_c^+ \\ \hline u_c^- \end{array}\right) = \left(\begin{array}{cc|c} g_{11} & g_{12} & g_{12} \\ g_{21} & g_{22} & g_{22} \\ \hline g_{31} & g_{32} & g_{32} \end{array}\right) \left(\begin{array}{c} \hat{\Omega}_1 \\ \tau_c^+ \\ \hline \tau_c^- \end{array}\right) \tag{9.18}$$

It is convenient to replace the displacements u_c^+ and u_c^- by the single crack opening displacement

$$\Delta u_c = u_c^+ - u_c^-, \tag{9.19}$$

and the tractions τ_c^+ and τ_c^- by the sum of tractions

$$\Sigma \tau_c = \tau_c^+ + \tau_c^-. \tag{9.20}$$

Thus the reduced system of equations can be rewritten as

$$\left(\begin{array}{cc|c} h_{11} & h_{12} & 0 \\ h_{21} & h_{22} & 0 \\ \hline h_{31} & h_{32} & 0 \end{array}\right) \left(\begin{array}{c} \Omega_1 \\ \Delta u_c \\ \hline 0 \end{array}\right) = \left(\begin{array}{cc|c} g_{11} & g_{12} & 0 \\ g_{21} & g_{22} & 0 \\ \hline g_{31} & g_{32} & 0 \end{array}\right) \left(\begin{array}{c} \hat{\Omega}_1 \\ \Sigma \tau_c \\ \hline 0 \end{array}\right). \tag{9.21}$$

The usual boundary condition is the derivative quantity (e.g. traction, flux) and thus only the hypersingular equation should be employed. It therefore allows to solve the smaller 2×2 block system for the unknowns $(\Omega_1, \Delta u_c)$.

As explained above, it suffices to discretize the upper crack surface Γ_c^+. Thus, the BIE and HBIE written for an interior point P take the following form:

$$u_k(P) = \int_{\Gamma_b} [U_{kj}(P,Q)\,\tau_j(Q) - T_{kj}(P,Q)\,u_j(Q)]\,dQ$$
$$+ \int_{\Gamma_c^+} [U_{kj}(P,Q)\,\Sigma\tau_j(Q) - T_{kj}(P,Q)\,\Delta u_j(Q)]\,dQ\,, \qquad (9.22)$$

$$\sigma_{k\ell}(P) = \int_{\Gamma_b} [D_{\ell km}(P,Q)\,\tau_m(Q) - S_{\ell km}(P,Q)\,u_m(Q)]\,dQ$$
$$+ \int_{\Gamma_c^+} [D_{\ell km}(P,Q)\,\Sigma\tau_m(Q) - S_{\ell km}(P,Q)\,\Delta u_m(Q)]\,dQ\,. \quad (9.23)$$

However, since the crack surfaces are usually symmetrically loaded, i.e. $\tau_c^- = -\tau_c^+$, one gets

$$u_k(P) = \int_{\Gamma_b} [U_{kj}(P,Q)\,\tau_j(Q) - T_{kj}(P,Q)\,u_j(Q)]\,dQ$$
$$- \int_{\Gamma_c^+} T_{kj}(P,Q)\,\Delta u_j(Q)\,dQ\,, \qquad (9.24)$$

$$\sigma_{k\ell}(P) = \int_{\Gamma_b} [D_{\ell km}(P,Q)\,\tau_m(Q) - S_{\ell km}(P,Q)\,u_m(Q)]\,dQ$$
$$- \int_{\Gamma_c^+} S_{\ell km}(P,Q)\,\Delta u_m(Q)\,dQ\,. \qquad (9.25)$$

Previous boundary element solutions of fracture mechanics problems in terms of displacement discontinuities have been presented by Crouch and his co-workers [60, 61]. In the Galerkin approximation for the non-crack boundary Γ_b, the limit of (9.24) and (9.25) is taken as $P \to \Gamma_{b(u)}$ and $\Gamma_{b(\tau)}$, respectively. Since tractions are prescribed on the crack surface Γ_c^+, only Eq.(9.25) is written for source points on Γ_c^+ and, following the Galerkin approximation, the limit of (9.25) as $P \to \Gamma_c^+$ is considered. Converting the stress equation (9.25) into a traction equation through the identity $\tau_k(P) = \sigma_{\ell k}(P)\,n_\ell(P)$, with $n_\ell(P)$ being the outward normal at P and discretizing, the following system is obtained from Eqs. (9.24) and (9.25) in block matrix form:

$$\begin{bmatrix} H_{bb} & H_{bc} \\ H_{cb} & H_{cc} \end{bmatrix} \begin{Bmatrix} u_b \\ \Delta u_c \end{Bmatrix} = \begin{bmatrix} G_{bb} & 0 \\ G_{cb} & G_{cc} \end{bmatrix} \begin{Bmatrix} \tau_b \\ -\tau_c^+ \end{Bmatrix}\,, \qquad (9.26)$$

where the subscripts b and c denote the contribution of the non-crack boundary and upper crack surface respectively. With traction free cracks ($\tau_c = 0$), the system of equations reduces to

$$\begin{bmatrix} H_{bb} & H_{bc} \\ H_{cb} & H_{cc} \end{bmatrix} \begin{Bmatrix} u_b \\ \Delta u_c \end{Bmatrix} = \begin{bmatrix} G_{bb} & 0 \\ G_{cb} & G_{cc} \end{bmatrix} \begin{Bmatrix} \tau_b \\ 0 \end{Bmatrix}\,. \qquad (9.27)$$

The vector τ_b is a mixture of known traction τ_{bv} and unknown traction τ_*, similarly u_b is a mixture of known displacement u_{bv} and unknown displacement u_*. Eq.(9.27)

178 SYMMETRIC GALERKIN

can be written in terms of the known and unknown boundary displacement and traction values as

$$
\begin{bmatrix} H_{b_u b_u} & H_{b_u b_\tau} & H_{b_u c} \\ H_{b_\tau b_u} & H_{b_\tau b_\tau} & H_{b_\tau c} \\ H_{cb_u} & H_{cb_\tau} & H_{cc} \end{bmatrix} \begin{Bmatrix} u_{bv} \\ u_* \\ \Delta u_c \end{Bmatrix} = \begin{bmatrix} G_{b_u b_u} & G_{b_u b_\tau} & 0 \\ G_{b_\tau b_u} & G_{b_\tau b_\tau} & 0 \\ G_{cb_u} & G_{cb_\tau} & G_{cc} \end{bmatrix} \begin{Bmatrix} \tau_* \\ \tau_{bv} \\ 0 \end{Bmatrix},
$$
(9.28)

where, the subscripts b_u, b_τ and c represents the terms corresponding to the non-crack boundary with prescribed displacements $\Gamma_{b(u)}$, non-crack boundary with prescribed tractions $\Gamma_{b(\tau)}$ and the crack surface Γ_c^+, respectively. By rearranging Eq.(9.28) into the form $[A]\{x\} = \{b\}$, and multiplying the HBIEs by -1, we get the system of the matrix,

$$
\begin{bmatrix} -G_{b_u b_u} & H_{b_u b_\tau} & H_{b_u c} \\ G_{b_\tau b_u} & -H_{b_\tau b_\tau} & -H_{b_\tau c} \\ G_{cb_u} & -H_{cb_\tau} & -H_{cc} \end{bmatrix} \begin{Bmatrix} \tau_* \\ u_* \\ \Delta u_c \end{Bmatrix} = \begin{Bmatrix} -H_{b_u b_u} u_{bv} + G_{b_u b_\tau} \tau_{bv} \\ H_{b_\tau b_u} u_{bv} - G_{b_\tau b_\tau} \tau_{bv} \\ H_{cb_u} u_{bv} - G_{cb_\tau} \tau_{bv} \end{Bmatrix}.
$$
(9.29)

The final coefficient matrix of this system is symmetric due to the symmetric properties of the kernel tensors [118, 223]. Further details can be found in references [118, 136, 165, 223, 257].

9.4 ON COMPUTATIONAL METHODS FOR EVALUATING FRACTURE PARAMETERS

In the BEM literature, the displacement correlation technique (DCT) is widely used to evaluate SIFs due to its simplicity in numerical implementation [13]. The general expression of SIFs by means of the DCT technique are given by

$$
K_I = \frac{\mu}{\kappa + 1} \lim_{r \to 0} \sqrt{\frac{2\pi}{r}} \Delta u_2
$$

$$
K_{II} = \frac{\mu}{\kappa + 1} \lim_{r \to 0} \sqrt{\frac{2\pi}{r}} \Delta u_1
$$
(9.30)

where Δu_k is the crack opening displacement in the coordinate system associated with the crack tip under consideration, μ is shear modulus and ν is Poisson's ratio,

$$
\kappa = \begin{cases} (3 - \nu)/(1 + \nu) & \text{plane stress} \\ (3 - 4\nu) & \text{plane strain.} \end{cases}
$$
(9.31)

Perhaps the most accurate and elegant method for computing fracture parameters is the path-independent integral based techniques. Among the path independent integral methods, Aliabadi [2] applied the J-integral to mixed-mode crack problems by decoupling the J into its symmetrical and anti-symmetrical portions. Sladek and Sladek [258] used the conservation integral method in thermoelasticity problems to calculate the T-stress, and the J-integral to calculate the SIFs. Denda [75] implemented the interaction integral for mixed-mode analysis of multiple cracks in anisotropic solids using a dislocation and point force approach

(Lekhnitskii-Eshelby-Stroh formalism). Wen and Aliabadi [286] proposed a different contour integral based on Westergaard's solution and Betti's reciprocal theorem to calculate the SIFs. Reviews of the application of the BEM in fracture can be found in references [3, 4].

In order to calculate the T-stress, researchers have used several techniques such as the stress substitution method [159], the variational method [160], the Eshelby J-integral method [35, 146], the weight function method [248], the line spring method [283], the Betti-Rayleigh reciprocal theorem [48, 260], and the interaction integral method [48, 196]. Among these methods, the Eshelby J-integral method, the Betti-Rayleigh reciprocal theorem and the interaction integral method are based on path-independent integrals. In these methods, the fracture parameters can be calculated using data remote from the crack tip and, as a result, higher accuracy compared to local methods are generally achieved.

In order to solve mixed-mode crack problems, and moreover, to calculate the T-stress which plays significant role in crack growth direction, the interaction integral method is the ideal choice. The interaction integral method is based on conservation laws of elasticity and fundamental relationships in fracture mechanics [49]. SIFs pertaining to mixed-mode fracture problems can be easily obtained from this integral. The basis of the approach lies in the construction of a conservation integral for two superimposed states (actual and auxiliary) of a cracked elastic solid. The analysis requires integration along a suitably selected path surrounding the crack tip. The method was originally proposed by Chen and Shield [49], and later implemented numerically by Yau *et al.* [297] using the FEM applied to homogenous isotropic materials.

9.5 THE TWO-STATE INTERACTION INTEGRAL: M-INTEGRAL

The interaction integral or M-integral is derived from the path-independent J-integral [233] for two admissible states of a cracked elastic body. The formulation of the M-integral is presented here followed by techniques to calculate T-stress and SIFs. Finally, some aspects of the numerical scheme adopted for the implementation of the contour integral are discussed.

9.5.1 Basic Formulation

The path-independent J-integral [233] is defined as

$$J = \lim_{\Gamma \to 0} \int_{\Gamma} (\mathcal{W}\delta_{1j} - \sigma_{ij} \, u_{i,1}) \, n_j \, d\Gamma, \tag{9.32}$$

where $\mathcal{W}$ is the strain energy density given by

$$\mathcal{W} = \int_0^{\varepsilon_{kl}} \sigma_{ij} d\varepsilon_{ij} \tag{9.33}$$

and n_j is the outward normal vector to the contour Γ, as shown in Figure 9.2.

If two independent admissible fields are considered where the displacements, strains and stresses of the actual fields and the auxiliary fields are denoted by $(\boldsymbol{u}, \boldsymbol{\varepsilon}, \boldsymbol{\sigma})$ and $(\boldsymbol{u}^{aux}, \ \boldsymbol{\varepsilon}^{aux}, \ \boldsymbol{\sigma}^{aux})$, respectively, then the J-integral of the

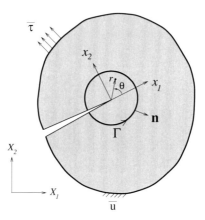

Figure 9.2 Schematic of integration path and coordinate systems. Notice the Cartesian (x_1, x_2) and polar (r, θ) coordinate systems at the crack tip. The notation $\bar{u}$ denotes prescribed displacements and $\bar{\tau}$ denotes prescribed tractions.

superimposed fields (actual and auxiliary) can be written as:

$$J^s = \int_\Gamma \left\{ \frac{1}{2}(\sigma_{ik} + \sigma_{ik}^{aux})(\varepsilon_{ik} + \varepsilon_{ik}^{aux})\delta_{1j} - (\sigma_{ij} + \sigma_{ij}^{aux})(u_{i,1} + u_{i,1}^{aux}) \right\} n_j \, d\Gamma. \tag{9.34}$$

This integral can be conveniently decomposed into

$$J^s = J + J^{aux} + M, \tag{9.35}$$

where J is given by Eq.(9.32), J^{aux} is given by

$$J^{aux} = \int_\Gamma \left(\mathcal{W}^{aux} \delta_{1j} - \sigma_{ij}^{aux} u_{i,1}^{aux} \right) n_j \, d\Gamma \tag{9.36}$$

with

$$\mathcal{W}^{aux} = \int_0^{\varepsilon_{kl}^{aux}} \sigma_{ij}^{aux} d\varepsilon_{ij}^{aux}, \tag{9.37}$$

and M is the interaction integral involving the cross terms of actual and auxiliary fields, which is given by

$$M = \int_\Gamma \left\{ \frac{1}{2}(\sigma_{ik}\varepsilon_{ik}^{aux} + \sigma_{ik}^{aux}\varepsilon_{ik})\delta_{1j} - (\sigma_{ij}u_{i,1}^{aux} + \sigma_{ij}^{aux}u_{i,1}) \right\} n_j \, d\Gamma. \tag{9.38}$$

The M-integral deals with interaction terms only, and will be used directly for solving mixed-mode fracture mechanics problems.

9.5.2 Auxiliary Fields for T-stress

The auxiliary fields are judiciously chosen for the interaction integral depending on the nature of the problem to be solved. Since the T stress is a constant stress that is parallel to the crack, the auxiliary stress and displacement fields are chosen due to a point force f in the x_1 direction (locally), applied to the tip of a semi-infinite crack in an infinite homogeneous body, as shown in Figure 9.3(a). The auxiliary stresses are given by Michell's solution [182]:

$$\sigma_{11}^{aux} = -\frac{f}{\pi r}\cos^3\theta, \quad \sigma_{22}^{aux} = -\frac{f}{\pi r}\cos\theta\sin^2\theta, \quad \sigma_{12}^{aux} = -\frac{f}{\pi r}\cos^2\theta\sin\theta. \quad (9.39)$$

The corresponding auxiliary displacements are [278]:

$$\begin{aligned} u_1^{aux} &= -\frac{f(1+\kappa)}{8\pi\mu}\ln\frac{r}{d} - \frac{f}{4\pi\mu}\sin^2\theta \\ u_2^{aux} &= -\frac{f(\kappa-1)}{8\pi\mu}\theta + \frac{f}{4\pi\mu}\sin\theta\cos\theta \end{aligned} \quad (9.40)$$

where d is the coordinate of a fixed point on the x_1 axis (see Figure 9.3(a)) Expressions for displacement derivatives of auxiliary fields for determining the T-stress are given by

$$\begin{aligned} u_{1,1}^{aux} &= \frac{-f\cos\theta}{\pi E'r}\left(1 - \frac{\sin^2\theta}{1-\nu}\right) \\ u_{2,1}^{aux} &= \frac{+f\sin\theta}{\pi E'r}\left(1 - \frac{\cos^2\theta}{1-\nu}\right) \\ u_{1,2}^{aux} &= \frac{-f\sin\theta}{\pi E'r}\left(1 + \frac{\cos^2\theta}{1-\nu}\right) \\ u_{2,2}^{aux} &= \frac{-f\cos\theta}{\pi E'r}\left(1 - \frac{\cos^2\theta}{1-\nu}\right) \end{aligned} \quad (9.41)$$

where E' is given by Eq.(9.54).

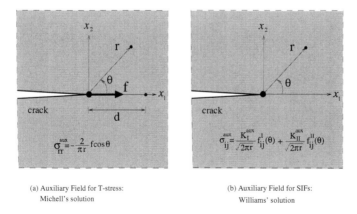

(a) Auxiliary Field for T-stress: Michell's solution

(b) Auxiliary Field for SIFs: Williams' solution

Figure 9.3 The loading configuration of the auxiliary fields: (a) a point force applied to the crack in an infinite plate for T-stress computation [182]; (b) Williams' asymptotic solution for SIF computation [289].

182 SYMMETRIC GALERKIN

9.5.3 Determination of T-stress

By considering the auxiliary field in Eq.(9.39), a simple expression for the T-stress in terms of the interaction integral (M), the point force for the auxiliary field (f), and material properties (E, ν) can be obtained. Since the modulus of elasticity is same for both the actual and the auxiliary states, the stresses are

$$\sigma_{ij} = C_{ijkl}\varepsilon_{kl} \qquad \text{and} \qquad \sigma_{ij}^{aux} = C_{ijkl}\varepsilon_{kl}^{aux}, \tag{9.42}$$

where C_{ijkl} is the constitutive tensor. Then from Eq.(9.42),

$$\sigma_{ij}\varepsilon_{kl}^{aux} = \sigma_{ij}^{aux}\varepsilon_{kl}. \tag{9.43}$$

Therefore, Eq.(9.38) can be rewritten as

$$M = \int_{\Gamma} \left\{ (\sigma_{ik}\varepsilon_{ik}^{aux}\delta_{1j} - (\sigma_{ij}u_{i,1}^{aux} + \sigma_{ij}^{aux}u_{i,1}) \right\} n_j \; d\Gamma. \tag{9.44}$$

The M-integral is path-independent, and thus any arbitrary path can be chosen to evaluate the integral. The actual stress field is composed of singular terms, T-stress term and higher order terms (see Eqs. 9.1 and 9.3). Considering a circular integration path, if Γ shrinks to zero $(r \to 0)$, then the contribution of the higher order terms on the M-integral tends to zero. The coefficients of the singular terms $O(r^{-1/2})$ after the integration over θ from $-\pi$ to $+\pi$ in Eq.(9.44) sum to zero. The contribution of the singular terms in the M-integral for calculating the T-stress is evaluated. Let

$$M = \int_{\Gamma} \left\{ (\sigma_{ik}\varepsilon_{ik}^{aux}n_1 - (\sigma_{ij}u_{i,1}^{aux} + \sigma_{ij}^{aux}u_{i,1})n_j \right\} d\Gamma. \tag{9.45}$$

By selecting a circular integration path, the coefficients of the singular terms $O(r^{-1/2})$ from the integration over θ from $-\pi$ to $+\pi$ in Eq.(9.45) are evaluated. After integration, the first, second and the third term of Eq.(9.45) yield

$$\int_{-\pi}^{\pi} \sigma_{ik}\varepsilon_{ik}^{aux}n_1 d\theta = -\frac{1}{210} \frac{f\sqrt{2}K_I \left(49\,\kappa - 27 \right)}{\pi\sqrt{r\pi}\mu}, \tag{9.46}$$

$$-\int_{-\pi}^{\pi} \sigma_{ij}u_{i,1}^{aux}n_j d\theta = -\frac{1}{105} \frac{f\sqrt{2}K_I \left(7\,\kappa + 3 \right)}{\pi\sqrt{r\pi}\mu}, \tag{9.47}$$

and

$$-\int_{-\pi}^{\pi} \sigma_{ij}^{aux}u_{i,1}n_j d\theta = \frac{1}{10} \frac{f\sqrt{2}K_I \left(3\,\kappa - 1 \right)}{\pi\sqrt{r\pi}\mu}, \tag{9.48}$$

respectively. These three terms add to zero.

As a result, the only contribution to the M-integral comes from the T-stress term of the stress field. Hence, the only stress to be considered is in the direction parallel to the crack, i.e.

$$\sigma_{ij} = T\delta_{1i}\delta_{1j}, \tag{9.49}$$

or $\sigma_{11} = T$. By means of Eq.(9.49), the stress-strain and strain-displacement relationships are

$$u_{1,1} = \frac{T}{E'} \qquad \text{and} \qquad u_{1,1}^{aux} = \varepsilon_{11}^{aux}. \tag{9.50}$$

Thus the first two terms of Eq.(9.44) cancel out and we get

$$M = -\lim_{\Gamma \to 0} \int_{\Gamma} \sigma_{ij}^{aux} n_j u_{i,1} \, d\Gamma = -\frac{T}{E'} \lim_{\Gamma \to 0} \int_{\Gamma} \sigma_{ij}^{aux} n_j \, d\Gamma. \tag{9.51}$$

In the auxiliary state, the force f is in equilibrium (see Figure 9.3(a)), thus

$$f = -\lim_{\Gamma \to 0} \int_{\Gamma} \sigma_{ij}^{aux} n_j \, d\Gamma, \tag{9.52}$$

and by substituting back into Eq.(9.51), we obtain

$$T = \frac{E'}{f} M \tag{9.53}$$

where

$$E' = \begin{cases} E & \text{plane stress} \\ E/(1 - \nu^2) & \text{plane strain.} \end{cases} \tag{9.54}$$

By calculating the M-integral from Eq.(9.44) and plugging the calculated value in Eq.(9.53), the T-stress can be readily obtained.

9.5.4 Auxiliary Fields for SIFs

The mixed-mode stress intensity factor can be extracted from the interaction integral, Eq.(9.38), through an appropriate definition of auxiliary fields. Local Cartesian and polar coordinates originate from the crack tip (see Figure 9.2). According to Figure 9.3(b), the auxiliary stress fields (expressed in polar coordinates) are given by

$$\sigma_{ij}^{aux} = \frac{K_I^{aux}}{\sqrt{2\pi r}} f_{ij}^I(\theta) + \frac{K_{II}^{aux}}{\sqrt{2\pi r}} f_{ij}^{II}(\theta), \; (i,j = 1,2) \tag{9.55}$$

where the angular functions $f_{ij}(\theta)$ are same as before Eq.(9.4). The corresponding auxiliary displacement fields are given by:

$$u_i^{aux} = \frac{K_I^{aux}}{\mu} \sqrt{\frac{r}{2\pi}} g_i^I(\theta) + \frac{K_{II}^{aux}}{\mu} \sqrt{\frac{r}{2\pi}} g_i^{II}(\theta), \; (i = 1,2) \tag{9.56}$$

where μ is the shear modulus, and K_I^{aux} and K_{II}^{aux} are the auxiliary mode I and mode II SIFs, respectively. The angular functions $g_{ij}(\theta)$ are given by

$$g_1^I(\theta) = \frac{1}{4} \left[(2\kappa - 1) \cos \frac{\theta}{2} - \cos \frac{3\theta}{2} \right],$$

$$g_1^{II}(\theta) = \frac{1}{4} \left[(2\kappa + 3) \sin \frac{\theta}{2} + \sin \frac{3\theta}{2} \right],$$

$$g_2^I(\theta) = \frac{1}{4} \left[(2\kappa + 1) \sin \frac{\theta}{2} - \sin \frac{3\theta}{2} \right],$$

$$g_2^{II}(\theta) = -\frac{1}{4} \left[(2\kappa - 3) \cos \frac{\theta}{2} + \cos \frac{3\theta}{2} \right]. \tag{9.57}$$

184   SYMMETRIC GALERKIN

These angular functions can be found in many references on fracture mechanics, e.g., the textbook by Anderson [5] or Hills *et al.* [135]. Expressions for displacement derivatives of Auxiliary fields for SIFs are given by

$$u_{1,1}^{aux} = \lambda \left[K_I^{aux} \left((2\kappa - 3) \cos \frac{\theta}{2} + \cos \frac{5\theta}{2} \right) - K_{II}^{aux} \left((2\kappa + 1) \sin \frac{\theta}{2} - \sin \frac{5\theta}{2} \right) \right]$$

$$u_{2,1}^{aux} = \lambda \left[K_I^{aux} \left((2\kappa + 3) \sin \frac{\theta}{2} + \sin \frac{5\theta}{2} \right) - K_{II}^{aux} \left((2\kappa + 5) \cos \frac{\theta}{2} + \cos \frac{5\theta}{2} \right) \right]$$

$$u_{1,2}^{aux} = \lambda \left[K_I^{aux} \left(-(2\kappa + 3) \sin \frac{\theta}{2} + \sin \frac{5\theta}{2} \right) - K_{II}^{aux} \left((2\kappa - 1) \cos \frac{\theta}{2} - \cos \frac{5\theta}{2} \right) \right]$$

$$u_{2,2}^{aux} = \lambda \left[K_I^{aux} \left((2\kappa - 1) \cos \frac{\theta}{2} + \cos \frac{5\theta}{2} \right) + K_{II}^{aux} \left(-(2\kappa - 5) \cos \frac{\theta}{2} + \sin \frac{5\theta}{2} \right) \right]$$

where

$$\lambda = \frac{1}{8\mu\sqrt{2\pi r}}$$

9.5.5   Determination of SIFs

The relationship among the J-integral and the mode I and mode II stress intensity factors (K_I and K_{II}) is established as:

$$J = \frac{K_I^2 + K_{II}^2}{E'} \tag{9.58}$$

where E' is given by Eq.(9.54). By superimposing the actual and auxiliary fields, and using Eq.(9.58), one obtains

$$J^s = \frac{(K_I + K_I^{aux})^2 + (K_{II} + K_{II}^{aux})^2}{E'} = J^{aux} + J + M \tag{9.59}$$

where

$$J^{aux} = \frac{(K_I^{aux})^2 + (K_{II}^{aux})^2}{E'} \tag{9.60}$$

and

$$M = \frac{2}{E'}(K_I K_I^{aux} + K_{II} K_{II}^{aux}). \tag{9.61}$$

The mode I stress intensity factor (K_I) is computed by assigning the SIFs of the auxiliary field to $K_I^{aux} = 1.0$ and $K_{II}^{aux} = 0.0$ in Eq.(9.61), i.e.

$$K_I = \frac{E'}{2} M \quad (K_I^{aux} = 1.0, K_{II}^{aux} = 0.0). \tag{9.62}$$

Similarly, the mode II stress intensity factor (K_{II}) can be obtained by assigning $K_I^{aux} = 0.0$ and $K_{II}^{aux} = 1.0$ in Eq.(9.61), i.e.

$$K_{II} = \frac{E'}{2} M \quad (K_I^{aux} = 0.0, K_{II}^{aux} = 1.0). \tag{9.63}$$

The M-integral in Eqs. (9.62) and (9.63) is evaluated by means of Eq.(9.38) and the auxiliary fields given by Eqs. (9.55) and (9.56).

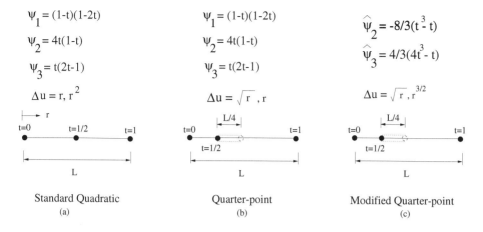

Figure 9.4 The shape functions, positions of the nodes and order of the crack opening displacement function for standard quadratic, quarter-point and the modified quarter-point element are shown.

9.5.6 Crack-tip elements

In fracture analysis, a quarter-point element at the the crack tip accounts for the $\sqrt{r}$ displacement behaviour of the crack tip, where r is the distance from the source point to the tip. The quarter-point element is formed from the standard quadratic element by simply moving the mid-node coordinates three fourths of the way towards the crack tip [15, 134] (see Figure 9.4). The modified quarter-point (MQP) crack-tip element [121] is based on the standard quarter-point element, but altered to account for a constraint on the linear term as suggested by Gray and Paulino [119]. This constraint is implemented in the modified shape functions by including a cubic term. Use of this element has been shown to greatly improve the accuracy in computed values for SIFs by means of local techniques [121]. The algorithm makes use of all the crack-tip elements shown in Figure 9.4 .

The standard quadratic shape functions are defined in terms of the intrinsic coordinate $t \in [0, 1]$ by

$$\psi_1(t) = (1-t)(1-2t), \qquad \psi_2(t) = 4t(1-t), \qquad \psi_3(t) = t(2t-1) \quad (9.64)$$

and thus the boundary interpolation is

$$\Gamma(t) = \sum_{j=1}^{3} (x_j \psi_j(t),\ y_j \psi_j(t)) \ . \qquad (9.65)$$

For the standard quarter-point crack-tip element, shown in Figure 9.4, the mid-side node is moved to the quarter-point position. The effect of this is that $t \approx \sqrt{(r)}$, which provides the singular behavior at the tip. For the modified quarter-point, the displacement discontinuity is given by

$$\Delta u(t) = \sum_{j=2}^{3} \left(\Delta u_1^j \hat{\psi}_j(t),\ \Delta u_2^j \hat{\psi}_j(t) \right) , \qquad (9.66)$$

where the new shape functions $\hat{\psi}$ are

$$\begin{aligned}\hat{\psi}_2(t) &= 4t(1-t) - 4t(1-t)(1-2t)/3 = -8(t^3-t)/3 \;, \\ \hat{\psi}_3(t) &= t(2t-1) + 2t(1-t)(1-2t)/3 = (4t^3-t)/3 \;. \end{aligned} \quad (9.67)$$

As the intrinsic coordinate t is proportional to $\sqrt{r}$, the MQP element is seen to give crack opening displacements which are of the form $\Delta u_k = A\sqrt{r} + Br^{3/2}$; the absence of the linear term in r is consistent with the proof by Gray and Paulino [119]. It should be noted that the shape function $\hat{\psi}_1(t)$ is not defined, as it multiplies the crack opening displacement at the tip which is known to be zero.

9.5.7 Numerical implementation of the M-integral

The accuracy of the computation of the M-integral depends on the integration points of the path and the method of integration. Integrals are evaluated along the circular path centered at the crack tip as shown in Figure 9.5.

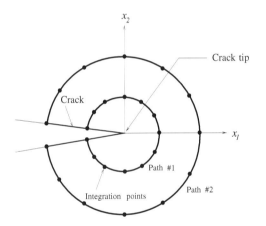

Figure 9.5 Integration contours for evaluation of the M-integral.

The integration along the contour path can be performed by using simple trapezoidal rule or Gaussian quadrature formula. The contour integrals can be written as follows

$$M = \int_{\Gamma_i} F_N(x,y) d\Gamma_i, \quad (x,y) \in \Gamma \quad (9.68)$$

where F_N denotes the integrand of Eq.(9.38) and Γ_i is the integration path of interest. The following formula is used for the trapezoidal rule,

$$\int_{\Gamma_i} F_N(x,y) d\Gamma_i = \frac{\pi}{m} \sum_{n=1}^{2m} F_N\left(h\cos\frac{\pi n}{m}, h\sin\frac{\pi n}{m}\right) \quad (9.69)$$

where h is the radius of the circular path. Alternatively Eq.(9.68) in polar (r,ω) coordinates is given by

$$M = h \int_{-\pi}^{\pi} F_N(r,\omega) d\omega. \quad (9.70)$$

NUMERICAL EXAMPLES **187**

In order to use the Gaussian-Legendre quadrature rule, Eq.(9.70) is written as

$$M = h\pi \int_{-1}^{1} F_N(r, \pi s)ds. \tag{9.71}$$

Convergence is achieved by using 30 or more integration points for the trapezoidal rule and 8 or more integration points for the Gauss-Legendre rule. For the numerical examples presented in this chapter, the Gaussian integration scheme is used.

9.6 NUMERICAL EXAMPLES

The performance of the SGBEM for fracture analysis and the M-integral formulation is examined by means of numerical examples. In order to assess the various features of the method, the following examples are presented:

1. Infinite plate with an interior inclined crack,

2. Slanted edge crack in a finite plate,

3. Multiple interacting cracks,

4. Fracture specimen configurations,

    (a) Single edge notch tension specimen (SENT),

    (b) Single edge notch bending specimen (SENB),

    (c) Center cracked tension (CCT) or Middle crack specimen (MT),

    (d) Double edge notched tension (DENT).

The first example is an inclined central crack in an infinite plate. This problem has analytical solution for both T-stress and SIFs. An edge crack in a finite plate is analysed in the second problem and compared with reference solutions. The third problem is an interesting problem consisting of two interacting cracks in a finite plate. The last example investigates the benchmark examples as used for laboratory experiments and provides solution for the T-stress, SIF and the associated biaxiality ratios. Unless otherwise stated, in all the examples we use modified quarter-point element at the crack-tips.

9.6.1 Infinite plate with an interior inclined crack

Consider a plate containing a single interior crack of length $2a$ oriented at an angle θ with the horizontal direction as shown in Figure 9.6. The plate is loaded with a uniform far-field traction $\sigma = 1$ applied symmetrically in the vertical direction and $\lambda\sigma$ in the horizontal direction, where λ is the lateral load ratio. The crack length is $2a = 2$ and the plate dimensions are $2H = 2W = 100$, which can be considered as an infinite domain. T-stress, K_I and K_{II} are calculated for various values of θ where $0 < \theta < \pi/2$. The boundary of the plate is discretized with only 2 quadratic elements on each side, and the crack is discretized with 4 elements. The number of Gauss points used for integration is $n = 8$. The Young's modulus is taken as

$E = 1.0$ (consistent units) and Poisson's ratio is $\nu = 0.3$. The exact solutions of the stress intensity factors and the T-stress for this problem [261] are

$$\begin{aligned}
K_I &= \sigma(\lambda \sin^2\theta + \cos^2\theta)\sqrt{\pi a} \\
K_{II} &= \sigma(1-\lambda)\cos\theta\sin\theta\sqrt{\pi a} \\
T &= -(1-\lambda)\sigma\cos 2\theta.
\end{aligned} \qquad (9.72)$$

The results of T-stress, normalized K_I and K_{II} for the right crack-tip and the corresponding analytical solution for $\lambda = 0$ (uniaxially loaded) and $\lambda = 0.5$ (biaxially loaded) are presented in Table 9.1 and Table 9.2, respectively. The results show good agreement between numerical (SGBEM) and analytical (Eq.(9.72)) results.

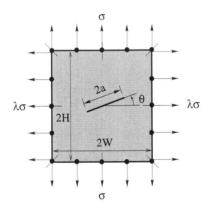

Figure 9.6 A single interior inclined crack in a plate subject to biaxial loading. The outer boundary of the plate is discretized with only 2 quadratic elements on each edge.

Table 9.1 T-stress and normalized SIFs as functions of the crack angle θ for the lateral load ratio $\lambda = 0$.

Angle θ	SGBEM (M integral) $K_I/\sqrt{\pi a}$	$K_{II}/\sqrt{\pi a}$	T	Analytical solution $K_I/\sqrt{\pi a}$	$K_{II}/\sqrt{\pi a}$	T
0°	1.0002	0.0000	-1.0013	1.0000	0.0000	-1.0000
15°	0.9332	0.2502	-0.8672	0.9330	0.2500	-0.8660
30°	0.7502	0.4334	-0.5009	0.7500	0.4330	-0.5000
45°	0.5001	0.5004	-0.0006	0.5000	0.5000	0.0000
60°	0.2500	0.4333	0.4997	0.2500	0.4330	0.5000
75°	0.0670	0.2502	0.8659	0.0670	0.2500	0.8660
90°	0.0000	0.0000	1.0000	0.0000	0.0000	1.0000

The influences of several parameters i.e., the radius of the integration contour r, the number of integration points n, crack discretization m, and the crack-tip elements on the calculated SIFs and the T-stress are studied.

Table 9.2 T-stress and normalized SIFs as function of the crack angle θ for the lateral load ratio $\lambda = 0.5$.

Angle θ	SGBEM (M integral) $K_I/\sqrt{\pi a}$	$K_{II}/\sqrt{\pi a}$	T	Analytical solution $K_I/\sqrt{\pi a}$	$K_{II}/\sqrt{\pi a}$	T
0°	1.0003	0.0000	-0.5011	1.0000	0.0000	-0.5000
15°	0.9668	0.1251	-0.4341	0.9665	0.1250	-0.4330
30°	0.8752	0.2167	-0.2509	0.8750	0.2165	-0.2500
45°	0.7502	0.2502	0.0007	0.7500	0.2500	0.0000
60°	0.6251	0.2167	0.2494	0.6250	0.2165	0.2500
75°	0.5336	0.1251	0.4324	0.5335	0.1250	0.4330
90°	0.5001	0.0000	0.4994	0.5000	0.0000	0.5000

Effect of the radius of the integration contour. The M-integral is evaluated along different circular integration paths as shown in Figure 9.5. The K_I, K_{II} and T-stress results are obtained for several circular paths at the right crack tip with the radius of the circular contour r ranging from $r = 0.025a$ (near crack tip) to $r = 1.5a$ (closer to opposite crack tip). According to Figure 9.6, the crack angle is $\theta = 30°$, and the lateral load ratio is $\lambda = 0$. The outer boundary of the plate is discretized with only 2 quadratic elements on each side. The crack is discretized with 10 elements since smaller values of r/a are used. The number of Gauss points for integration is $n = 8$. A plot of normalized K_I, K_{II} and T-stress versus r/a is plotted in Figure 9.7, which verifies the path independence of the integral. However, notice that the T-stress results are more sensitive to the radius of integration (r/a) than the SIFs.

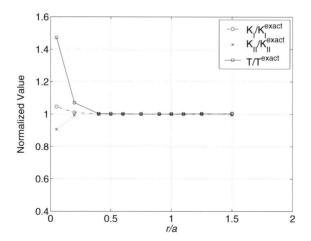

Figure 9.7 Variation of normalized SIFs and T-stress with different values of r/a. Parameters adopted: $\theta = 30°, \lambda = 0, m = 10, n = 8$.

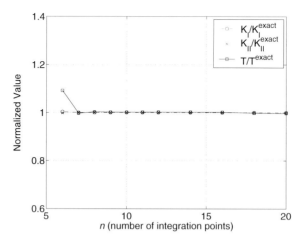

Figure 9.8 Normalized SIFs and T-stress versus number of integration points (n) on the contour. Parameters adopted: $\theta = 30°, \lambda = 0.5, m = 10, r/a = 1.0$.

Effect of the number of integration points in the contour. A convergence study with respect to the number of Gauss integration points (n) is carried out. The SIFs and T-stress are calculated and normalized values are plotted against n in Figure 9.8 for n ranging from 6 to 20. According to Figure 9.6, the crack angle is $\theta = 30°$ and the lateral load ratio is $\lambda = 0.5$. The outer boundary of the plate is discretized with only 2 quadratic elements on each side, the crack is discretized with 10 elements and $r/a = 1.0$. Figure 9.8 shows that the results converge when n is equal to or greater than 8. The T-stress results are more sensitive to the number of integration points (n) than the SIFs. For higher values of n (e.g., $n > 20$), stress results at integration points very close to the crack face are required. Accurate stress evaluation close to the boundary requires appropriate treatment of near-singular integrals. In the implementation, no special treatment has been considered for this purpose. Instead, by increasing the number of elements to discretize the crack (m), accurate results are obtained when n is large. However, for all the other problems presented in this chapter the number of integration points n used was between 8 and 12, which proved sufficient.

Effect of crack discretization. A convergence study on crack discretization with the number of the elements on the crack m ranging from 2 to 14 elements is done. The crack is oriented at $\theta = 30°$ and $\lambda = 0.5$. The outer boundary of the plate is discretized with only 2 quadratic elements on each side. The number of Gauss points is $n = 8$. The normalized values of K_I/K_I^{exact}, K_{II}/K_{II}^{exact} and T/T^{exact} are presented in Table 9.3, which shows that discretizing the crack with 4 elements is sufficient for the problem.

Effect of crack-tip elements. The influence of the type of crack-tip elements on the SIFs and the T-stress results is studied. The crack angle is $\theta = 30°$ and the lateral load ratio is $\lambda = 0.5$. The outer boundary of the plate is discretized with only 2 quadratic elements on each side and the crack is discretized with 6 elements. The number of Gauss points is $n - 8$ and $r/a - 1.0$. Table 9.4 shows normalized SIFs

Table 9.3 Effect of the crack discretization. The parameter m denotes the number of elements on the crack surface.

m	K_I/K_I^{exact}	K_{II}/K_{II}^{exact}	T/T^{exact}
2	0.9982	0.9992	1.0621
3	0.9913	1.0075	1.0774
4	1.0002	1.0009	1.0043
5	1.0001	1.0010	1.0035
6	1.0002	1.0009	1.0036
7	1.0003	1.0014	1.0033
8	1.0003	1.0005	1.0034
9	1.0003	0.9997	1.0034
10	1.0003	1.0009	1.0034
11	1.0003	0.9992	1.0034
12	1.0003	1.0013	1.0034
14	1.0003	1.0004	1.0034

and the T-stress obtained by using standard quadratic elements (Figure 9.4(a)), quarter-point elements (Figure 9.4(b)) and modified quarter-point elements (Figure 9.4(c)) as crack-tip elements. As expected, there is no significant difference in the results, especially for the latter two elements. Since the M-integral is computed away from the crack tip, the details of the local crack tip interpolation do not have much influence on the results. However, such details are relevant for a local method like the DCT [121].

Table 9.4 Effect of crack-tip elements on normalized SIFs and T-stress.

Element type	$K_I/\sqrt{\pi a}$	$K_{II}/\sqrt{\pi a}$	T-stress
Quadratic (no quarter-point)	0.8737	0.2164	-0.2509
Quarter-point	0.8752	0.2167	-0.2509
Modified Quarter-point	0.8752	0.2167	-0.2509
Analytical solution	0.8750	0.2165	-0.2500

9.6.2 Slanted edge crack in a finite plate

Figure 9.9 shows a slanted edge crack in a finite plate loaded with a uniform traction $\sigma = 1$ applied symmetrically at the ends. The crack length is $a/W = 0.4\sqrt{2}$ and the plate dimensions are $H = 2W = 1$ (consistent units). T-stress, K_I and K_{II} are calculated for the crack angle $\theta = 45°$. The outer boundary of the plate is discretized using 20 quadratic elements on the left side and 10 quadratic elements on rest of the sides. The crack is discretized using 6 elements ($m = 6$). The number of Gauss points is $n = 10$, and $r/a = 0.5$. The Young's modulus is taken as $E = 1.0$ (consistent units) and Poisson's ratio is $\nu = 0.3$. Kim and Paulino solved this problem previously using FEM by the J_k-integral [148] and the interaction integral [150]. Table 9.5 shows a comparison of the SGBEM results with those obtained in references [148, 150], which indicates good agreement.

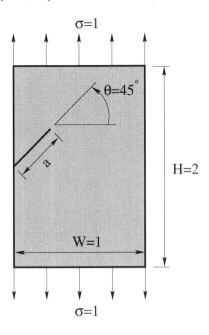

Figure 9.9 Slanted edge crack in a plate.

Table 9.5 Comparison of normalized SIFs and T-stress.

Method	$K_I/\sqrt{\pi a}$	$K_{II}/\sqrt{\pi a}$	T-stress
M-integral (SGBEM)	1.446	0.615	0.775
J_k-integral (FEM [148])	1.451	0.604	0.787
M-integral (FEM [149])	1.446	0.615	0.764

9.6.3 Multiple interacting cracks

Figure 9.10 shows two cracks of length $2a$ oriented with an angle θ_i ($\theta_1 = 30°, \theta_2 = 60°$) in a finite two-dimensional plate. The distance from the origin of the coordinate system (see Figure 9.10) to the two crack tips which are closer to the origin is 1.0. Kim and Paulino [148] have provided finite element solution and Shbeeb et al. [253] have provided semi-analytical solutions using integral equation method for this problem. The applied load is $\sigma = 1.0$, the crack length is $2a = 2$, the plate dimensions are given by $H/W = 1.0, W = 20$, and the material properties are $E = 1.0, \nu = 0.0$ (consistent units). The number of integration point is $n = 12$ and $r/a = 1$. The outer boundary is discretized with 10 quadratic elements, and each of the cracks is discretized with 10 elements ($m = 10$). Table 9.6 shows a comparison of the normalized SIFs at crack tips for the lower crack oriented at an angle $\theta = 30°$ computed by the interaction integral (M) with those obtained using the J_k-integral [148] and the integral equation method [253]. Here a^- and a^+ refer to the left and right crack tips respectively in Table 9.6. The SGBEM results agree very well with those by Kim and Paulino [148] and Shbeeb et al. [253].

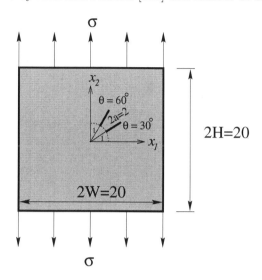

Figure 9.10 Two interacting cracks in a plate.

Table 9.6 Comparison of the Normalized SIFs[a] for the lower crack among various methods.

Method	$K_I(a^-)$	$K_{II}(a^-)$	$K_I(a^+)$	$K_{II}(a^+)$
M integral (SGBEM)	0.601	0.430	0.808	0.433
J_k integral (FEM [148])	0.603	0.431	0.801	0.431
Integral Equation ([253])	0.59	0.43	0.78	0.42

[a] Note that all the SIFs are normalized by $\sqrt{\pi a}$.

9.6.4 Various fracture specimen configurations

This final example investigates the T-stress and the SIFs for various benchmark fracture specimens, i.e. center cracked tension (CCT) or middle crack tension specimen (MT), the single edge notched tension specimen (SENT), the single edge notched bending specimen (SENB), and the double edge notched tension specimen (DENT) as shown in Figure 9.11. In order to understand the behaviour of the M-integral a plot of the integrand along the integration path ω from $-\pi$ to $+\pi$ for the SENT specimen with $a/W = 0.5$ is depicted in Figure 9.12.

The analyses were carried out using plane strain conditions with Young's modulus $E = 1.0$ and Poisson's ratio $\nu = 0.3$. The applied load is $\sigma = 1$ for the different load configurations of Figure 9.11 (consistent units). The crack is discretized using

194 SYMMETRIC GALERKIN

10 elements ($m = 10$). The number of Gauss integration point is $n = 10$. The outer boundary of the CCT specimen is discretized with 10 quadratic elements on each side, while for the rest of the specimens (SENT, SENB and DENT) the outer boundary is discretized with 50 quadratic elements on the left edge and 30 quadratic elements on rest of the edges. Only half of the DENT specimen was analysed due to its symmetry. Table 9.7 shows good agreement between SGBEM results and those available in the literature. Figure 9.13 illustrates the variation of biaxiality ratio ($B = T\sqrt{\pi a}/K_I$) versus the ratio of crack length to width a/W for various specimens ($H/W = 12$) and compares the results published by Fett et al. [89] using the boundary collocation method, and by Paulino and Kim [212] using the FEM. Notice that, for the SENB the SGBEM solution is closer to that by Fett et al. [89], while, for the DENT, the SGBEM solution is closer to that by Paulino and Kim [212]. In general, all the solutions (SGBEM, Fett et al. [89], and Paulino and Kim [212]) show very good agreement in Figure 9.13. The sign of the biaxiality ratio changes from negative to positive as a/W increases in SENT and SENB, while the sign remains the same for CCT and DENT specimens.

Table 9.7 Normalized T-stress, biaxiality ratio (B), and normalized mode I SIF for various fracture specimens.

Fracture specimen	Sources	T/σ	$B = T\sqrt{\pi a}/K_I$	$K_I/(\sigma\sqrt{\pi a})$
CCT or MT (a/W=0.3, H/W=1.0)	SGBEM	-1.1554	-1.0286	1.1232
	Chen et al. [48]	-1.1554	-1.0286	1.1232
	Fett [89]	-1.1557	-1.0279	-
	Leevers & Radon [160]	-	-1.0255	-
	Cardew et al. [35]	-	-1.026	-
SENT (a/W=0.3, H/W=12)	SGBEM	-0.6105	-0.3679	1.6597
	Paulino and Kim [212]	-0.6139	-0.3700	1.6594
	Chen et al. [48]	-0.6103	-0.3677	1.6598
	Fett [89]	-0.6141	-0.3664	-
	Sham [248]	-0.6142	-0.3707	1.6570
SENT (a/W=0.5, H/W=12)	SGBEM	-0.4184	-0.1481	2.8241
	Paulino and Kim [212]	-0.4309	-0.1481	2.8237
	Chen et al. [48]	-0.4217	-0.1493	2.8246
	Fett [89]	-0.4182	-0.1481	-
	Sham [248]	-0.4314	-0.1529	2.8210
SENB (a/W=0.3, H/W=12)	SGBEM	-0.0800	-0.0712	1.1235
	Chen et al. [48]	-0.0792	-0.0704	1.1241
	Fett [89]	-0.0771	-0.0671	-
	Sham [248]	-0.0824	-0.0734	1.1220
SENB (a/W=0.5, H/W=12)	SGBEM	0.3986	0.2662	1.4973
	Chen et al. [48]	0.3975	0.2655	1.4972
	Fett [89]	0.3921	0.2620	-
	Sham [248]	0.3911	0.2616	1.4951
DENT (a/W=0.3, H/W=12)	SGBEM	-0.5326	-0.4780	1.1143
	Paulino and Kim [212]	-0.5384	-0.4444	1.2115
	Fett [89]	-0.5319	-0.4720	-
DENT (a/W=0.5, H/W=12)	SGBEM	-0.5521	-0.4725	1.1685
	Paulino and Kim [212]	-0.5597	-0.4454	1.2567
	Fett [89]	-0.5216	-0.4396	-

NUMERICAL EXAMPLES   **195**

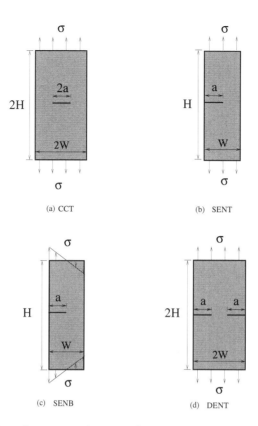

Figure 9.11 Various fracture specimen configurations. The thickness of each specimen is t.

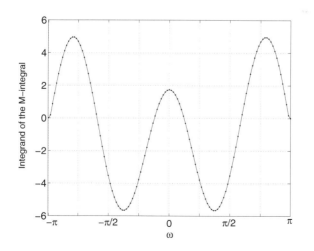

Figure 9.12 Variation of integrand of the M-integral along the integration path ω with $r/a = 0.5$ where a is the crack length. Parameters adopted: $m = 10, n = 10$.

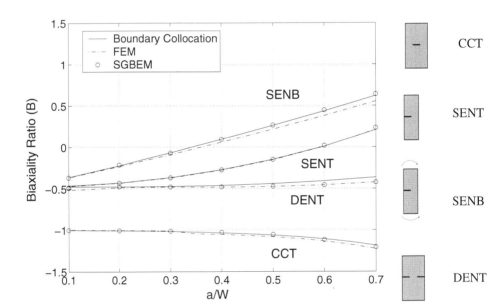

Figure 9.13 Biaxiality ratio versus a/W for various fracture specimens.

CHAPTER 10

NONHOMOGENOUS MEDIA

Synopsis: In this chapter we demonstrate that the same recipe for the symmetric Galerkin formulations for homogenoues media can be successfully used for problems in nonhomogeneous media. A symmetric Galerkin formulation and implementation for heat conduction in a three dimensional functionally graded material is presented. Special emphasis is given to the treatment of complicated hypersingular integrals that arise. A transient implementation using the Laplace transform Galerkin boundary element method is also presented and test examples are provided to verify the numerical implementation.

10.1 INTRODUCTION

Advancements in material processing technology have enabled the design and manufacture of new material systems that can withstand very high temperatures and large temperature gradients. Functionally graded materials or FGMs are a new generation of composites where the volume fraction of the FGM constituents vary gradually, giving a non-uniform microstructure with continuously graded macro-properties such as heat conductivity, specific heat, density, etc. Typically, in an FGM, one face of a structural component is an engineering ceramic that can resist severe thermal loading and the other face is a metal which has excellent structural strength. FGMs consisting of heat-resisting ceramic and fracture-resisting metal can improve the properties of thermal barrier systems because cracking and delamination, which are often observed in conventional layered systems, are reduced

The Symmetric Galerkin Boundary Element Method. By Sutradhar, Paulino and Gray **197**
ISBN 978-3-540-68770-2 ©2008 Springer-Verlag Berlin Heidelberg

198 NONHOMOGENOUS MEDIA

by proper smooth transition of material properties. Ceramic based FGMs have also been used for thermal protection [38]. FGMs are being developed as thermal barrier materials for combustion chambers, gas vanes, air vanes, nose cones, fuel valve sheets and piston crowns which undergo high-temperature gradient and high-thermal cycles in addition to wear. A comprehensive treatment of the science and technology of FGMs can be found, for example, in the books by Miyamoto et al. [184], Suresh and Mortensen [266], and the review article by Paulino et al. [211].

In the context of BEM, problems in nonhomogeneous media has been previously studied by Cheng [53,54], Ang *et al.* [7], Bialecki and Kuhn [20], Shaw [250], Gray *et al.* [114] and Dumont *et al.* [83], among others. The majority of these works have emphasized obtaining the Green's function. Cheng [53] presented a direct Green's function approach for Darcy's flow with spatially variable permeability. He also presented the Green's function for a class of permeability variations whose square root of hydraulic conductivity satisfies the Laplace and the Helmholtz equations in one to three dimensions. Shaw [251] presented a two dimensional fundamental solution involving axisymmetric material variation and also showed the inter-relationship between the fundamental solutions for different heterogeneous potential, wave and advective-diffusion problems. Lafe and Cheng [156] used a pertubation boundary element for steady state ground water flow for heterogeneous aquifers where the governing equation is decomposed into a Laplace equation and a sequence of Poisson's equations with known right-hand sides. Harrouni *et al.* [86, 87] used a global interpolation based dual reciprocity boundary element method (DRBEM) for Darcy's flow in heterogenous media where the governing equation is transformed into a Poisson-type equation with modified boundary conditions. Kassab and Divo [76, 78] introduced a technique for heat conduction in heterogeneous media based on a fundamental solution that is a locally radially symmetric response to a non-symmetric forcing function. Shaw and Manolis [252] employed a conformal mapping technique to solve heat conduction problems in graded materials.

Multiple techniques have been used to deal with the numerical implementation, *e.g.* the *iterative scheme* involving domain integrals and iterations, the *domain scheme* and the *direct Green's function scheme* [53]. The domain technique (*e.g.*, DRBEM) requires use of domain integrals or radial basis functions [86, 274]. The iterative and the domain scheme decrease the inherent efficiency of the BEM as the boundary-only nature of the method is lost. Another simple technique is the multi-zone approach [157], where the conductivity is assummed to be constant over several zones. Bialecki and Kuhn [20] presented a multizone approach where the material property was modeled as constant in certain zones in the layered media. This approach at times become inefficient because in order to capture the continuous variation of the material property a large number of sub-regions or zones are necessary. Sutradhar and Paulino [267] presented a *transformation approach*, called the 'simple BEM', for potential theory problems in nonhomogeneous media where nonhomogeneous problems are transformed into known problems in homogeneous media. The method leads to a pure boundary-only formulation.

10.2 STEADY STATE HEAT CONDUCTION

Steady state isotropic heat conduction in a solid is governed by the equation

$$\nabla \bullet (\mathrm{k}(x, y, z)\nabla \phi) = 0 \ . \tag{10.1}$$

STEADY STATE HEAT CONDUCTION **199**

where • denotes the inner product, $\phi = \phi(x, y, z)$ is the temperature function, $k(x, y, z)$ is the thermal conductivity which can be a function of the Cartesian coordinates. Let the FGM be defined by the thermal conductivity that varies exponentially in one Cartesian coordinate, i.e.

$$k(x, y, z) = k(z) = k_0 e^{2\beta z} , \tag{10.2}$$

where β denotes the material nonhomogeneity parameter. Substituting this material expression into Eq.(10.1), one obtains

$$\nabla^2 \phi + 2\beta \phi_z = 0, \tag{10.3}$$

where ϕ_z is the derivative of ϕ with respect to z, i.e.

$$\phi_z \equiv \partial \phi / \partial z. \tag{10.4}$$

10.2.1 On the FGM Green's function

The Green's function is the solution to the adjoint equation with a delta function force, namely

$$\nabla^2 G(P, Q) - 2\beta G_z(P, Q) = -\delta(Q - P) , \tag{10.5}$$

where δ denotes the Dirac delta function, P and Q denote the source point and the field point, respectively, and G_z is the derivative of G with respect to z. The solution of this equation is derived [114, 163] as

$$G(P, Q) = \frac{e^{\beta(-r + R_z)}}{4\pi r} \tag{10.6}$$

where

$$R_z = z_Q - z_P \quad \text{and} \quad r = \|\mathbf{R}\| = \|\mathbf{Q} - \mathbf{P}\|. \tag{10.7}$$

The Green's function for the nonhomogeneous problem is essential for developing a boundary-only integral equation formulation. Note that the Green's function for an FGM can be rewritten as [214]

$$G(P, Q) = \frac{1}{4\pi r} + \frac{e^{\beta(-r + R_z)} - 1}{4\pi r}. \tag{10.8}$$

The first term of Eq.(10.8) is the Green's function for the Laplace equation in homogeneous media. The second term is bounded in the limit as $r \to 0$ and is a consequence of the grading; when β tends to zero (material is homogeneous), this graded term vanishes. This form shows that the singularity for the FGM Green's function is precisely the same as for the homogeneous. It will therefore not be surprising that the divergent terms in the hypersingular integral are the same as for the Laplace equation [112].

10.2.2 Symmetric Galerkin Formulation

The governing interior limit BIE corresponding to Eq.(10.3) is

$$\phi(P) + \int_\Sigma \phi(Q) \left(\frac{\partial}{\partial n} G(P, Q) - 2\beta n_z G(P, Q) \right) dQ = \int_\Sigma G(P, Q) \frac{\partial}{\partial n} \phi(Q) \, dQ, \tag{10.9}$$

which differs in form from the usual integral statements by the presence of the additional term multiplying $\phi(Q)$, i.e. $[-2\beta n_z G(P,Q)]$, due to the material gradation.

In the SGBEM, the symmetry comes from the symmetry properties of the kernel functions [22,25]. For the homogeneous Laplace equation, the fundamental solution is symmetric, but the FGM Green's function, Eq.(10.6), is not. In order to get symmetric kernels, the FGM boundary integral equations are re-formulated in terms of physical variables (flux instead of normal derivative), as described below. Thus, to obtain a symmetric matrix, the equations are written in terms of the surface flux,

$$\mathcal{F}(Q) = -k(z_Q)\frac{\partial}{\partial n}\phi(Q). \tag{10.10}$$

The boundary integral equation (BIE) for surface temperature $\phi(P)$ on the boundary Σ (see Figure 10.1) is therefore

$$\phi(P) + \int_{\Sigma} F(P,Q)\phi(Q)\,dQ = \int_{\Sigma} G_S(P,Q)\mathcal{F}(Q)\,dQ\,, \tag{10.11}$$

and the kernel functions are

$$
\begin{aligned}
G_S(P,Q) &= -\frac{G(P,Q)}{k(z_Q)} = -\frac{1}{4k_0\pi}\frac{e^{\beta(-r-z_Q-z_P)}}{r} \\
F(P,Q) &= \frac{\partial}{\partial n}G(P,Q) - 2\beta n_z G(P,Q) \\
&= -\frac{e^{\beta(-r+R_z)}}{4\pi}\left(\frac{\mathbf{n}\cdot\mathbf{R}}{r^3} + \beta\frac{\mathbf{n}\cdot\mathbf{R}}{r^2} + \beta\frac{n_z}{r}\right).
\end{aligned}
\tag{10.12}
$$

Notice that $G_S(P,Q)$, unlike $G(P,Q)$, is symmetric with respect to P and Q. This is one of the conditions needed for symmetry. The hypersingular boundary integral equation (HBIE) is obtained by differentiating Eq.(10.11) with respect to the source P in the direction $\mathbf{N}$, which is normal to the boundary at P. In this case, however, it needs to be multiplied by $-k(z_P)$ to obtain the corresponding equation for surface flux, i.e.

$$\mathcal{F}(P) + \int_{\Sigma} W(P,Q)\phi(Q)\,dQ = \int_{\Sigma} S(P,Q)\mathcal{F}(Q)\,dQ\,. \tag{10.13}$$

The kernel functions are computed as

$$
\begin{aligned}
S(P,Q) &= -k(z_P)\frac{\partial}{\partial N}G_S(P,Q) \\
&= \frac{e^{\beta(-r-R_z)}}{4\pi}\left(\frac{\mathbf{N}\cdot\mathbf{R}}{r^3} + \beta\frac{\mathbf{N}\cdot\mathbf{R}}{r^2} - \beta\frac{N_z}{r}\right)
\end{aligned}
\tag{10.14}
$$

and

$$
\begin{aligned}
W(P,Q) &= -k(z_P)\frac{\partial}{\partial N}F(P,Q) \\
&= \frac{k_0}{4\pi}e^{\beta(-r+z_Q+z_P)}\left(3\frac{(\mathbf{n}\cdot\mathbf{R})(\mathbf{N}\cdot\mathbf{R})}{r^5} + 3\beta\frac{(\mathbf{n}\cdot\mathbf{R})(\mathbf{N}\cdot\mathbf{R})}{r^4}\right. \\
&\quad + \frac{\beta^2(\mathbf{n}\cdot\mathbf{R})(\mathbf{N}\cdot\mathbf{R}) - \beta(N_z\mathbf{n} - n_z\mathbf{N})\cdot\mathbf{R} - \mathbf{n}\cdot\mathbf{N}}{r^3} \\
&\quad \left. - \beta\frac{\beta(N_z\mathbf{n} - n_z\mathbf{N})\cdot\mathbf{R} + \mathbf{n}\cdot\mathbf{N}}{r^2} - \beta^2\frac{N_z n_z}{r}\right).
\end{aligned}
\tag{10.15}
$$

STEADY STATE HEAT CONDUCTION

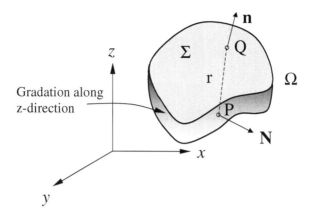

Figure 10.1 Illustration of a generic body with boundary Σ and domain Ω. The present SGBEM relies on a boundary-only formulation for FGMs. The source point is P (normal **N**) and the field point is Q (normal **n**).

The three symmetry requirements for the kernel functions are now fulfilled, *i.e.*

$$G_S(P,Q) = G_S(Q,P), \quad W(P,Q) = W(Q,P), \quad S(P,Q) = F(Q,P) \quad (10.16)$$

Interchanging Q and P implies replacing $\mathbf{N}(P)$ with $\mathbf{n}(Q)$ and changing the sign of $\mathbf{R}$, and thus all the conditions necessary for symmetry are seen to hold.

The direct limit procedure is employed to define and evaluate the singular integrals. If the limit is taken with the source point P approaching the boundary from *outside* the domain, then the "free terms" $\phi(P)$ in the BIE (Eq.(10.11)) and $\mathcal{F}(P)$ in the HBIE (Eq.(10.13)) are not present. Thus, the exterior limit BIE and HBIE take the form

$$\mathcal{P}_s(P) \equiv \int_\Sigma F(P,Q)\phi(Q)\,dQ - \int_\Sigma G_S(P,Q)\mathcal{F}(Q)\,dQ = 0,$$

$$\mathcal{F}_s(P) \equiv \int_\Sigma W(P,Q)\phi(Q)\,dQ - \int_\Sigma S(P,Q)\mathcal{F}(Q)\,dQ = 0, \quad (10.17)$$

where $\mathcal{P}_s$ denotes the equation for the surface temperature (potential) and $\mathcal{F}_s$ denotes the equation for the surface flux. The free terms are automatically incorporated in the "*exterior limit*" evaluation of the $F(P,Q)$ and $S(P,Q)$ integrals. Thus, a separate computation of these free terms is avoided, and they are obtained as a natural outcome of the direct limit procedure in Chapter 4.

The surface temperature and surface flux are approximated in terms of values at element nodes Q_j and shape functions $\psi_j(Q)$, *i.e.*,

$$\phi(Q) = \sum_j \phi(Q_j)\psi_j(Q), \quad \mathcal{F}(Q) = \sum_j \mathcal{F}(Q_j)\psi_j(Q). \quad (10.18)$$

In a Galerkin approximation, Eq.(10.17) is enforced in an average sense, with the shape functions employed as the weighting functions. Therefore, the Galerkin boundary integral equations take the form

$$\int_{\Sigma} \psi_k(P)\mathcal{P}_s(P)\,\mathrm{d}P = 0 \tag{10.19}$$

$$\int_{\Sigma} \psi_k(P)\mathcal{F}_s(P)\,\mathrm{d}P = 0 \ . \tag{10.20}$$

After discretization, the set of equations can be written in block-matrix form as $[H]\{\phi\} = [G]\{\mathcal{F}\}$, and in block-matrix these equations become

$$\begin{bmatrix} H_{11} & H_{12} \\ H_{21} & H_{22} \end{bmatrix} \begin{Bmatrix} \phi_{bv} \\ \phi_u \end{Bmatrix} = \begin{bmatrix} G_{11} & G_{12} \\ G_{21} & G_{22} \end{bmatrix} \begin{Bmatrix} \mathcal{F}_u \\ \mathcal{F}_{bv} \end{Bmatrix} . \tag{10.21}$$

Symmetry of the coefficient matrix for a general mixed boundary value problem is achieved by the following simple arrangement. The BIE is employed on the Dirichlet surface, and the HBIE equation is used on the Neumann surface. The first row represents the BIE written on the Dirichlet surface, and the second row represents the HBIE written on the Neumann surface. Similarly, the first and the second columns arise from integrating over Dirichlet and Neumann surfaces, respectively. The subscripts in the matrix therefore denote known boundary values (bv) and unknown (u) quantities. Rearranging Eq.(10.21) into the form $[\mathbf{A}]\{x\} = \{b\}$, one obtains

$$\begin{bmatrix} -G_{11} & H_{12} \\ G_{21} & -H_{22} \end{bmatrix} \begin{Bmatrix} \mathcal{F}_u \\ \phi_u \end{Bmatrix} = \begin{Bmatrix} H_{11}\phi_{bv} + G_{12}\mathcal{F}_{bv} \\ H_{21}\phi_{bv} - G_{22}\mathcal{F}_{bv} \end{Bmatrix} . \tag{10.22}$$

The symmetry of the coefficient matrix, $G_{11} = G_{11}^T$, $H_{22} = H_{22}^T$, and $H_{12} = G_{21}^T$, now follows from the properties of the kernel functions (see Eq.(10.16)).

10.2.3   Treatment of Singular and Hypersingular Integrals

This section discusses the evaluation of hypersingular integrals that arises in the Galerkin boundary element method (GBEM) (e.g. symmetric and non-symmetric) formulation for FGMs. The procedures for treating the hypersingular integral are applicable to other less singular integrals.

The two key features of the singular integration are first, the definition of the integrals as limits from the exterior of the domain, and second, the combination of analytical and numerical evaluation procedures. The boundary limit provides a consistent scheme for defining all singular integrals, weakly singular, strongly singular, and hypersingular, resulting in direct evaluation algorithms. The key task for the direct evaluation will be to isolate the divergent terms and to show that they cancel. Symbolic computation is exploited to simplify the work involved in carrying out the limit process and analytic integration.

In this chapter, the analysis for a linear element will be considered in detail, as this forms the basis for handling higher order interpolations. An equilateral triangle parameter space $\{\eta, \xi\}$, where

$$-1 \leq \eta \leq 1, \quad 0 \leq \xi \leq \sqrt{3}(1 - |\eta|) \tag{10.23}$$

will be employed (see Figure 10.2). This choice of parameter space is convenient for executing the coincident integration, as will be explained in the next section. The three linear shape functions are

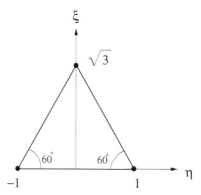

Figure 10.2 Isoparametric equilateral triangular linear element in $\{\eta, \xi\}$ space, where $-1 \leq \eta \leq 1$, $0 \leq \xi \leq \sqrt{3}(1 - |\eta|)$.

$$\psi_1(\eta, \xi) = \frac{\sqrt{3}(1-\eta) - \xi}{2\sqrt{3}}, \quad \psi_2(\eta, \xi) = \frac{\sqrt{3}(1+\eta) - \xi}{2\sqrt{3}}, \quad \psi_3(\eta, \xi) = \frac{\xi}{\sqrt{3}}. \tag{10.24}$$

10.3 EVALUATION OF SINGULAR DOUBLE INTEGRALS

The non-singular integrals can be evaluated using standard Gaussian quadrature formulas. In the direct limit approach for evaluating the singular integrals, the integrals for the coincident and the edge-adjacent cases are forced to be finite by moving the source P off the boundary in the direction $\mathbf{N}$ at a distance of ϵ. The next step is to employ polar coordinate transformations and then integrate analytically with a *fixed distance from the singularity*. After the exact integration, the limit $\epsilon \to 0$ is considered. It will be demonstrated that the coincident and the edge-adjacent hypersingular integrals are separately divergent, producing terms of the form $\log(\epsilon)$. However, the divergent terms from the coincident case can be shown to cancel out with the divergent terms from the edge-adjacent case, and therefore the divergent terms are removed exactly in this approach. Taking the limit $\epsilon \to 0$ back to the boundary results in finite expressions, thus giving a well behaved integral. Once the divergent terms have been identified and removed, the remaining terms of the integral can be evaluated using standard numerical quadrature. The discussion here about $\log(\epsilon)$ singularity, etc. applies only to the hypersingular equation. The direct approach has been designed to handle this worst case, but applies equally to the less-singular integrals.

Compared to the simple Laplace equation treated in Chapter 4, the challenge here is to work with the complicated hypersingular kernel function $W(P, Q)$ defined in Eq.(10.15). In particular, the exponential in this function precludes a complete analytic integration as in Chapter 4, and thus additional procedures are required.

204 NONHOMOGENOUS MEDIA

10.3.1 Coincident Integration

The details of the procedure to evaluate the hypersingular integrals involving the kernel $W(P,Q)$ are described in this section. However, the integration of the kernels G_S, $S(P,Q)$ or $F(P,Q)$ can be handled in exactly the same manner, with the added simplification that no divergent terms appear in the limit $\varepsilon \to 0$. When the source point P and the field point Q lie within the same element E, $E_P = E_Q = E$, the coincident integral to be evaluated is

$$\int_E \psi_k(P) \int_E \phi(Q) W(P,Q) \, \mathrm{d}Q \, \mathrm{d}P$$

$$= \sum_{j=1}^{3} \phi(Q_j) \int_E \psi_k(P) \int_E \psi_j(Q) W(P,Q) \, \mathrm{d}Q \, \mathrm{d}P \qquad (10.25)$$

where E is defined by nodes P_k, $1 \le k \le 3$. Let the parametric variables for the outer P integration be denoted by (η, ξ), and that for Q by (η^*, ξ^*). Transfering the integral to the parametric space ($\mathrm{d}Q \to J_Q \mathrm{d}\xi^* \mathrm{d}\eta^*$ and $\mathrm{d}P \to J_P \mathrm{d}\xi \mathrm{d}\eta$) introduces the Jacobians J_Q and J_P. For coincident integration considering linear elements, the J_Q and J_P are equal and constant. The Jacobians can be conveniently incorporated into the hypersingular kernels, i.e.

$$J_P^2 W(P,Q) = \frac{\kappa_0}{4\pi} e^{\beta(-r+z_Q+z_P)} \left(3\frac{(J_P \mathbf{N} \cdot \mathbf{R})^2}{r^5} + 3\beta\frac{(J_P \mathbf{N} \cdot \mathbf{R})^2}{r^4} \right.$$
$$\left. + \frac{\beta^2 (J_P \mathbf{N} \cdot \mathbf{R})^2 - J_P^2}{r^3} - \beta\frac{J_P^2}{r^2} - \beta^2 \frac{(J_P N_z)^2}{r} \right) \quad (10.26)$$

First Polar Coordinate transformation $\{\eta^*, \xi^*\} \to \{\rho, \theta\}$. For the inner Q integration, the first step is to define a polar coordinate system centered at $P = (\eta, \xi)$,

$$\eta^* - \eta = \rho \cos(\theta), \qquad \xi^* - \xi = \rho \sin(\theta) \qquad (10.27)$$

as shown in Figure 10.3. Polar coordinate transformations centered at the singularity are particularly effective, as the Jacobian of the transformation, $\rho \, \mathrm{d}\rho$, reduces the order of the singularity. This aspect will be used in all the singular integrations.

The upper limit of ρ ($0 < \rho < \rho_L(\theta)$) is different for the three edges of the triangle and consequently, the (ρ, θ) integration is split into three sub-triangles (see Figure 10.3). It suffices to consider the calculation for the lower sub-triangle (shaded portion of the triangle in Figure 10.3). By exploiting the symmetry of the equilateral parametric space, the remaining sub-triangles are handled by rotating the element and employing the formulas for the lower sub-triangle associated with the edge $\xi^* = 0$.

For the lower sub-triangle, the integration limits are

$$0 \le \rho \le \rho_L \quad \text{and} \quad \Theta_1 \le \theta \le \Theta_2 \qquad (10.28)$$

where

$$\rho_L = -\frac{\xi}{\sin(\theta)}, \qquad \Theta_1 = -\frac{\pi}{2} - \tan^{-1}\left(\frac{1+\eta}{\xi}\right), \qquad \Theta_2 = -\frac{\pi}{2} + \tan^{-1}\left(\frac{1-\eta}{\xi}\right).$$
$$(10.29)$$

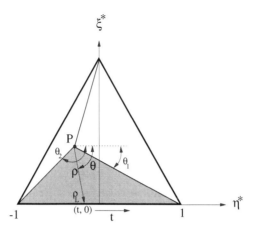

Figure 10.3 First polar coordinate transformation, $\{\eta^*, \xi^*\} \to \{\rho, \theta\}$, for the coincident integration. The variable t eventually replaces θ. Note that $P = P(\eta, \xi)$.

In the limit to the boundary approach as P is moved off the boundary in the direction of the source normal $\mathbf{N}$ at distance of ε, P is replaced by $P + \varepsilon \mathbf{N}$ for the exterior boundary limit (see Figure 10.4), therefore, the distance $r = \|\mathbf{Q} - \mathbf{P}\|$ takes the simple form

$$r^2(\rho, \theta) = \varepsilon^2 + a^2(\theta) \rho^2 , \qquad (10.30)$$

where

$$a^2(\theta) = a_{cc} \cos^2(\theta) + a_{cs} \cos(\theta) \sin(\theta) + a_{ss} \sin^2(\theta). \qquad (10.31)$$

The three coefficients a_{cc}, a_{cs}, a_{ss}, depend solely upon the coordinates of the element nodes (a^2 is a positive quantity), i.e.

$$\begin{aligned}
a_{cc} &= 1/4[(x_2 - x_1)^2 + (y_2 - y_1)^2 + (z_2 - z_1)^2], \\
a_{cs} &= 1/(2\sqrt{3}) \left[(x_2 - x_1)(x_1 + x_2 - 2x_3) + (y_2 - y_1)(y_1 + y_2 - 2y_3) \right. \\
&\quad \left. + (z_2 - z_1)(z_1 + z_1 - 2z_3) \right], \\
a_{ss} &= 1/12[(y_1 + y_2 - 2y_3)^2 + (x_1 + x_2 - 2x_3)^2 + (z_1 + z_2 - 2z_3)^2].
\end{aligned} \qquad (10.32)$$

Here $P_k = (x_k, y_k, z_k)$, $(k = 1, 2, 3)$ are the (x, y, z) coordinates of the element nodes.

The term $J_P \mathbf{N} \cdot \mathbf{R}$ in the kernel (see expression (10.26)) becomes $-\varepsilon J_P$ as P is moved to $P + \varepsilon \mathbf{N}$. The shape function of P, $\psi_j(P)$ is a function of η and ξ. With the polar coordinate transformation centered on P (Eq.(10.27)), the shape function $\psi_j(Q)$ is a linear function of ρ, i.e.

$$\psi_j(Q) = c_{j,0}(\eta, \xi) + c_{j,1}(\eta, \xi, \theta)\rho = \sum_{m=0}^{1} c_{j,m} \rho^m \qquad (10.33)$$

$$\psi_j(P) = c_{j,0}(\eta, \xi). \qquad (10.34)$$

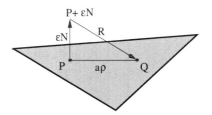

Figure 10.4 The source P is moved off the surface boundary in the direction of $\mathbf{N}$ at a distance of ϵ.

As a result of the polar transformation, $(z_Q - z_P)$ can be written as $\alpha \rho$ where α is a function of θ and z coordinates of the element vertices, i.e.,

$$\alpha = \frac{1}{2}\cos(\theta) + \frac{1}{2\sqrt{3}}(2z_3 - z_1 - z_2)\sin(\theta) \tag{10.35}$$

Since $e^{2\beta z_P}$ is independent of ρ, it is taken outside of the ρ integral, and the exponential term in Eq.(10.26) can be written as

$$e^{\beta(-r+z_Q+z_P)} = e^{\beta(-r+z_Q-z_P+2z_P)}$$
$$= e^{2\beta z_P} e^{\beta(-\sqrt{\epsilon^2+a^2\rho^2}+\alpha\rho)}. \tag{10.36}$$

Employing the boundary limit procedure and expressing the kernel function in polar coordinates, one obtains Eq.(10.25) as

$$\frac{1}{4\pi}\sum_{m=0}^{1}\int_{-1}^{1}d\eta\int_{0}^{\sqrt{3}(1-|\eta|)}\psi_k(\eta,\xi)\,d\xi\int_{\Theta_1}^{\Theta_2}c_{j,m}\,d\theta\int_{0}^{\rho_L}\rho^{m+1}$$

$$e^{\beta(-\sqrt{\epsilon^2+a^2\rho^2}+\alpha\rho)}\left(\frac{3\epsilon^2 J_P^2}{(\epsilon^2+a^2\rho^2)^{5/2}} - \frac{J_P^2}{(\epsilon^2+a^2\rho^2)^{3/2}} + \frac{3\beta\epsilon^2 J_P^2}{(\epsilon^2+a^2\rho^2)^2}\right.$$

$$\left.+\frac{\beta^2\epsilon^2 J_P^2}{(\epsilon^2+a^2\rho^2)^{3/2}} - \frac{\beta J_P^2}{\epsilon^2+a^2\rho^2} - \frac{\beta^2 J_P^{z\,2}}{\sqrt{\epsilon^2+a^2\rho^2}}\right)d\rho. \tag{10.37}$$

Here J_P^z denotes $J_P N_z$, which is the Jacobian J_P multiplied by the z component of the normal at P, i.e. N_z.

First Taylor expansion of the exponential function. Due to the exponential term in the kernel, it is not possible to integrate the entire expression analytically. Our goal for analytic integration is to explicitly identify the divergent terms in the integral. This can be done by employing a Taylor expansion of the exponential term. Expanding up to the first two terms is sufficient to identify the divergent terms, i.e.

$$e^{\beta\left(-\sqrt{\epsilon^2+a^2\rho^2}+\alpha\rho\right)} = 1 + \beta\left(-\sqrt{\epsilon^2+a^2\rho^2}+\alpha\rho\right) + \mathcal{O}(\rho^2). \tag{10.38}$$

As the remainder of the expansion is of order ρ^2, this expansion leads to a sufficiently well behaved expression for the remainder of the integral so that numerical quadrature can be safely used.

First analytical integration (on ρ). Incorporating the Taylor expansion, the integral to be evaluated analytically is

$$
\sum_{m=0}^{1} \int_0^{\rho_L} \rho^{m+1} \left(1 + \beta \left(-\sqrt{\epsilon^2 + a^2 \rho^2} + \alpha \rho\right)\right) \left(\frac{3\epsilon^2 J_P{}^2}{(\epsilon^2 + a^2 \rho^2)^{5/2}}\right.
$$

$$
\left. - \frac{J_P{}^2}{(\epsilon^2 + a^2 \rho^2)^{3/2}} + \frac{3\beta\,\epsilon^2 J_P{}^2}{(\epsilon^2 + a^2 \rho^2)^2} + \frac{\beta^2 \epsilon^2 J_P{}^2}{(\epsilon^2 + a^2 \rho^2)^{3/2}} - \frac{\beta J_P{}^2}{\epsilon^2 + a^2 \rho^2} - \frac{\beta^2 (J_P^z)^2}{\sqrt{\epsilon^2 + a^2 \rho^2}}\right) d\rho. \tag{10.39}
$$

For $m = 0$ (see Eq.(10.33)), the exact analytical integration results in

$$
F_0 = \frac{\beta^3}{2} (J_P^z)^2 \left(1 - \frac{\alpha}{a}\right) \rho_L^2 + \frac{\beta^2}{a} \left[\left(1 - \frac{\alpha}{a}\right) J_P{}^2 - (J_P^z)^2\right] \rho_L
$$

$$
+ 2\beta J_P{}^2 \frac{\alpha}{a^3} + \beta J_P{}^2 \frac{\alpha}{a^3} \log(\epsilon) + \frac{J_P{}^2}{a^3 \rho_L}. \tag{10.40}
$$

All terms are well behaved at $\epsilon = 0$ except for the last two. However the expression

$$
\frac{\alpha \beta J_P{}^2 \log(\varepsilon)}{a^3} \tag{10.41}
$$

is *not* the divergent term that is being sought. It is easily seen that this term cancels out in the subsequent integration over θ. As the term does not contain ρ_L, a complete integration over $0 \le \theta \le 2\pi$ can be considered. Note that α is a linear function of $\cos(\theta)$ and $\sin(\theta)$, and thus satisfies $\alpha(\pi + \theta) = -\alpha(\theta)$. From Eq.(10.31), $a(\pi + \theta) = a(\theta)$, and from Eq.(10.33), $c_{j,m}$ is independent of θ for $m = 0$. Hence, the subsequent integration of the $\log(\varepsilon)$ term on expression (10.41) results in zero, i.e.

$$
-\log(\varepsilon) c_{j,0}(\eta, \xi) J_P{}^2 \int_0^{2\pi} \frac{\alpha \beta}{a^3} d\theta = 0. \tag{10.42}
$$

For $m = 1$, the analytical integration becomes,

$$
F_1 = \frac{\beta^3}{2} (J_P^z)^2 \left(1 - \frac{\alpha}{a}\right) \rho_L^3 + \frac{\beta^2}{2a} \left[\left(1 - \frac{\alpha}{a}\right) J_P{}^2 - (J_P^z)^2\right] \rho_L^2
$$

$$
+ 2\frac{J_P{}^2}{a^3} + \frac{J_P{}^2}{a^3} \log(\epsilon) - J_P{}^2 \frac{\log(a \rho_L)}{a^3}. \tag{10.43}
$$

A divergent term similar to expression (10.41) also appears in this case. This term is also seen to cancel out in the subsequent integration over θ. The coefficient $c_{j,1}(\eta, \xi, \theta)$ is linear in $\cos(\theta)$ and $\sin(\theta)$, and therefore satisfies $c_{j,1}(\eta, \xi, \pi + \theta) = -c_{j,1}(\eta, \xi, \theta)$. Thus

$$
-\log(\varepsilon) J_P{}^2 \int_0^{2\pi} \frac{c_{j,1}(\eta, \xi, \theta)}{a^3} d\theta = 0. \tag{10.44}
$$

This first analytic integration is not sufficient to display the divergent term, and the subsequent integration on θ will not pose any problem. It is the next integration on ξ, (cf. Eq.(10.37)) which has to be dealt with analytically. The analytic integration over ρ produces results that behave as $1/\rho_L$ (see the last term in Eq.(10.40)) and,

208 NONHOMOGENOUS MEDIA

from Eq.(10.29), $\rho_L = -\xi/\sin(\theta)$. This term is capable of producing a $\log(\varepsilon)$ contribution upon ξ integration with the lower limit of $\xi = 0$. Therefore it is necessary to interchange the order of the integration on θ and ξ in order to identify the divergent term through analytical integration.

Variable Tranformation $\{\theta\} \rightarrow \{t\}$. As the limits of the θ integration (i.e. Θ_1 and Θ_2) depend on ξ and η, the integration on the variable θ and ξ is not interchangeable. To circumvent this problem, a new variable t $(-1 \leq t \leq 1)$ is introduced via

$$\theta = -\frac{\pi}{2} + \tan^{-1}\left(\frac{t - \eta}{\xi}\right), \qquad \frac{d\theta}{dt} = \frac{\xi}{\xi^2 + (t - \eta)^2}, \qquad (10.45)$$

which also results in $\rho_L = \left(\xi^2 + (t - \eta)^2\right)^{1/2}$. As depicted in Figure 10.3, t is the 'end-point' $(t, 0)$ of ρ on the ξ^*-axis.

Interchanging the order of integration and tranforming the variable from $\theta \rightarrow t$, the integral (cf. Eq.(10.39)) for $m = 0$ can be written as,

$$\frac{J_P^2}{4\pi} \int_{-1}^{1} d\eta \int_{-1}^{1} dt \int_{0}^{\sqrt{3}(1-|\eta|)} \psi_k(\eta, \xi)\, c_{j,0}\, F_0(\rho_L)\, d\xi . \qquad (10.46)$$

Second Polar Coordinate transformation $\{t, \xi\} \rightarrow \{\Lambda, \Psi\}$. From Eq.(10.45), the singularity is now at $t = \eta$, $\xi = 0$, and another polar coordinate transformation $\{\Lambda, \Psi\}$, replacing $\{t, \xi\}$, is employed (see Figure 10.5),

$$t = \Lambda \cos(\Psi) + \eta, \qquad \xi = \Lambda \sin(\Psi) . \qquad (10.47)$$

The goal is to integrate Λ analytically. With the two changes of variables, $\theta \rightarrow t$ and $\{t, \xi\} \rightarrow \{\Lambda, \Psi\}$, $\cos(\theta)$ becomes $\cos(\Psi)$ and $\sin(\theta)$ becomes $-\sin(\Psi)$. Thus, $a(\theta)$, Eq.(10.31), becomes simply $a(\Psi)$ and is a constant as far as the Λ integration is concerned. As shown in Figure 10.5, the $\{t, \xi\}$ domain is a rectangle, and integrating over $\{\Lambda, \Psi\}$ will necessitate a decomposition into three subdomains

$$0 \leq \Psi \leq \Psi_1, \Psi_1 \leq \Psi \leq \pi - \Psi_2, \quad \text{and} \quad \pi - \Psi_2 \leq \Psi \leq \pi, \qquad (10.48)$$

where

$$\Psi_1 = \tan^{-1}\left(\frac{\sqrt{3}(1 - |\eta|)}{1 - \eta}\right), \qquad \Psi_2 = \tan^{-1}\left(\frac{\sqrt{3}(1 - |\eta|)}{1 + \eta}\right) \qquad (10.49)$$

Second Taylor expansion of the exponential function. With this final coordinate transformation, the P shape functions are linear in Λ, as are the coefficients $c_{j,m}$ from the Q shape functions. Part of the exponential term, $e^{2\beta z_P}$, which was previously kept outside of the ρ integral (cf. Eq.(10.36)), is a function of Λ, and this term has to be included in the integration on Λ. In order to integrate analytically, a Taylor expansion of the exponential term is once again necessary. Note that z_P has a constant part which is independent of Λ,

$$z_P = z_P^0 + \Lambda z_P^1. \qquad (10.50)$$

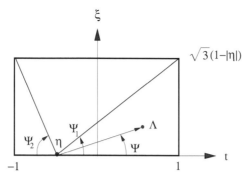

Figure 10.5 Geometry of the second polar coordinate transformation, $\{t, \xi\} \to \{\Lambda, \Psi\}$, for the coincident integration.

Thus the term $e^{2\beta z_P^0}$ can be kept out of the Λ integration. The rest can be expanded up to three terms in order to obtain the necessary divergent term,

$$e^{2\beta\Lambda z_P^1} = 1 + 2\beta\Lambda z_P^1 + 2\beta^2\Lambda^2(z_P^1)^2 + \mathcal{O}(\Lambda^3). \tag{10.51}$$

Theoretically, two terms in the above expansion are sufficient, however, an additional term was considered for convergence of numerical results. A discussion of additional terms in the Taylor expansion employed in HBIEs has been presented by Gray and Paulino [119].

Second analytical integration (on Λ). The product of the shape functions of P and Q produces terms of Λ up to order 2. The integrals (cf. Eq.(10.46)) to be evaluated are therefore of the form

$$-\frac{J_P^2}{4\pi} \int_{-1}^{1} d\eta \int \sin(\Psi)\, d\Psi \int_{0}^{\cdot} \Lambda^s f(\Lambda)\, d\Lambda \tag{10.52}$$

for $s = 0, 1, 2$ where $f(\Lambda)$ is a function of Λ. The missing limits of the Λ and Ψ integrals depend upon the particular sub-triangle in Figure 10.5 being considered. The Λ integrations for $s = 1$ and $s = 2$ are straightforward. For $s = 0$, a finite contribution plus a divergent term of the form

$$L_{kj}^c = \log(\varepsilon)\, \frac{J_P^2}{4\pi} \int_{-1}^{1} \hat{\psi}_k^0\, \hat{\psi}_j^0\, d\eta \int_{0}^{\pi} \frac{\sin(\Psi)}{a^3}\, d\Psi, \tag{10.53}$$

is found. Here, $\hat{\psi}_l^0$ are the shape functions evaluated at $\Lambda = \rho = 0$, as

$$\hat{\psi}_1^0 = \frac{1-\eta}{2}, \quad \hat{\psi}_2^0 = \frac{1+\eta}{2}, \quad \hat{\psi}_3^0 = 0. \tag{10.54}$$

Note that as $a = a(\Psi)$ is independent of η, Eq.(10.53) simplifies to

$$L_{kj}^c = \log(\varepsilon)\, \frac{J_P^2}{4\pi}\, \frac{1+\delta_{kj}}{3} \int_{0}^{\pi} \frac{\sin(\Psi)}{a^3}\, d\Psi, \tag{10.55}$$

where δ_{kj} is the usual Kronecker delta function and $1 \leq k, j \leq 2$. For $m = 1$, following the same procedure as above does not produce any divergent terms.

210 NONHOMOGENOUS MEDIA

The divergent term L_{kj}^c is precisely the same as that obtained from the hypersingular *homogeneous* Laplace equation as in section 4.2.6 in Chapter 4. This comes as no surprise because the new feature in the FGM kernels is the exponential term. In the Taylor expansion,

$$e^{\beta r} = 1 + (\beta r) + \frac{(\beta r)^2}{2} + \frac{(\beta r)^3}{6} + ..., \tag{10.56}$$

the leading constant is the most singular, the subsequent terms actually help to kill off the singularity. Thus the divergence comes from the first term of the expansion, which is exactly the same term for the Laplace equation. The important consequence of this observation is that it will not be necessary to prove that Eq.(10.55) cancels with the corresponding divergence from the adjacent edge integration (obtained below). The proof in section 4.2.6 in Chapter 4 suffices to demonstrate this point.

10.3.2 Edge Adjacent Integration

In this case an edge is shared between the two elements. Orient the elements so that the shared edge is defined by $\xi = 0$ in E_P, and $\xi^* = 0$ for E_Q, and the singularity occurs when $\eta = -\eta^*$.

First Polar Coordinate transformation $\{\eta^*, \xi^*\} \to \{\rho, \theta\}$. The first step is to employ polar coordinates for the Q integration [112],

$$\eta^* = \rho \cos(\theta) - \eta, \quad \xi^* = \rho \sin(\theta) \tag{10.57}$$

As shown in Figure 10.6(a), the θ integration must be split into two pieces (for simplicity the integrands are omitted, but it will be useful to retain the Jacobians of the transformations)

$$\int_{-1}^{1} d\eta \int_{0}^{\sqrt{3}(1-|\eta|)} d\xi \left[\int_{0}^{\Theta_1(\eta)} d\theta \int_{0}^{L_1(\theta)} \rho \, d\rho + \int_{\Theta_1(\eta)}^{\pi} d\theta \int_{0}^{L_2(\theta)} \rho \, d\rho \right], \tag{10.58}$$

where

$$L_1(\theta) = \frac{\sqrt{3}(1+\eta)}{\sin(\theta) + \sqrt{3}\cos(\theta)}, \quad L_2(\theta) = \frac{\sqrt{3}(1-\eta)}{\sin(\theta) - \sqrt{3}\cos(\theta)}. \tag{10.59}$$

The break-point in θ,

$$\theta_1(\eta) = \frac{\pi}{2} - \tan^{-1}\left(\frac{\eta}{\sqrt{3}}\right) \tag{10.60}$$

is only a function of η. The integrations can therefore be rearranged as follows,

$$\int_{-1}^{1} d\eta \int_{0}^{\theta_1(\eta)} d\theta \int_{0}^{\sqrt{3}(1-|\eta|)} d\xi \int_{0}^{L_1(\theta)} \rho \, d\rho$$

$$+ \int_{-1}^{1} d\eta \int_{\theta_1(\eta)}^{\pi} d\theta \int_{0}^{\sqrt{3}(1-|\eta|)} d\xi \int_{0}^{L_2(\theta)} \rho \, d\rho . \tag{10.61}$$

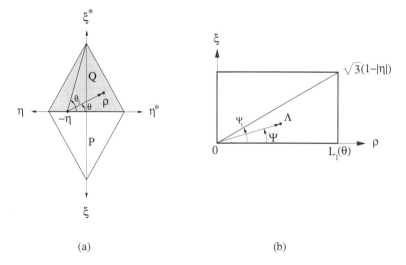

Figure 10.6 (a) Polar coordinate transformation employed in the Q element, $\{\eta^*, \xi^*\} \to \{\rho, \theta\}$; (b) Second polar coordinate transformation $\{\rho, \xi\} \to \{\Lambda, \Psi\}$ for the edge-adjacent integration.

Second Polar Coordinate transformation $\{\rho, \xi\} \to \{\Lambda, \Psi\}$. As the singularity occurs at $\rho = \xi = 0$, a second polar coordinate transformation is introduced

$$\rho = \Lambda \cos(\Psi), \quad \xi = \Lambda \sin(\Psi). \tag{10.62}$$

The Ψ integration must also be taken in two parts (Figure 10.6(b)), resulting in the four integrals

$$\int_{-1}^{1} d\eta \int_{0}^{\Theta_1(\eta)} d\theta \left[\int_{0}^{\Psi_1} d\Psi \int_{0}^{\lambda_{11}} \cos(\Psi)\Lambda^2 \, d\Lambda + \int_{\Psi_1}^{\pi/2} d\Psi \int_{0}^{\lambda_{12}} \cos(\Psi)\Lambda^2 \, d\Lambda \right] +$$

$$\int_{-1}^{1} d\eta \int_{\Theta_1(\eta)}^{\pi} d\theta \left[\int_{0}^{\Psi_2} d\Psi \int_{0}^{\lambda_{21}} \cos(\Psi)\Lambda^2 \, d\Lambda + \int_{\Psi_2}^{\pi/2} d\Psi \int_{0}^{\lambda_{22}} \cos(\Psi)\Lambda^2 \, d\Lambda \right]. \tag{10.63}$$

The formulas for the Λ limits are simply

$$\lambda_{n1} = L_n(\theta)/\cos(\Psi), \quad \lambda_{n2} = L_n(\theta)/\sin(\Psi), \tag{10.64}$$

for $n = 1, 2$. The distance function takes the form

$$r^2 = \varepsilon^2 + \varepsilon a_1 \Lambda + a_2 \Lambda^2, \tag{10.65}$$

but this quadratic expression (in the denominators) can be integrated exactly. The Λ^2 factor from the two polar transformations sufficiently reduces the order of the singularity such that one analytic integration (over Λ) will produce the $\log(\varepsilon)$ term.

Taylor Expansion of the exponential function. The Λ integral can not be evaluated analytically unless a Taylor expansion is once again utilized for the exponential

212   NONHOMOGENOUS MEDIA

term. To extract the divergent terms from this integral, a one term expansion will be enough (same arguments as for the coincident integration). Similarly to Eq.(10.36) and Eq.(10.51), the exponential term for this case can be written as,

$$
\begin{aligned}
e^{\beta(-r+z_Q+z_P)} &= e^{\beta(-r+z_Q-z_P+2z_P)} \\
&= e^{\beta(-\sqrt{\epsilon^2+a^2\Lambda^2}+\alpha\Lambda+2(z_P^0+\Lambda z_P^1))} \\
&= e^{2\beta z_P^0} e^{\beta(-\sqrt{\epsilon^2+a^2\Lambda^2}+\alpha\Lambda+2\Lambda z_P^1)}.
\end{aligned}
\tag{10.66}
$$

The term $e^{2\beta z_P^0}$ is taken outside of the Λ integration since it is independent of Λ. A one term Taylor expansion is employed in the remaining exponential term, i.e.

$$
e^{\beta(-\sqrt{\epsilon^2+a^2\Lambda^2}+\alpha\Lambda+2\Lambda z_P^1)} = 1 + \mathcal{O}(\Lambda).
\tag{10.67}
$$

The rest of the expansion is of the order of Λ. This expansion along with the Λ^2 from the Jacobian of the polar transformation will be sufficient to provide a smooth function for numerical integration. This follows from the observation that $\Lambda = 0$ encapsulates all three conditions for $r = 0$, namely $\xi = \xi^* = 0$, and $\eta = -\eta^*$. Thus, as in the coincident algorithm, the exact integration is with respect "*to the distance from the singularity*".

Analytical Integration on Λ.   The Λ integration results in a finite quantity plus a divergent contribution, which is given by

$$
L_{kj}^e = \log(\varepsilon)\, \frac{1}{4\pi} \int_{-1}^{1} \hat{\psi}_k^0\, \hat{\psi}_j^0\, d\eta \int_0^{\pi} d\theta \int_0^{\pi/2} \cos(\Psi) \left(\frac{3 J_{PN}^1 J_{Qn}^1}{a_2^{5/2}} - \frac{J_{Qn}\boldsymbol{\cdot} J_{PN}}{a_2^{3/2}} \right) d\Psi \,,
\tag{10.68}
$$

where J_{Qn}^1 and J_{PN}^1 are the coefficients of Λ in $J_Q\mathbf{n}\boldsymbol{\cdot}\mathbf{R}$ and $J_P\mathbf{N}\boldsymbol{\cdot}\mathbf{R}$ respectively,

$$
J_P\mathbf{N}\boldsymbol{\cdot}\mathbf{R} = J_{PN}^1\Lambda - J_P\varepsilon, \qquad J_Q\mathbf{n}\boldsymbol{\cdot}\mathbf{R} = J_{Qn}^1\Lambda - J_Q\mathbf{n}\boldsymbol{\cdot}\mathbf{N}.
\tag{10.69}
$$

As in Eq.(10.54), $\hat{\psi}_k^0$ and $\hat{\psi}_j^0$ denote the shape functions evaluated at $\Lambda = 0$. This expression, Eq.(10.68), and the expression for L_{kj}^c, Eq.(10.55), cancel one another. As noted above, the proof of such cancellation [112] is precisely the same as that for Laplace equation explained in Chapter 4.

10.3.3   Vertex Adjacent Integration

In this case the singularity is limited to a single point in the four dimensional integration. Orient the P and Q elements so that the singular point is $\eta = -1$ and $\eta^* = -1$.

Two Polar Coordinate transformations ($\{\eta^*, \xi^*\} \to \{\rho_q, \theta_q\}$, $\{\eta, \xi\} \to \{\rho_p, \theta_p\}$). Two separate polar coordinates are first introduced in each element (see Figure 10.7(a)),

$$
\begin{aligned}
\eta^* &= \rho_q \cos(\theta_q) - 1, \qquad \xi^* = \rho_q \sin(\theta_q) \\
\eta &= \rho_p \cos(\theta_p) - 1, \qquad \xi = \rho_p \sin(\theta_p).
\end{aligned}
\tag{10.70}
$$

This results in an integral of the form

$$
\int_0^{\pi/3} d\theta_p \int_0^{L_p(\theta_p)} \rho_p\, d\rho_p \int_0^{\pi/3} d\theta_q \int_0^{L_q(\theta_q)} \rho_q\, d\rho_q,
\tag{10.71}
$$

EVALUATION OF SINGULAR DOUBLE INTEGRALS   213

where

$$L_p(\theta_p) = 2\sqrt{3}/\left[\sin(\theta_p) + \sqrt{3}\cos(\theta_p)\right], \quad (10.72)$$

$$L_q(\theta_q) = 2\sqrt{3}/\left[\sin(\theta_q) + \sqrt{3}\cos(\theta_q)\right]. \quad (10.73)$$

As the vertex adjacent integration will not produce a divergent term, we omit the kernel function and just keep track of the Jacobians in the subsequent expressions. Now interchanging the order of the integration between ρ_p and θ_q, one obtains

$$\int_0^{\pi/3} d\theta_p \int_0^{\pi/3} d\theta_q \int_0^{L_p(\theta_p)} \rho_p \, d\rho_p \int_0^{L_q(\theta_q)} \rho_q \, d\rho_q.$$

Third Polar Coordinate transformations $\{\rho_p, \rho_q\} \to \{\Lambda, \Psi\}$. The only singularity is at the common vertex $\rho_p = \rho_q = 0$, and thus a third polar coordinate transformation is performed

$$\rho_p = \Lambda\cos(\Psi), \quad \rho_q = \Lambda\sin(\Psi).$$

As illustrated in Figure 10.7(b), the $\{\rho_p, \rho_q\}$ domain is a rectangle, and thus the Ψ integration must be taken in two pieces. The combined Jacobian in this case is $\cos(\Psi)\sin(\Psi)\Lambda^3$, and thus Eq.(10.71) becomes

$$\int_0^{\pi/3} d\theta_p \int_0^{\pi/3} d\theta_q \left[\int_0^{\Psi_1} \cos(\Psi)\sin(\Psi)\,d\Psi \int_0^{L_1(\Psi)} \Lambda^3\,d\Lambda + \right. \quad (10.74)$$

$$\left. \int_{\Psi_1}^{\pi/2} \cos(\Psi)\sin(\Psi)\,d\Psi \int_0^{L_2(\Psi)} \Lambda^3\,d\Lambda \right],$$

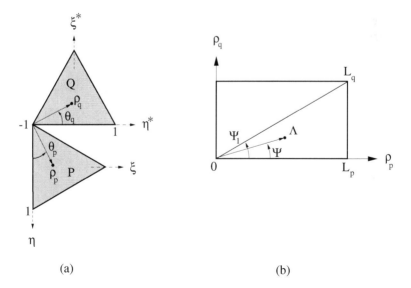

(a) (b)

Figure 10.7   (a) Initial polar coordinate transformation employed in both P and Q elements; (b) Final polar coordinate transformation $\{\rho_p, \rho_q\} \to \{\Lambda, \Psi\}$ for the vertex adjacent integration.

where

$$L_1(\Psi) = L_P(\theta_p)/\cos(\Psi) \quad \text{and} \quad L_2(\Psi) = L_Q(\theta_q)/\sin(\Psi). \tag{10.75}$$

With the Λ^3 factor mulitplying the kernel function, it is possible to immediately set $\varepsilon = 0$, and the distance function is then $r^2 = a^2\Lambda^2$ (the coefficient being a function of all three angles and nodal coordinates). It is then immediately apparent that this integral is finite. Although numerical evaluation with Gaussian quadrature could be employed for the entire integration we prefer to execute the Λ integral semi-analytically in order to achieve better accuracy.

10.3.4 Numerical Example

A simple FGM cube with constant temperature on two planes is considered first. As the analytical solution of this problem can be obtained, this problem is suitable for a convergence study. The problem of interest is shown in Figure 10.8. The top surface of the cube at $[z = 1]$ is maintained at a temperature of $T = 100$ while the bottom face at $[z = 0]$ is zero. The remaining four faces are insulated (zero normal flux).

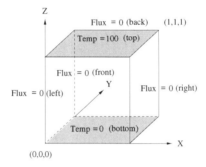

Figure 10.8 Geometry and boundary conditions of the FGM unit cube problem with constant temperature on two planes. The faces with prescribed temperature are shaded.

The thermal conductivity is

$$k(x, y, z) = k_0 e^{2\beta z} = 5e^{2\beta z}, \tag{10.76}$$

and various values of β will be considered. The analytical solution for the temperature is

$$\phi(x, y, z) = T \frac{1 - e^{-2\beta z}}{1 - e^{-2\beta L}}, \tag{10.77}$$

where L is the dimension of the cube (in the z-direction), and the analytical solution for the flux (on $z = 1$) is

$$q(x, y, z) = -k(x, y, z) \frac{2\beta T e^{-2\beta z}}{1 - e^{-2\beta L}} = -k_0 \frac{2\beta T}{1 - e^{-2\beta L}} \tag{10.78}$$

The cube is discretized with 432 boundary elements and 294 nodes. The computed temperature variation in the z direction is plotted with different values of β (material nonhomogeneity parameter) and compared with the analytical solution in

EVALUATION OF SINGULAR DOUBLE INTEGRALS 215

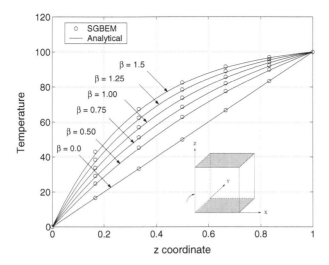

Figure 10.9 Temperature profile in z direction for the FGM cube discretized with 432 elements and 294 nodes.

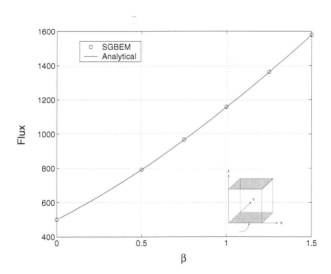

Figure 10.10 Variation of flux on the face $z = 0$ for different values of the material nonhomogeneity parameter β.

Figure 10.9. For this problem the flux on each face of the unit cube is constant. Figure 10.10 shows the variation of the flux on the face $z = 0$ with respect to the change of the material nonhomogeneity parameter β, and compares the results with the analytical flux solution. For the face $z = 1$, the flux has the same variation as on the face $z = 0$, but with opposite sign (cf. Eq.(10.78)).

216 NONHOMOGENOUS MEDIA

10.4 TRANSIENT HEAT CONDUCTION IN FGMS

Next we present a transient formulation for heat conduction in FGMs. The Green's function for three-dimensional transient heat conduction (diffusion equation) for functionally graded materials (FGMs) is derived. The thermal conductivity and heat capacitance both vary exponentially in one coordinate. In the process of solving this diffusion problem numerically, a Laplace transform approach is used to eliminate the dependence on time. The fundamental solution in Laplace space is derived and the boundary integral equation formulation for the Laplace Transform Boundary Element Method (LTBEM) is obtained. The numerical implementation is performed using a Galerkin approximation, and the time-dependence is restored by numerical inversion of the Laplace transform.

Transient heat conduction problems are usually solved using either the time domain approach or the Laplace transform domain approach. In the *time domain approach*, a time marching scheme associated with the BEM solution at each time step is used, and solutions are found directly in the time domain. The time dependent fundamental solution is used to transform the differential system into a boundary integral equation. The numerical solution of the boundary integral requires both space and time discretization. Early works using the time-domain approach include those by Chang *et al.* [42], Shaw [249], Curran *et al.* [66], Wrobel and Brebbia [293], and many others. Recent works involve that of Lesnic *et al.* [162], Coda and Venturini [57,58], Pasquetti and Caruso [228], Wrobel *et al.* [207], Divo and Kassab [77], etc.

By employing a time-dependent fundamental solution together with recent developments in techniques for converting volume integrals into a (series of) boundary integrals, the diffusion problem can be solved by means of finite-differencing in time and BEM discretization for the spatial variables. The volume integral can be converted through either a set of local interpolation functions – known as the dual reciprocity method, as presented by Brebbia and Wrobel [294] – or a hierarchy of higher order fundamental solutions – known as the multiple reciprocity method, as presented by Nowak [202]. A drawback of time-marching schemes is that they can be numerically inefficient.

An alternative is to employ a *transform space* approach, wherein the time dependent derivative is eliminated in favor of an (algebraic) transform variable. However, once the differential system is solved in transform space, reconstituting the solution in the time domain requires an inverse transform. Although this approach is simple and attractive, the accuracy depends upon an efficient and accurate numerical inverse transform. For diffusion problems, Laplace transform (LT) seems to be the best choice. The first such formulation utilizing the Laplace transform approach was proposed by Rizzo and Shippy [234] for solution of heat conduction problems in solids. Later Liggett and Liu [166] extended the method to unsteady flows in confined aquifers. Early Laplace inversion methods were not efficient, as they employed a type of curve fitting process for which the behavior of the solution had to be known a priori. However, with the advancement of techniques for inverse LT, this approach has received renewed attention. Moridis and Reddell [186] successfully used their Laplace transform boundary element method (LTBEM) for diffusion type problems. Cheng *et al.* [52] also used the BEM to solve axisymmetric diffusion problems in the LT space. Zhu *et al.* [302] and Zhu and Satravaha [301] extended the work to the Laplace transform dual reciprocity method (LTDRM) for solving nonlinear diffusion equations with a source term and temperature-dependent diffusivity. Goldberg and Chen [104] used the Method of fundamental solutions (MFS) in Laplace space for both diffusion and Helmholtz equations. Maillet *et al.* [177] recently used this approach to solve heat transfer problems by the quadrupole method. Some details of both approaches (i.e. time domain and transformed space) in transient problems can be found in Reference [302]. Similar problems for different applications have been presented by Cheng [53] for ground water flow in heterogeneous media, and by Wu and Lee [295] and Lacerda *et al.* [154] for acoustic propagation with a mean flow.

The Green's function for the three-dimensional (3D) FGM transient diffusion equation is derived using an exponential variable transform; the boundary integral equation based upon this Green's function is then solved numerically. The exponential transform technique has been previously used by Carslaw and Jaeger [39] to obtain analytical solutions for various problems. Moreover, Li and Evans [163], Onishi and Ikeuchi [137], Ramachandran [229] and, more recently, Singh and Tanaka [171] have used this transform to solve advection-diffusion problems.

10.4.1 Basic Equations

The transient diffusion equation is given by

$$\nabla \cdot (k\nabla\phi) = c\rho\frac{\partial\phi}{\partial t} \tag{10.79}$$

where $\phi = \phi(x, y, z; t)$ is the temperature function, c is the specific heat, ρ is the density, and k is the thermal conductivity. We assume that the thermal conductivity varies exponentially in one cartesian coordinate, i.e.

$$k(x, y, z) = k_0 e^{2\beta z} \tag{10.80}$$

in which β is the nonhomogeneity parameter. The specific heat is also graded with the same functional variation as the conductivity,

$$c(x, y, z) = c_0 e^{2\beta z} \tag{10.81}$$

Substituting these material expressions into Eq.(10.79), one obtains

$$\nabla^2\phi + 2\beta\phi_z = \frac{1}{\alpha}\frac{\partial\phi}{\partial t} \tag{10.82}$$

where $\alpha = k_0/(c_0\rho)$ and ϕ_z denotes the derivative of ϕ with respect to z (i.e. $\phi_z \equiv \partial\phi/\partial z$).

Two types of boundary conditions are prescribed. The Dirichlet condition for the unknown potential ϕ is

$$\phi(x, y, z; t) = \overline{\phi}(x, y, z; t) \tag{10.83}$$

on boundary Σ_1 and the Neumann condition for its flux is

$$q(x, y, z; t) = -k(\cdot)\frac{\partial\phi(x, y, z; t)}{\partial n} = \overline{q}(x, y, z; t) \tag{10.84}$$

on boundary Σ_2, where $\mathbf{n}$ is the unit outward normal to Σ_2. Here a bar over the quantity of interest means that it assumes a prescribed value. For a well-posed

218 NONHOMOGENOUS MEDIA

problem $\Sigma_1 \cup \Sigma_2 = \Sigma$ with Σ being the entire boundary. As the problem is time dependent, in addition to these boundary conditions, an initial condition at a specific time t_0 must also be prescribed. A zero initial temperature distribution has been considered in all the examples in this chapter, i.e.

$$\phi(x, y, z; t_0) = \phi_0(x, y, z) = 0 \tag{10.85}$$

A nonzero initial temperature distribution may be solved with the dual reciprocity method [206].

10.4.2 Green's Function

The Green's function for Eq.(10.82) can be derived by employing the substitution

$$\phi = e^{-\beta z - \beta^2 \alpha t} u \ . \tag{10.86}$$

Thus, the derivatives in Eq.(10.82) can be expressed in terms of u, as follows:

$$\frac{\partial \phi}{\partial z} = -\beta e^{-\beta z - \beta^2 \alpha t} u + e^{-\beta z - \beta^2 \alpha t} \frac{\partial u}{\partial z} \tag{10.87}$$

$$\frac{\partial^2 \phi}{\partial z^2} = \beta^2 e^{-\beta z - \beta^2 \alpha t} u - 2\beta e^{-\beta z - \beta^2 \alpha t} \frac{\partial u}{\partial z} + e^{-\beta z - \beta^2 \alpha t} \frac{\partial^2 u}{\partial z^2} \tag{10.88}$$

$$\frac{1}{\alpha} \frac{\partial \phi}{\partial t} = \frac{1}{\alpha}(-\beta^2 \alpha e^{-\beta z - \beta^2 \alpha t}) u + \frac{e^{-\beta z - \beta^2 \alpha t}}{\alpha} \frac{\partial u}{\partial t} \tag{10.89}$$

Substituting Eqs. 10.87, 10.88 and 10.89 into Eq.(10.82), one obtains

$$\nabla^2 u = \frac{1}{\alpha} \frac{\partial u}{\partial t} \tag{10.90}$$

which is the standard diffusion equation for a homogeneous material problem. The time dependent fundamental solution for this equation is known [29], and is given by

$$u^* = \frac{1}{(4\pi\alpha\tau)^{\frac{3}{2}}} e^{-\frac{r^2}{4\alpha\tau}} \tag{10.91}$$

where $\tau = t_F - t$. Note that the function u^* represents the temperature field at time t_F produced by an instantaneous source of heat at point $P(x_p, y_p, z_p)$ and time t. The 3D fundamental solution to the FGM diffusion equation can be obtained by backsubstitution (using Eq.(10.86)) as

$$\phi^* = \frac{1}{(4\pi\alpha\tau)^{\frac{3}{2}}} e^{-\beta(z - z_p) - \beta^2 \alpha\tau - \frac{r^2}{4\alpha\tau}} \tag{10.92}$$

10.4.3 Laplace Transform BEM (LTBEM) Formulation

Let the Laplace transform (LT) of ϕ be denoted by

$$\tilde{\phi}(Q,s) = \int_0^\infty \phi(Q,t)e^{-st}dt \tag{10.93}$$

Thus, in LT space, the differential equation (10.42) becomes

$$\nabla^2\tilde{\phi} + 2\beta\tilde{\phi}_z - \frac{s}{\alpha}\tilde{\phi} = 0 \tag{10.94}$$

where $\phi_0 = 0$ (at $t = 0$) is considered (see Eq.(10.85)).

Following the usual practice, the corresponding boundary integral statement can be obtained by 'orthogonalizing' this equation against an arbitrary (for now) function $f(x,y,z) = f(Q)$, i.e., integrating over a bounded volume V

$$\int_V f(Q)\left(\nabla^2\tilde{\phi} + 2\beta\tilde{\phi}_z - \frac{s}{\alpha}\tilde{\phi}\right)dV_Q = 0 \tag{10.95}$$

According to Green's second identity, if the two functions ϕ and λ have continuous first and second derivatives in V, then

$$\int_V (\phi\nabla^2\lambda - \lambda\nabla^2\phi)dV = \int_\Sigma \left(\phi\frac{\partial\lambda}{\partial n} - \lambda\frac{\partial\phi}{\partial n}\right)dS \tag{10.96}$$

Using this relation and denoting the boundary of V by Σ, the first term of Eq.(10.95) becomes,

$$\int_V f(Q)\nabla^2\tilde{\phi}\,dV_Q = \int_V \tilde{\phi}(Q)\nabla^2 f(Q)dV_Q + \int_\Sigma \left(f(Q)\frac{\partial}{\partial n}\tilde{\phi}(Q) - \tilde{\phi}(Q)\frac{\partial}{\partial n}f(Q)\right)dS_Q. \tag{10.97}$$

Integrating by parts the second term of Eq.(10.95) we obtain,

$$\int_V 2\beta f(Q)\tilde{\phi}_z dV_Q = \int_\Sigma 2\beta f(Q)n_z\tilde{\phi}(Q)dS_Q - \int_V 2\beta\frac{\partial f}{\partial z}\tilde{\phi}(Q)dV_Q \tag{10.98}$$

and using Eq.(10.97) and Eq.(10.98) in Eq.(10.95), we get after simplification,

$$0 = \int_\Sigma \left(f(Q)\frac{\partial}{\partial n}\tilde{\phi}(Q) - \tilde{\phi}(Q)\frac{\partial}{\partial n}f(Q) + 2\beta n_z(Q)\tilde{\phi}(Q)f(Q)\right)dS_Q$$

$$+ \int_V \tilde{\phi}(Q)\left(\nabla^2 f(Q) - 2\beta f_z(Q) - \frac{s}{\alpha}f(Q)\right)dV_Q \tag{10.99}$$

where $f_z = \partial f/\partial z$, and $\mathbf{n}(Q) = (n_x, n_y, n_z)$ is the unit outward normal on Σ.

If we select $f(Q) = G(P,Q)$ as a Green's function, then the Green's function equation is (cf. Eq.(10.82))

$$\nabla^2 G(P,Q) - 2\beta G_z(Q) - \frac{s}{\alpha}G(P,Q) = -\delta(Q - P), \tag{10.100}$$

220 NONHOMOGENOUS MEDIA

where δ is the Dirac Delta function. Thus the source point volume integral in Eq.(10.99) becomes $-\tilde{\phi}(P)$. By means of Eq.(10.100), Eq.(10.99) can be rewritten as

$$\tilde{\phi}(P) + \int_{\Sigma} \left(\frac{\partial}{\partial n} G(P,Q) - 2\beta n_z G(P,Q) \right) \tilde{\phi}(Q) dS_Q = \int_{\Sigma} G(P,Q) \frac{\partial}{\partial n} \tilde{\phi}(Q) dS_Q. \tag{10.101}$$

In order to obtain the Green's function in Laplace space, Eq.(10.100) is modified by using the substitution

$$G = e^{\beta z} v. \tag{10.102}$$

In this case, the differential equation for the LT space is

$$\nabla^2 v - (\beta^2 + \frac{s}{\alpha}) v = 0. \tag{10.103}$$

This equation is the modified Helmholtz equation, whose Green's function is known. Thus the Green's function in 3D LT space is

$$v = \frac{1}{4\pi r} e^{-\sqrt{\beta^2 + \frac{s}{\alpha}} \, r}. \tag{10.104}$$

By back substitution (see Eq.(10.102)), we obtain

$$G(P,Q,s) = \frac{1}{4\pi r} e^{\beta z} e^{-\sqrt{\beta^2 + \frac{s}{\alpha}} \, r}. \tag{10.105}$$

The boundary conditions, Eq.(10.83) and Eq.(10.84), must also be transformed into Laplace space, i.e.

$$\tilde{\phi}(Q,s) = \int_0^\infty \overline{\phi}(Q,t) e^{-st} dt, \qquad \tilde{q}(Q,s) = \int_0^\infty \overline{q}(Q,t) e^{-st} dt \tag{10.106}$$

respectively. For constant boundary conditions the above equations reduce to (see Brebbia, Telles and Wrobel [29], p. 143)

$$\tilde{\phi}(Q,s) = \frac{\overline{\phi}(Q,t)}{s}, \qquad \tilde{q}(Q,s) = \frac{\overline{q}(Q,t)}{s} \tag{10.107}$$

respectively.

The modified kernel functions, in terms of the Laplace variable s, are

$$G(P,Q,s) = \frac{1}{4\pi r} e^{\beta(z_Q - z_P) - \sqrt{\beta^2 + \frac{s}{\alpha}} \, r} \tag{10.108}$$

and

$$\frac{\partial}{\partial n} G(P,Q,s) - 2\beta n_z G(P,Q,s)$$

$$= \frac{e^{\beta R_z - \sqrt{\beta^2 + \frac{s}{\alpha}} \, r}}{4\pi} \left(-\frac{1}{r^2} \frac{\mathbf{n} \cdot \mathbf{R}}{r} - \frac{1}{r} \sqrt{\beta^2 + \frac{s}{\alpha}} \frac{\mathbf{n} \cdot \mathbf{R}}{r} + \frac{1}{r} \beta n_z - \frac{2\beta n_z}{r} \right)$$

$$= -\frac{e^{\beta R_z - \sqrt{\beta^2 + \frac{s}{\alpha}} \, r}}{4\pi} \left(\frac{\mathbf{n} \cdot \mathbf{R}}{r^3} + \sqrt{\beta^2 + \frac{s}{\alpha}} \frac{\mathbf{n} \cdot \mathbf{R}}{r^2} + \frac{\beta n_z}{r} \right) \tag{10.109}$$

where $\mathbf{n}$ is the unit outward normal at a field point Q, n_z is the z component of $\mathbf{n}$, $\mathbf{R} = \mathbf{Q} - \mathbf{P}$, $R_z = z_Q - z_P$, and r is the norm of $\mathbf{R}$, i.e. $r = \|\mathbf{R}\| = \|\mathbf{Q} - \mathbf{P}\|$

10.4.4 Numerical Implementation of the 3D Galerkin BEM

The numerical methods employed in this work use standard Galerkin implementation techniques in conjunction with the LT method. A few aspects of the numerical methods are briefly reviewed in this section.

Division of the boundary into elements. Over each element, the variation of the geometry and the variables (potential and flux) is approximated by six noded isoparametric quadratic triangular elements (see Figure 10.11).

The geometry of an element can be defined by the coordinates of its six nodes using appropriate quadratic shape functions as follows

$$x_i(\xi, \eta) = \sum_{j=1}^{6} N_j(\xi, \eta)(x_i)_j \qquad (10.110)$$

In an isoparametric approximation, the same shape functions are used for the solution variables, as follows:

$$\phi_i(\xi, \eta) = \sum_{j=1}^{6} N_j(\xi, \eta)(\phi_i)_j$$

$$\frac{\partial \phi_i}{\partial n}(\xi, \eta) = \sum_{j=1}^{6} N_j(\xi, \eta) \left(\frac{\partial \phi_i}{\partial n}\right)_j. \qquad (10.111)$$

The shape functions can be explicitly written in terms of intrinsic coordinates ξ and η as follows (see Figure 10.11):

$$\begin{array}{ll} N_1(\xi, \eta) = (1 - \xi - \eta)(1 - 2\xi - 2\eta) & N_4(\xi, \eta) = 4\xi(1 - \xi - \eta) \\ N_2(\xi, \eta) = \xi(2\xi - 1) & N_5(\xi, \eta) = 4\xi\eta \\ N_3(\xi, \eta) = \eta(2\eta - 1) & N_6(\xi, \eta) = 4\eta(1 - \xi - \eta) \end{array} \qquad (10.112)$$

The intrinsic coordinate space is the right triangle with $\xi \geq 0$, $\eta \geq 0$ and $\xi + \eta \leq 1$.

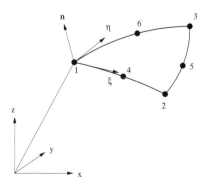

Figure 10.11 Isoparametric quadratic triangular element of 6 nodes. The intrinsic coordinate space is the right triangle in (ξ, η) space with $\xi \geq 0, \eta \geq 0$ and $\xi + \eta \leq 1$.

222 NONHOMOGENOUS MEDIA

Galerkin Boundary Integral Equation. Define

$$\mathcal{B}(P) \equiv \phi(P) + \int_\Sigma \left(\frac{\partial}{\partial n} G(P, Q) - 2\beta n_z G(P, Q) \right) \phi(Q) dS_Q - \int_\Sigma G(P, Q) \frac{\partial \phi}{\partial n}(Q) dS_Q$$

and thus for an exact solution $\mathcal{B}(P) \equiv 0$.

In a Galerkin approximation, the error in the approximate solution is orthogonalized against the shape functions, *i.e.*, the shape functions are the weighting functions and $\mathcal{B}(P) = 0$ is enforced in the 'weak sense'

$$\int_\Sigma N_k(P)\mathcal{B}(P)dP = 0 \tag{10.113}$$

After replacing the boundary and the boundary functions by their interpolated approximations, a set of linear algebraic equations emerges,

$$[H]\{\phi\} = [G] \left\{ \frac{\partial \phi}{\partial n} \right\}. \tag{10.114}$$

In computing the matrix elements, the singular integrals must of course be evaluated differently. The treatment of the singularity are analogous to the procedure described in Chapter 4.

10.4.5 Numerical Inversion of the Laplace Transform

In the LTBEM approach, the numerical inversion of the LT is a key issue. The LT technique has been efficiently applied in conjunction with different numerical methods such as finite difference and finite element methods for the solution of ground water flow and solute transport problems (Sudicky [265], Moridis and Reddell [186]), and heat conduction problems (Chen and Chen [50], Chen and Lin [51]). In these papers different Laplace inversion algorithms such as those of Talbot [272], Dubner and Abate [81], Durbin [84], Crump [62], and Stehfest [263, 264] were used. The advantages and deficiencies of some algorithm were pointed out in Maillet *et al.* [177]. Davies and Martin [68] made a critical study of the various algorithms. Later Duffy [82] examined three popular methods for numerical inversion of the Laplace transform, i.e. direct integration [82], Week's [285] method and Talbot's method [272].

As LT inversion is an ill-posed problem, small truncation errors can be greatly magnified in the inversion process, leading to poor numerical results. In recent times, Moridis and Reddell [186] showed that Stehfest's algorithm poses no such problems and high accuracy may be achieved. Subsequently Zhu *et al.* [302], and Satravaha and Zhu [301] had similar success using numerical inversion of LT in BEM problems. Maillet *et al.* [177] critically reviewed the Stehfest's algorithm and pointed out its advantages and disadvantages. For the present study, a computer code has been written following Stehfest's algorithm [263, 264].

The Stehfest's algorithm originates from Gaver [100]. If $\tilde{P}(s)$ is the Laplace Transform of $F(t)$ then an approximate value F_a of the inverse $F(t)$ for a specific time $t = T$ is given by

$$F_a = \frac{\ln 2}{T} \sum_{i=1}^N V_i \tilde{P} \left(\frac{\ln 2}{T} i \right) \tag{10.115}$$

where

$$V_i = (-1)^{N/2+i} \sum_{k=\frac{i+1}{2}}^{min(i,N/2)} \frac{k^{N/2}(2k)!}{(N/2-k)!k!(k-1)!(i-k)!(2k-i)!} \qquad (10.116)$$

Equations (10.115) and (10.116) correspond to the final form used in the numerical implementation.

When inverting a function from its Laplace transform, one should compare the results for different N, to verify whether the function is smooth enough, to observe the accuracy, and to determine an optimum value of N. Originally Stehfest suggested $N = 10$ for single precision arithmetic, however, Moridis and Reddell [186], Sutradhar and Paulino [202] and Zhu et $al.$ [302] found no significant change in their results for $6 \leq N \leq 10$. In the present calculations, $N = 10$ was adopted.

Most of the methods for the numerical inversion of the LT require the use of complex values of the LT parameter, and as a result the use of complex arithmetic leads to additional storage and an increase in computation time. The disadvantage of using complex arithmetic has been overcome in Stehfest's method. It uses only real arithmetic and thus produces significant reduction in storage together with an increased efficiency in computation time.

10.4.6 Numerical Examples

As noted above, the integral equation is numerically approximated via the non-symmetric Galerkin BEM Method. Standard 6-node isoparametric quadratic triangular elements are used to interpolate the boundary geometry and boundary functions for the physical variables. For all the examples, $N = 10$ is used for the Laplace inversion algorithm using the Stehfest's Method.

Consider a unit cube with initial and boundary conditions given by

$$\begin{aligned}
\phi(x, y, 0; t) &= 0 \\
\phi(x, y, 1; t) &= 100 \\
\phi(x, y, z; 0) &= 0, \qquad (10.117)
\end{aligned}$$

and the remaining faces are insulated. The thermal conductivity and the specific heat are taken to be

$$\begin{aligned}
k(x, y, z) &= k_0 e^{2\beta z} = 5e^{3z} & (10.118) \\
c(x, y, z) &= c_0 e^{2\beta z} = 1e^{3z} & (10.119)
\end{aligned}$$

The analytical solution for temperature is,

$$\begin{aligned}
\phi(x, y, z; t) &= \phi_s(x, y, z) + \phi_t(x, y, z; t) \\
&= T \frac{1 - e^{-2\beta z}}{1 - e^{-2\beta L}} + \sum_{n=1}^{\infty} B_n \sin \frac{n\pi z}{L} e^{-\beta z} e^{-\left(\frac{n^2\pi^2}{L^2} + \beta^2\right)\alpha t} \qquad (10.120)
\end{aligned}$$

where L is the dimension of the cube (in the z-direction) and the analytical solution for flux is,

$$\begin{aligned} q(x,y,z;t) &= -k(x,y,z)\frac{\partial \phi}{\partial z} \\ &= -k(x,y,z)\left[\frac{2\beta T e^{-2\beta z}}{1-e^{-2\beta L}}\right. \\ &\left. + \sum_{n=1}^{\infty} B_n e^{-\beta z} e^{-\left(\frac{n^2\pi^2}{L^2}+\beta^2\right)\alpha t}\left(\frac{n\pi}{L}\cos\frac{n\pi z}{L} - \beta\sin\frac{n\pi z}{L}\right)\right] \end{aligned}$$
(10.121)

where

$$B_n = -\frac{2Te^{\beta L}}{\beta^2 L^2 + n^2\pi^2}\left[\beta L \sin n\pi \frac{1+e^{-2\beta L}}{1-e^{-2\beta L}} - n\pi \cos n\pi\right] \quad (10.122)$$

The Galerkin BEM mesh has 10 elements along each edge totaling to 200 elements for each face and 1200 elements for the cube. Numerical solutions for the temperature profile at different times are shown in Figure 10.13. Notice that the temperature variation matches the analytical solution. Figure 10.14 shows the change of flux with time. At the top face the flux rapidly approaches the steady state flux, while on the bottom face where the temperature is zero, the flux gently approaches to the steady state flux. It is worth observing that the flux from the Galerkin BEM matches the analytical solution within plotting accuracy.

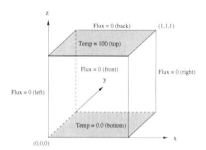

Figure 10.12 Geometry and boundary conditions of the FGM cube problem with constant temperature on two planes. The faces with prescribed temperature are shaded.

10.5 CONCLUDING REMARKS

Symmetric Galerkin boundary element analysis can be successfully applied to FGMs. For exponentially graded (nonhomogeneous) materials, the FGM Green's function is determined and a boundary-only formulation is obtained. The numerical results presented in this chapter indicate that it is feasible to implement the complicated FGM Green's function (and its derivatives) in a standard boundary integral (symmetric Galerkin) approximation, and accurate results are obtained. In particular, the present SGBEM for FGMs can also handle crack geometries [25] as the hypersingular equation has been successfully implemented. Also it was shown that using

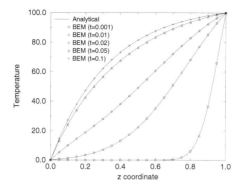

Figure 10.13 Temperature profile in z direction at different time levels for the FGM cube problem with constant temperature on two planes.

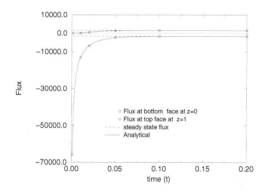

Figure 10.14 Change of flux with time for the FGM cube with constant temperature on two planes.

a Laplace Transform BEM the transient heat conduction problem can be solved as well. The principal computational difficulty with the LTBEM is the numerical inversion of the transform, which was handled accurately and efficiently by Stehfest's algorithm. In these Green's function based approach, each material variation requires different fundamental solution to be derived and consequently, new computer codes to be developed. In order to alleviate this constraint a "simple" Galerkin boundary element method is proposed by Sutradhar [267] where the nonhomogeneous problems can be transformed to known homogeneous problems for a class of variations (quadratic, exponential and trigonometric) of thermal conductivity. The material property can have a functional variation in one, two and three dimensions. Recycling existing codes for homogeneous media, the problems in nonhomogeneous media can be solved maintaining a pure boundary only formulation. This method can be used for any problem governed by potential theory. Galerkin boundary element method formulations for steady state and transient heat conduction, and fracture problems involving multiple interacting cracks in three-dimensional graded material systems can be obtained [213, 268, 269].

CHAPTER 11

BEAN: BOUNDARY ELEMENT ANALYSIS PROGRAM

Synopsis: This chapter is devoted to a description of the educational MATLAB computer code Boundary Element ANalysis, including its graphical user interface. BEAN is a user-friendly adaptive symmetric Galerkin BEM code to solve the two-dimensional Laplace Equation. This chapter outlines both the specific procedures to set up 2-D Laplace problems and the steps to utilize BEAN's post-processing capabilities. The book web-site contains additional related material *i.e.* example library and a tutorial video.

11.1 INTRODUCTION

BEAN is an adaptive symmetric Galerkin BEM code to solve 2D Laplace Equation problems. It can solve single and multi-region problems of arbitrary geometry. BEAN is written closely following the algorithm and notations used in the book. The self-adaptive meshing strategy introduced in Chapter 8 is used in the formulation. More examples and a tutorial video are included in the book website at `http://www.ghpaulino.com/SGBEM_book`. This chapter is a very brief user manual for the BEAN computer code. In the following sections, the graphical user interface and details on how to setup a problem are explained.

The Symmetric Galerkin Boundary Element Method. By Sutradhar, Paulino and Gray
ISBN 978-3-540-68770-2   ©2008 Springer-Verlag Berlin Heidelberg

11.2 MAIN CONTROL WINDOW: BEAN

BEAN is the control window for the boundary analysis program. Pre-processing procedures and analysis are performed through this window.

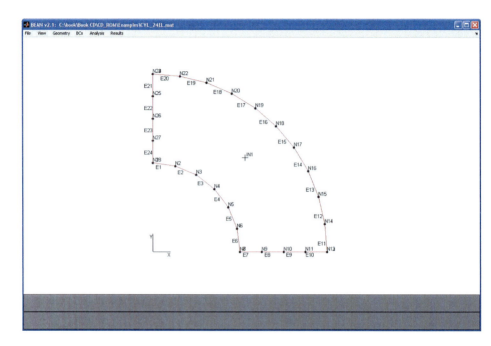

Figure 11.1 Snapshot of BEAN

11.2.1 Menu

The following describes the menus found in BEAN.

11.2.2 File

Standard file options including open, save as, save, new, and quit.

11.2.3 Geometry

This menu allows the User to set up the geometry for a 2-D Laplace problem. The options described here are accessed through BEAN's Geometry menu.

Define Boundary Nodes
   The user must enter numbers for both the x and y coordinates. Blanks are not accepted. Select APPLY to create the boundary node. Nodes will appear on screen as dots with appropriate node labels.

MAIN CONTROL WINDOW: BEAN **229**

Remove Boundary Nodes

Removes existing boundary nodes. Use the mouse to highlight unattached node(s) and press APPLY to remove. Otherwise, press ALL UNATTACHED to choose all free nodes. Pressing CLEAR will deselect all currently selected nodes.

Define Elements

Defines an element by its first node (ith node) and last node (jth node). After highlighting first node, highlight last node and select APPLY to define the element. User can toggle between which node (i.e. first or last) is currently being defined by selecting either the Node i radio button or the Node j radio button, respectively. Elements will appear on screen as dot-dashed lines with appropriate element labels.

Subdivide Elements

Subdivides element(s) into a specified number of elements using a specified ratio. Use mouse to highlight element(s) or select ALL to choose all elements. Enter a RATIO in the form of m:n indicating that the last division's length is (n/m) that of the first division. DIVISIONS indicate how many elements into which the original element is to be divided. Select APPLY to generate subdivisions.

Remove Elements

Removes existing element(s). Highlight an element to be removed or select ALL to highlight all existing elements. Select APPLY to remove selected elements.

Renumber Nodes and Elements

Renumbers the existing mesh in a standard format acceptable to the analysis routine. Select double nodes (if any) by highlighting desired nodes. Select APPLY to renumber the mesh. BEAN will select the furthest node from the origin that is located in quadrant three as the start node (Node 1) and will number in a counter-clockwise fashion for external boundary nodes and a clockwise fashion for internal boundary nodes (i.e. holes). Double nodes will be labeled twice. Elements will now change from dot-dashed to uniformly dashed lines. After the procedure is complete, the number of elements should equal the number of nodes less the number of double nodes.

Define Interior Node

Defines a node within the interior (i.e. the domain) of the boundary so that the potential and flux can be calculated at that specified point. As with defining a boundary node, indicate the x and y coordinates of the interior node in the corresponding sections. Blanks are not accepted. Select APPLY to create interior node. Interior nodes will appear on screen as plus signs with appropriate interior node labels.

Remove Interior Node

Removes existing interior nodes. Select interior node(s) by highlighting node or by selecting ALL. Select APPLY to remove highlighted nodes.

11.2.4 Boundary Conditions (BCs)

Define Boundary Node Conditions

Defines boundary conditions and boundary values for boundary nodes. Highlight desired element(s), select a boundary condition from pop-up menu, and enter its value for the chosen element(s). Each element is associated with one boundary

230 BEAN: BOUNDARY ELEMENT ANALYSIS PROGRAM

condition (i.e. either potential or flux) and each node of the element has its own boundary value associated to the boundary condition of the element. Select APPLY to attach boundary conditions and boundary values to the highlighted elements. Elements that have boundary conditions associated with them will be highlighted yellow.

11.2.5 Analysis

Menu in which User selects the type of analysis to be performed..

Laplace
Performs BEM using symmetric Galerkin method. Select the maximum number of iterations for the adaptive mesh procedure (maximum 6 iterations). If 6 iterations is selected and more iterations are required, then by selecting Analysis on the final mesh, another 6 iterations can be done. Select APPLY to begin the analysis. A separate window will appear that will display each iterative mesh, starting with the initial mesh and ending with the final mesh. Once the analysis is complete, the final mesh will be redrawn in the main window.

11.2.6 Results

Menus accessing BEAN's post-processing capabilities.

Condition Number of System Matrix
Displays the condition number of the system matrix. The condition number measures the loss of significant digits of accuracy that can result when solving ill-conditioned matrices. Generally, the higher the condition number, the less accurate the results will be.

Elapsed CPU Time
Displays total amount of computer time needed to complete analysis, which includes both the BE procedure and the adaptive meshing procedure.

View Output
Displays output of analysis onto the MATLAB window. Output consists of boundary node number, it's x-coordinate, y-coordinate, potential, and flux. If any interior nodes were specified, it's node number, x-coordinate, y-coordinate, potential, x-gradient, and y-gradient will be displayed.

Line Plot
Calls the plotting program, BEANPlot. Refer to Section 11.3 for further details.

Contour Plot
Calls the contour program BEANContour. Refer to Section 11.4 for further details.

View Mesh Adaptation
Displays the meshes used in the adaptive mesh procedure if analysis has been performed.

11.3 BEANPLOT

BEANPlot displays the output from BEAN in a 2-D plot. It plots either the potential or flux against the arc length of the chosen section of boundary.

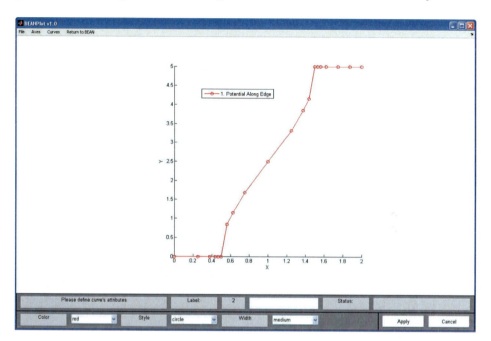

Figure 11.2 Snapshot of BEAN plot

11.3.1 Menus

The following are descriptions of the menus found in BEANPlot.

Print Photo
  Allows User to print the current figure including all its respective graphics.

Axes
  Options to control the appearance of the axes in the window.

X-Attributes
  Adjusts the appearance of the X-axis. LABEL provides a label for the axis. The default is "X". MIN and MAX allow the User to control the minimum and maximum values on the X-axis that are shown in the window. The User can control how many ticks are shown on the axis by specifying how many DIVISIONS are wanted. The default is 10 divisions. Select APPLY to apply changes.

Y-Attributes
  Adjusts the appearance of the Y-axis.

Fit Axes
  Adjusts both the X and Y-axis properties so that the entire plot is shown in window.

232 BEAN: BOUNDARY ELEMENT ANALYSIS PROGRAM

11.3.2 Curves

Menu that sets up the properties for the actual curve.

Define X-Data

Defines the X values for the current plot. START NODE indicates the node at which the User wants to get the first Y value. END NODE indicates the node at which the User wants to get the last Y value. The X-axis will be the distance along the boundary from the START NODE to the END NODE (i.e. the arc length between START NODE and END NODE). Select APPLY to define the X-data according to these choices.

Define Y-Data

Defines the Y values for the current plot. Y values can either be POTENTIAL or FLUX. Select either through the pop-up menu and select APPLY to define Y-data.

Generate Curve

Generates a curve according to the X-data, Y-data, and the selected curve attributes. The User can provide a LABEL and choose the COLOR, line STYLE, and line WIDTH through the pop-up menus, for the current curve. Select APPLY to generate the curve with these properties.

Remove Curve

Removes an existing curve. Select the number of the curve through the arrow buttons and select APPLY to delete the curve.

Return to BEAN

Brings the main window, BEAN, to the front.

11.4 BEANCONTOUR

BEANPlot converts analysis data from BEAN to a contour plot.

11.4.1 Menus

The following are descriptions of the menus found in BEANContour.

File

Print Photo

Allows User to print current figure including all its respective graphics.

Properties

Menu with options to control the properties of the contour.

Color

Adjusts the COLORMAP and number of LEVELS used in the contour. COLORMAP changes the spectrum that will be used in the contour. LEVELS corresponds to the number of isolines that will appear in the contour. The higher the LEVELS, the wider the range of colors that will be used in the spectrum.

Contour

Menu that contains options to set up the data used to create the contour.

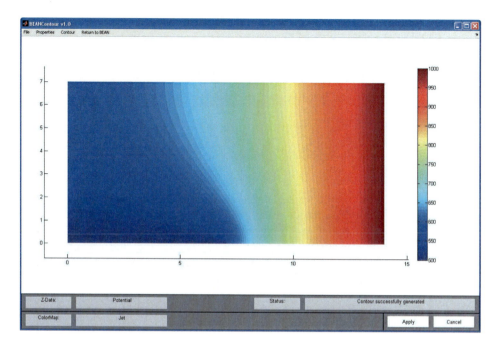

Figure 11.3 Snapshot of BEAN Contour

Define Z-Data
Defines Z-data for current plot. Select either Potential, X-Gradient, or Y-Gradient from the pop-up menu. Select APPLY to define Z-data.

Generate Contour
Generates the contour, given the specified properties.

Remove Contour
Deletes the current contour

Return to BEAN
Brings the main window, BEAN, to the front.

BEANContour Notes

Grid Data
Currently, the points at which the potential or flux are sampled cannot be manually adjusted. These points are selected by the combination of the X and Y coordinates of the boundary nodes. It takes all unique X and Y coordinates and creates a grid based on these points and calculates the appropriate value at that point. Consequently, the fineness of this grid is dependent on the fineness of the boundary mesh. This generally leads to a grid that is non-uniformly spaced, which in turn will be interpolated to determine the values within the interior. The User should take this into account when interpreting these results.

Geometry
In the current version, BEANContour can generate contours for regions that are bounded only by vertical and horizontal boundaries. This means shapes such

234 BEAN: BOUNDARY ELEMENT ANALYSIS PROGRAM

as circles, triangles, or polygons other than rectangles and squares, will not be contoured correctly. The analysis may still be performed using BEAN but a contour currently cannot be generated using BEANContour. An algorithm for arbitrary shapes and domains is currently being developed.

11.5 GENERAL INSTRUCTIONS

The following are step by step instructions for setting up a 2-D Laplace problem.

Setting up Geometry

Set up as many boundary nodes as necessary to effectively describe the geometry and connect with elements. These elements should appear as dot-dashed lines. At this point, it is not necessary to define nodes/elements in any specific order. For efficiency, use the subdivision option as often as possible. Internal boundaries (i.e. holes) should also be created at this time using the same procedure for creating external boundaries. Once User has discretized the mesh, renumber the mesh. After renumbering, the mesh will be difficult (but not impossible) to alter! If at all possible, make sure the geometry is correct before continuing! Renumber the mesh to assure that BEAN appropriately analyzes the boundary. At this point, any required double nodes can be designated. Double nodes are needed at points where the geometry changes and/or where there are discontinuous boundary conditions. An acceptable mesh is numbered counter-clockwise on the external boundary and clockwise on the internal boundary(s), with double nodes at any locations as described above. An acceptable mesh will be indicated by elements with a uniformly dashed linestyle. Interior nodes can be defined only after the mesh has been correctly renumbered to assure that these nodes lie in the domain specified by the boundary. Interior nodes are not required to run the analysis. However, interior nodes allow the User the option to calculate the potential/flux (for Laplace problems) at points other than the boundaries. At this point, the setup of the geometry is complete. Note the order of the nodes and elements. Node i is the first node of an element and Node j is the last node of an element. Nodes and elements are ordered counterclockwise on the external boundary and clockwise on the internal boundaries. Remember this orientation to correctly define boundary conditions and values.

Defining Boundary Conditions and Values

In a Laplace problem, the User has two choices of boundary conditions for each element: potential or flux. Only one boundary condition can be assigned to a specific element. At each node of an element, the User needs to assign a value for the corresponding boundary condition. If the values at both end nodes are equal, then the value along the element is assumed to be constant. If the values at the end nodes are different, then the boundary value is assumed to vary linearly along the element. Once these conditions are defined for a specific element, that element will be highlighted yellow.

At Elements Attached to Only Single Nodes

The User needs to remember that Node i of this type of element is the same as Node j of the previous element. Similarly, Node j of the current element corresponds to Node i of the next element. Therefore, the boundary condition, as well as its corresponding value, must be equal at these nodes. There can not be any discontinuities in the boundary conditions and boundary values at single nodes. If required, place double nodes at discontinuous points.

At Elements Attached to Double Nodes

Double nodes do not carry the same restrictions as single nodes. Double nodes indicate points that allow discontinuities in boundary conditions (or values) and/or changes in geometry. Therefore double nodes allow the User to define different boundary conditions to adjacent elements. However, note that if the potential is defined for both elements, the corresponding value at the double node must be equal since potential is a scalar value (i.e. temperature).

Running Analysis

When the previous steps have been taken correctly, the mesh will be colored yellow. This indicates that the problem is ready to be analyzed using BEAN. The User can choose how many iterations the adaptive mesh routine will perform (up to six iterations). One iteration means the current mesh will be analyzed without any adaptation. Once the analysis is started, a separate window will appear displaying each iterative mesh with its current number of elements. This status window will allow the User to follow BEAN as it refines each mesh. Note that analysis time depends on both the speed of the computer and the size of the mesh. When the analysis is complete, the final mesh will be redrawn onto the main window. The User can now access BEAN's post-processing features to visualize the results.

Post-Processing Functions The User has several options to view the output of the analysis.

Analysis Output

A printout of the analysis consists of the boundary node coordinates, specified boundary conditions, and corresponding potential and flux values. If any internal nodes were specified, the same information is provided with the exception of flux, which is calculated in both the X and Y-directions.

Line Plots

The User may access BEANPlot to generate line plots along any part of the boundary of either the potential or flux versus the arc length. The plots are limited to only the boundaries; start and end nodes for arc must be those found on the boundaries! To see results for interior, use BEANContour.

Contour Plots

BEANContour generates contours using results from BEAN, allowing the User to visualize values within the boundary domain. However, BEANContour is currently limited to meshes composed of only horizontal and vertical lines. The User may try to contour these types of meshes, but it will result in a slightly distorted shape. With a finer mesh, however, the contour will approach the true shape.

11.6 TROUBLESHOOTING

This section is dedicated to explaining error messages that may appear in BEAN and what to do in those situations.

236 BEAN: BOUNDARY ELEMENT ANALYSIS PROGRAM

11.6.1 Error Message Meanings

Not a Valid Number

Most text inputs needed in BEAN require real numbers. Any characters other than those that constitute real numbers will be rejected.

Node Already Exists at Those Coordinates

Boundary Nodes

While setting up initial mesh, only one boundary node can be specified at any given coordinate. This ensures that the renumbering routine of BEAN will work. If the User is trying to create a double node, this must be specified at the beginning of the renumbering routine. This is the only time at which two nodes can be placed on the same coordinates.

Interior Nodes

Duplicate interior nodes do not adversely affect the analysis. They do not provide any special purpose, either, except to give results at a specific point twice. Therefore, duplicate interior nodes are of no use.

No Nodes/Elements Selected

No nodes/elements were highlighted. Select a node/element by clicking it with the mouse.

Element Already Exists with Those End Nodes

An element can not be defined if another element exists with the selected end nodes. Direction (i.e. which node is designated as Node i or Node j), in this case, does not matter. Each element must be unique in terms of its end nodes.

Node(s) has Two Elements Already Attached to it

Each node can have at most two elements attached to it. The User will not be able to define an element with an end node that has two or more elements connected to it. Each node requires exactly two elements connected to it in order for the mesh to be renumbered.

Mesh is Not Correctly Defined

In order for a mesh to be renumbered, it must be completely closed. This means each node is connected to exactly two elements and that the boundary encloses a domain, which may include holes.

Mesh Must be Renumbered First

When defining an interior node, the mesh must be numbered correctly to ensure correct interior node definition. It is also to ensure that the steps outlined are being followed correctly. The User can not define boundary conditions and consequently complete the analysis without first properly renumbering the mesh.

Interior Node on Boundary

Interior nodes can not be placed on the boundary. This procedure can not be used to obtain values along the boundary, aside from those values calculated at boundary nodes. If this is required, define a boundary node at a coordinate instead.

Interior Node Outside of Boundary

Interior nodes clearly can not be placed outside of the boundary.

Inconsistent Boundary Conditions

Depending on the element, this error may mean two things:

For Elements Attached to Only Single Nodes: Make sure boundary conditions and boundary values are consistent with any attached elements. Remember that there can not be any discontinuities at single nodes. Adjacent elements must have the same boundary conditions as well as the same boundary value at shared nodes.

For Elements Attached to Double Nodes: If potential is specified as the boundary condition for both sides of the double node, the boundary value must be the same. Since potential is a scalar value (i.e. temperature), it can not be discontinuous at a point. In other words, the temperature can not be two different values at identical coordinates.

Boundary Conditions Need to be Defined

Analysis can not be completed without first defining boundary conditions and values for all elements. A mesh with acceptable boundary conditions and values will be colored yellow, indicating that the analysis can be performed.

Changes in mesh

Once the mesh has been renumbered, can BEAN incorporate changes to the geometry of the mesh? Yes! To do this, delete one of the elements that are attached to each existing double node. This will free up one of the double nodes and cause the elements to revert back to the dot-dashed Linestyle, indicating that the mesh needs to be renumbered again. Delete the extra nodes by either highlighting them or using the ALL UNATTACHED function. The geometry can now be altered in any fashion. However, the User will be required to redefine the boundary conditions. As one can see, this process is not particularly efficient. The best suggestion is to make sure the geometry is correct before renumbering.

11.7 SAMPLE PROBLEMS

The book website mentioned above, has different interesting examples. One example on heat conduction in a transformer coil is shown here. The geometry and the boundary conditions are shown in the Figure 11.4. The initial mesh, contour plot, mesh adaptation of the problem are shown in Figures 11.5, 11.6 and Figure 11.7, respectively.

238 BEAN: BOUNDARY ELEMENT ANALYSIS PROGRAM

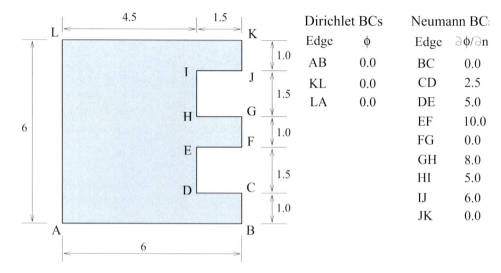

Figure 11.4 Geometry and boundary condition

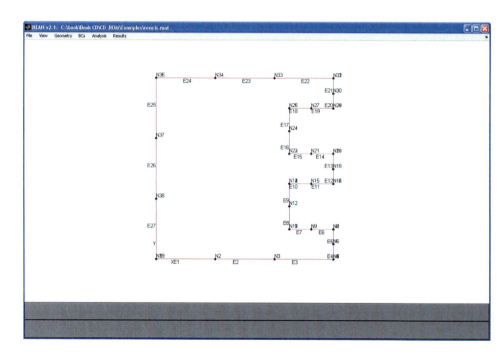

Figure 11.5 Initial mesh

SAMPLE PROBLEMS 239

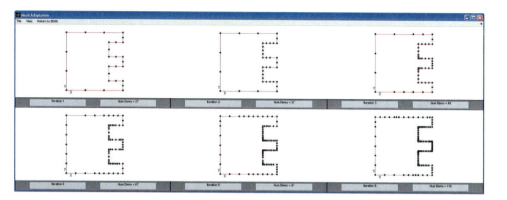

Figure 11.6 Mesh adaptation of the coil mesh

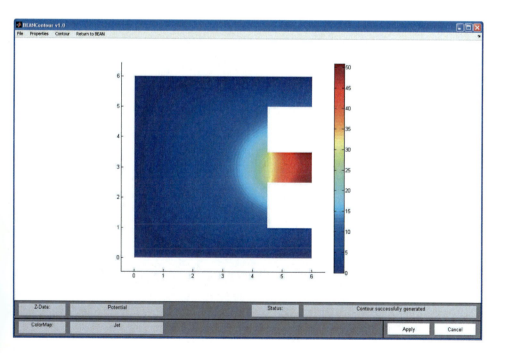

Figure 11.7 Contour plot of temperature in the coil

APPENDIX A

MATHEMATICAL PRELIMINARIES AND NOTATIONS

In this Appendix, a limited selection of mathematical results and formulas employed in the text are presented. Proofs and further details can be found in textbooks on advanced calculus, linear algebra and differential equations.

A.1 DIRAC DELTA FUNCTION

The mathematical representation of the concentrated force can be achieved with the aid of the Dirac Delta function $\delta(P, Q)$. The main feature of the Dirac delta function is that it is zero at all Q points, except at $P = Q$, where it becomes infinity. Thus it represents a point singularity at the source point P, i.e.

$$\delta(P, Q) = \begin{cases} \infty & \text{for } P = Q \\ 0 & \text{otherwise} \end{cases} \tag{A.1}$$

Technically the Dirac Delta function should only appear inside an integral, and its defining properties are

$$\int_V f(Q)\delta(P, Q)dV_Q = \begin{cases} f(P) & \forall P \in V \\ 0 & \forall P \notin V \end{cases} \tag{A.2}$$

The Symmetric Galerkin Boundary Element Method. By Sutradhar, Paulino and Gray **241**
ISBN 978-3-540-68770-2 ©2008 Springer-Verlag Berlin Heidelberg

242 MATHEMATICAL PRELIMINARIES AND NOTATIONS

$$\int_V f(Q)\frac{\partial}{\partial x_j}\delta(P,Q)dV_Q = \begin{cases} -\frac{\partial}{\partial x_j}f(P) & \forall P \in V \\ 0 & \forall P \notin V \end{cases} \qquad (A.3)$$

Application of δ-functions simplifies the derivation of BEM formulations.

A.2 KRONECKER DELTA FUNCTION

Kronecker Delta δ_{ij} represents the identity or unit matrix, i.e.,

$$\delta_{ij} = \begin{cases} 1 & \text{if } i = j \\ 0 & \text{if } i \neq j \end{cases} \text{ (Similar to the identity matrix)} \qquad (A.4)$$

Using this definition,

$$f_{,i}\delta_{ij} = f_{,j},$$
$$a_i b_j \delta_{ij} = a_i b_i,$$
$$A_{ij}\delta_{jk} = A_{ik},$$
$$\delta_{ij}\delta_{jk}\delta_{kl} = \delta_{il}.$$

if x_i is a set of N independent variables, then

$$\frac{\partial x_i}{\partial x_j} = \delta_{ij}$$

A.3 DERIVATIVE, GRADIENT, DIVERGENCE AND LAPLACIAN

The Cartesian components of the ∇ operator are the partial derivatives with respect to corresponding coordinates,

$$\nabla \equiv \left(\frac{\partial}{\partial x}, \frac{\partial}{\partial y}, \frac{\partial}{\partial z}\right). \qquad (A.5)$$

In two dimensions, only the x and y derivatives appear.

If ϕ is a scalar function, then its gradient $\nabla\phi$ is a vector defined as

$$\nabla\phi = \mathbf{e}_x\frac{\partial}{\partial x}\phi + \mathbf{e}_y\frac{\partial}{\partial y}\phi + \mathbf{e}_z\frac{\partial}{\partial z}\phi, \qquad (A.6)$$

where e_i are the unit Cartesian vectors.

The derivative in the direction of the normal $\mathbf{n}$ at e.g., point Q, for a scalar quantity ϕ is

$$\frac{\partial\phi}{\partial\mathbf{n}}(Q) = \nabla\phi \cdot \mathbf{n}(Q), \qquad (A.7)$$

The divergence of a vector is a scalar quantity defined as follows

$$\nabla\mathbf{u} = \frac{\partial\mathbf{u}_x}{\partial x} + \frac{\partial\mathbf{u}_y}{\partial y} + \frac{\partial\mathbf{u}_z}{\partial z} \qquad (A.8)$$

The Laplacian of a scalar function ϕ is a scalar quantity defined by the divergence of the gradient of the function as

$$\nabla \cdot (\nabla\phi) \equiv \nabla^2 = \frac{\partial^2 u}{\partial x^2} + \frac{\partial^2 u}{\partial y^2} + \frac{\partial^2 u}{\partial z^2} \qquad (A.9)$$

A.4 DIVERGENCE THEOREM

Let V be a volume that is bounded by the closed surface B, and $\mathbf{n}$ be the unit vector that is normal to B pointing outward. The Gauss Divergence theorem states that the volume integral of the divergence of any differentiable vector function $\mathbf{F}$ over V is equal to the flow rate of $\mathbf{F}$ across B. The Gauss theorem thus relates the volume integral to a surface integral.

$$\int_B \nabla \cdot \mathbf{F} \, dV = \int_{\partial B} \mathbf{F} \cdot \mathbf{n} \, ds = \tag{A.10}$$

$$or \int_B F_{i,i} \, dV = \int_{\partial B} F_i n_i$$

provided $\mathbf{F}$ and its derivatives are continuous in B and, on ∂B if ∂B is piecewise smooth. B may be multiply connected but the complete boundary must be included in the surface integral.

A.5 STOKES THEOREM

$$\oint_{\partial B} \mathbf{F} \cdot d\mathbf{r} = \int_B \mathbf{n} \cdot (\nabla \times \mathbf{F}) \, ds$$

where B is contained in a simple region and $\mathbf{F}$ is continuously differentiable.

If B is a plane area

$$\oint_{\partial B} P \, dx + Q \, dy = \int_B \left(\frac{\partial Q}{\partial x} - \frac{\partial P}{\partial y} \right) dx \, dy$$

with $\mathbf{F} = P\mathbf{i} + Q\mathbf{j}$

A.6 GREEN'S IDENTITIES

If u and v are scalar functions then $u\nabla v$ and $v\nabla u$ are two vector fields. So the divergence of these vector fields are

$$\nabla \cdot (u\nabla v) = \nabla u \cdot \nabla v + u\nabla^2 v \tag{A.11}$$

$$\nabla \cdot (v\nabla u) = \nabla v \cdot \nabla u + v\nabla^2 u \tag{A.12}$$

From the divergence theorem

$$\int_B \nabla \cdot \mathbf{F} \, dV = \int_{\partial B} \mathbf{F} \cdot \mathbf{n} \, ds \tag{A.13}$$

Plugging Eq.(A.12) into Eq.(A.13)

$$\int_B \left(\nabla u \cdot \nabla v + u\nabla^2 v \right) dV = \int_{\partial B} u \frac{\partial v}{\partial \mathbf{n}} \, ds \tag{A.14}$$

244   MATHEMATICAL PRELIMINARIES AND NOTATIONS

This is Green's 1st Identity.

Subtracting Eq.(A.12) from Eq.(A.11) and using Eq.(A.13), we get

$$\int_B \left(u\nabla^2 v - v\nabla^2 u \right) \, dV = \int_{\partial B} \left(u\frac{\partial v}{\partial \mathbf{n}} - v\frac{\partial u}{\partial \mathbf{n}} \right) \, ds \qquad \text{(A.15)}$$

This is Green's 2nd Identity.

A.7 FOURIER AND LAPLACE TRANSFORM

In many time dependent problem involving governing equations which are linear in the time variable and have time-independent coefficients it is possible to remove the time dependence in the equations by taking a Laplace ($\mathcal{L}$) or Fourier transform ($\mathcal{F}$). The resulting equations may then be solved in the transform space with the usual boundary element method. These equations are usually hyperbolic (or parabolic) in type. Once the equations are solved in the transformed space the original variable involving the time dependence may be recovered by inverting the transformation numerically. The type of transform used, either Laplace or Fourier will depend on the type of solution sought. As shown below, the Fourier Transform is also useful for deriving Green's functions.

If $f(t)$ is a piecewise differential function, the Fourier transform $\mathcal{F}(\omega)$ of $f(t)$ by

$$\mathcal{F}(\omega) \equiv \frac{1}{\sqrt{2\pi}} \int_{-\infty}^{\infty} f(t)e^{-i\omega t}dt \qquad \text{(A.16)}$$

The Laplace transform is an integral transform perhaps second only to the Fourier transform in its utility in solving physical problems. The Laplace transform is particularly useful in solving linear ordinary differential equations such as those arising in the analysis of electronic circuits. The Laplace transform is defined by

$$\mathcal{L}(\omega) \equiv \frac{1}{\sqrt{2\pi}} \int_0^{\infty} f(t)e^{-st}dt \qquad \text{(A.17)}$$

where $f(t)$ is a piecewise differential function.

A.8 FREE SPACE GREEN'S FUNCTION

In a region V with boundary B,

$$L(u) = f(x) \text{ in V} \qquad \text{(A.18)}$$

$$B(u) = 0 \text{ on B}, \qquad \text{(A.19)}$$

represent a linear second order partial differential equation and a linear boundary condition, respectively such that the problem has a uniques solution for each continuous f. Then $G(P, Q)$ is the Green's function for this problem if the unique solution is given by

$$u(\mathbf{x}) = \int_V G(P, Q)f(Q)dQ \qquad \text{(A.20)}$$

where dQ denotes the integration with respect to Q.

If $f(Q)$ is replaced by the Dirac delta function $\delta(Q, P)$, then the Green's function $G(P, Q)$ represents the free space Green's function or the fundamental solution for the problem

Laplace Green's Function. A derivation of this Green's function for the Laplace equation is given below. As with other Green's functions, this is accomplished using the Fourier Transform and its inverse

$$\hat{f}(\boldsymbol{\omega}) = \int_{\mathcal{R}^3} f(\mathbf{x}) e^{-i\boldsymbol{\omega}\bullet\mathbf{x}}\, d\mathbf{x} \tag{A.21}$$

$$f(\mathbf{x}) = \frac{1}{(2\pi)^3} \int_{\mathcal{R}^3} \hat{f}(\boldsymbol{\omega}) e^{i\boldsymbol{\omega}\bullet\mathbf{x}}\, d\boldsymbol{\omega}$$

Taking the Fourier Transform of $\nabla^2 G(P,Q) = -\delta(P,Q)$ and setting $\omega^2 = \|\boldsymbol{\omega}\|^2$, we obtain

$$-\omega^2 \hat{G}(\boldsymbol{\omega}, P) = -e^{-i\boldsymbol{\omega}\bullet P}$$

$$\hat{G}(\boldsymbol{\omega}, P) = \frac{e^{-i\boldsymbol{\omega}\bullet P}}{\omega^2} \tag{A.22}$$

$$G(P,Q) = \frac{1}{(2\pi)^3} \int_{\mathcal{R}^3} \frac{e^{i\boldsymbol{\omega}\bullet(Q-P)}}{\omega^2}\, d\boldsymbol{\omega}$$

To compute the Fourier integral, employ spherical coordinates with the z (polar) axis chosen as the $\mathbf{R}/r$ direction.

$$\begin{aligned} G(P,Q) &= \frac{1}{(2\pi)^3} \int_{\mathcal{R}^3} \frac{e^{i\boldsymbol{\omega}\bullet\mathbf{R}}}{\omega^2}\, d\boldsymbol{\omega} \\ &= \frac{1}{(2\pi)^3} \int_0^\infty d\rho \int_0^{2\pi} d\theta \int_0^\pi \sin(\psi) e^{i\rho r \cos(\psi)}\, d\psi \\ &= \frac{1}{2\pi^2} \frac{1}{r} \int_0^\infty \frac{\sin(r\rho)}{\rho}\, d\rho = \frac{1}{4\pi r} \end{aligned} \tag{A.23}$$

This last integral is a standard calculation from residue calculus,

$$\int_0^\infty \frac{\sin(r\rho)}{\rho}\, d\rho = \pi/2 \tag{A.24}$$

and we obtain the desired result for $G(P,Q)$.

APPENDIX B

GAUSSIAN INTEGRATION

Gaussian integration is employed due to its simplicity, versatility and accuracy. This appendix provides tables of Gauss points and weights for standard and logarithmic Gauss rules.

B.1 GAUSSIAN RULE FOR LOGARITHMIC SINGULARITIES

For the kernels that have a logarithmic singularity, a specific integration rule is required,

$$I = \int_0^1 \log \frac{1}{\xi} f(\xi) d\xi \approx \sum_{i=1}^n w_i f(\xi_i) \tag{B.1}$$

Table B.1 give the abscissae and weights of the Gauss points corresponding to the logarithmic case.

B.2 GAUSSIAN RULE FOR ONE-DIMENSIONAL NON-SINGULAR INTEGRATION

The integrals for this case can be written as,

$$I = \int_0^1 f(\xi) d\xi \approx \sum_{i=1}^n w_i f(\xi_i) \tag{B.2}$$

The Symmetric Galerkin Boundary Element Method. By Sutradhar, Paulino and Gray
ISBN 978-3-540-68770-2 ©2008 Springer-Verlag Berlin Heidelberg

Table B.1 Gauss point abscissae and weights for logarithmic singularity

n	ξ_i	w_i
2	0.1120088061669761	0.7185393190303845
	0.6022769081187381	0.2814606809696154
3	0.06389079308732544	0.5134045522323634
	0.3689970637156184	0.3919800412014877
	0.766880303938942	0.0946154065661483
4	0.04144848019938324	0.3834640681451353
	0.2452749143206022	0.3868753177747627
	0.5561654535602751	0.1904351269501432
	0.848982394532986	0.03922548712995894
5	0.02913447215197205	0.2978934717828955
	0.1739772133208974	0.3497762265132236
	0.4117025202849029	0.234488290044052
	0.6773141745828183	0.0989304595166356
	0.89477136103101	0.01891155214319462
6	0.02163400584411693	0.2387636625785478
	0.1295833911549506	0.3082865732739458
	0.3140204499147661	0.2453174265632108
	0.5386572173517997	0.1420087565664786
	0.7569153373774084	0.05545462232488041
	0.922668851372116	0.01016895869293513
8	0.01332024416089244	0.164416604728002
	0.07975042901389491	0.2375256100233057
	0.1978710293261864	0.2268419844319134
	0.354153994351925	0.1757540790060772
	0.5294585752348643	0.1129240302467932
	0.7018145299391673	0.05787221071771947
	0.849379320441094	0.02097907374214317
	0.953326450056343	0.003686407104036044
9	0.01086933608417545	0.1400684387481339
	0.06498366633800794	0.2097722052010308
	0.1622293980238825	0.211427149896601
	0.2937499039716641	0.1771562339380667
	0.4466318819056009	0.1277992280331758
	0.6054816627755208	0.07847890261203835
	0.7541101371585467	0.0390225049841783
	0.877265828834263	0.01386729555074604
	0.96225055941096	0.002408041036090773
10	0.00904263096219963	0.12095513195457
	0.05397126622250072	0.1863635425640733
	0.1353118246392511	0.1956608732777627
	0.2470524162871565	0.1735771421828997
	0.3802125396092744	0.135695672995467
	0.5237923179723384	0.0936467585378491
	0.6657752055148032	0.05578772735275126
	0.7941904160147613	0.02715981089692378
	0.898161091216429	0.00951518260454442
	0.9688479887196	0.001638157633217673
12	0.006548722279080035	0.09319269144393
	0.03894680956045022	0.1497518275763289
	0.0981502631060046	0.166557454364573
	0.1811385815906331	0.1596335594369941
	0.2832200676673157	0.1384248318647479
	0.398434435164983	0.1100165706360573
	0.5199526267791299	0.07996182177673273
	0.6405109167754819	0.0524069547809709
	0.7528650118926111	0.03007108900074863
	0.850240024421055	0.01424924540252916
	0.926749682988251	0.004899924710875609
	0.977756129778486	0.000834029009809656

Table B.2 give the abscissae and weights of the Gauss points corresponding to the non-singular case.

Table B.2 Gauss point abscissae and weights for one-dimensional non-singular integration

n	ξ_i	w_i
3	0.112701665379258	0.277777777777778
	0.500000000000000	0.444444444444444
	0.887298334620742	0.277777777777778
4	0.069431844202974	0.173927422568727
	0.330009478207572	0.326072577431273
	0.669990521792428	0.326072577431273
	0.930568155797026	0.173927422568727
5	0.046910077030668	0.118463442528095
	0.230765344947158	0.239314335249683
	0.500000000000000	0.284444444444444
	0.769234655052842	0.239314335249683
	0.953089922969332	0.118463442528095
6	0.033765242898424	0.085662246189585
	0.169395306766868	0.180380786524069
	0.380690406958402	0.233956967286346
	0.619309593041598	0.233956967286346
	0.830604693233132	0.180380786524069
	0.966234757101576	0.085662246189585
8	0.019855071751232	0.050614268145188
	0.101666761293187	0.111190517226687
	0.237233795041836	0.156853322938944
	0.408282678752175	0.181341891689181
	0.591717321247825	0.181341891689181
	0.762766204958165	0.156853322938944
	0.898333238706813	0.111190517226687
	0.980144928248768	0.050614268145188
12	0.009219682876640	0.023587668193256
	0.047941371814763	0.053469662997659
	0.115048662902848	0.080039164271673
	0.206341022856691	0.101583713361533
	0.316084250500910	0.116746268269177
	0.437383295744266	0.124573522906701
	0.562616704255734	0.124573522906701
	0.683915749499090	0.116746268269177
	0.793658977143309	0.101583713361533
	0.884951337097152	0.080039164271673
	0.952058628185237	0.053469662997659
	0.990780317123360	0.023587668193256
24	0.002406390001489	0.006170614899994
	0.012635722014345	0.014265694314467
	0.030862723998634	0.022138719408710
	0.056792236497800	0.029649292457718
	0.089999007013049	0.036673240705540
	0.129937904210723	0.043095080765977
	0.175953174031512	0.048809326052057
	0.227289264305580	0.053722135057983
	0.283103246186977	0.057752834026863
	0.342478660151918	0.060835236463902
	0.404440566263192	0.062918728173414
	0.467971553568697	0.063969097673376

Table B.2 (continued)

n	ξ_i	w_i
	0.532028446431303	0.063969097673376
	0.595559433736808	0.062918728173414
	0.657521339848082	0.060835236463902
	0.716896753813023	0.057752834026863
	0.772710735694420	0.053722135057983
	0.824046825968488	0.048809326052057
	0.870062095789277	0.043095080765977
	0.910000992986951	0.036673240705540
	0.943207763502200	0.029649292457718
	0.969137276001366	0.022138719408710
	0.987364277985655	0.014265694314467
	0.997593609998511	0.006170614899994

APPENDIX C
MAPLE CODES FOR TREATMENT OF
HYPERSINGULAR INTEGRAL

The following codes were implemented with Maple. Only a few basic Maple operations, integration and substitution, are employed, and thus it is likely that, with relatively minor changes, these scripts would work with other symbolic computation systems. The code shows the analytical integration to extract the divergent terms for the coincident case and the edge adjacent case. The naming of variables follows the notation in this book fairly closely, and it is therefore hoped that the codes are mostly self-explanatory. The Maple script employed to derive the results in Chapter 4 is exhibited and commented upon. The boundary limit parameter is the variable eps, a stand-in for ε employed above.

C.1 MAPLE SCRIPT: COINCIDENT

```
##
##    Coincident LINEAR  hypersingular
##
##
## define arrays
##
```

The Symmetric Galerkin Boundary Element Method. By Sutradhar, Paulino and Gray **251**
ISBN 978-3-540-68770-2   ©2008 Springer-Verlag Berlin Heidelberg

```
phi    := array(0..2,0..1);
Lterm := array(0..2,0..1):
##
##
## first integration
##
##
rh  := (eps^2 + a^2*rho^2)^(1/2);
JNR := -eps*jp;
##
## phi integral jnJN=jp^2  (JNR)*(jnR) = eps^2*jp^2
##
hyp  := jp^2/rh^3 - 3*JNR^2/rh^5;
for j from 0 to 1 do
## powers for Q shape function

 zz := int(rho^(j+1)*hyp,rho=0..QR);
 zz := subs(sqrt(eps^2)=eps,1/sqrt(eps^2)=1/eps,zz);
  if j=1 then
   zz := subs( ln(eps)=loge,eps=0,
               (a^2*QR^2)^(1/2)=a*QR,
               (a^2*QR^2)^(-1/2)=1/(a*QR),
               (a^2*QR^2)^(3/2)=a^3*QR^3,
               (a^2*QR^2)^(-3/2)=1/(a^3*QR^3),
          zz);
## this loge disappears after integrating theta (st,ct coeffs from shapes)
   zz := subs( loge=0,zz);
  fi;
 zz := subs(sqrt(a^2)=a,1/sqrt(a^2)=1/a,zz);
 zz := factor(expand(zz));
##
## change of variables: theta --> t
## polar coords: t-e=Lambda*cp  x=Lambda*sp
##
##  QR = sqrt(x^2+(t-e)^2) = Lambda
##  ct = (t-e)/sqrt(x^2+(t-e)^2) = cp
##  st = -x/sqrt(x^2+(t-e)^2) = -sp
 zz   := subs(QR=Lambda,zz);
## d(theta)/dt = x/(x^2+(t-e)^2) = sp/Lambda
## 1/Lambda cancels with Lambda d(Lambda)
 zz   := zz*sp;
 zz   := expand(zz);
 zz   := normal(zz);
##  new polar coords
##
## second integration
##  powers of rho for shape function product mp_k*mp_j
##
 for k from 0 to 2 do
  phi[k,j]   := int(Lambda^k*zz,Lambda=0..QL);
##
##  Note: 3 subtriangles for QR to divide rectangle
##  0<x<LE    -e<t<1-e      LE = sq3*(1-e) e>0    LE = sq3*(1+e) e<0
```

```
##  origin at (0,0)
##
phi[k,j]   := subs(sqrt(eps^2)=eps,1/sqrt(eps^2)=1/eps,
                   sqrt(a^2)=a,1/sqrt(a^2)=1/a,
               phi[k,j]);
phi[k,j]   := subs(ln(eps)=loge,phi[k,j]);
phi[k,j]   := collect(phi[k,j],loge):
Lterm[k,j] := subs(eps=0,coeff(phi[k,j],loge));
phi[k,j]   := coeff(phi[k,j],loge,0);
phi[k,j]   := subs(eps=0,phi[k,j]);
phi[k,j]   := subs((a^2*QL^2)^(1/2)=a*QL,
                   (a^2*QL^2)^(-1/2)=1/(a*QL),
               phi[k,j]);
phi[k,j]   := normal(expand(phi[k,j]));
 od;
od;
```

C.2 MAPLE SCRIPT: EDGE ADJACENT

```
##
##   Edge Adjacent LINEAR
##
##
## define arrays
##
phi   := array(0..2);
sng   := array(0..2);
##
rh  := (eps^2 + eps*a1*Lambda + a2*Lambda^2)^(1/2);
##
## kernel
##  Note: cos(psi) factor from jacobian product omitted
##
hyp := Lambda^2*(ndN/rh^3 - 3*(j1p*Lambda-jp*eps)*(j1q*Lambda-eps*ndN/jp)/rh^5);
## powers from shape function product
for ll from 0 to 2 do
 phi[ll] := Lambda^ll*hyp;^
 phi[ll] := int(phi[ll],Lambda=0..QL):
 phi[ll] := expand(phi[ll]);
 phi[ll] := subs( (eps^2)^(1/2)=eps,
                  (eps^2)^(3/2)=eps^3,
                  (eps^2)^(-1/2)=1/eps,
                  (eps^2)^(-3/2)=1/eps^3,
                  ln((a2)^(1/2)*eps*a1+2*eps*a2)=loge+logaa,
##  logaa = log(2*a2+a2s*a1)
             phi[ll]);
 phi[ll] := subs( eps=0,phi[ll]);
 phi[ll] := subs( (a2*QL^2)^(1/2)=a2s*QL,
                  (a2*QL^2)^(3/2)=a2s^3*QL^3,
                  (a2*QL^2)^(-1/2)=a2s^(-1)*QL^(-1),
                  (a2*QL^2)^(-3/2)=a2s^(-3)*QL^(-3),
```

254 MAPLE CODES FOR TREATMENT OF HYPERSINGULAR INTEGRAL

```
                  ln((a2)^(3/2)*QL+a2s*QL*a2)=logQ,
                  ln(a2s^3*QL+a2s*QL*a2)=logQ,
              phi[ll]);
## logQ = log(2*a2s^3*QL)
 phi[ll] := expand(phi[ll]):
 phi[ll] := collect(phi[ll],loge);
 sng[ll] := normal(coeff(phi[ll],loge,1));
 phi[ll] := normal(coeff(phi[ll],loge,0));
 phi[ll] := subs(logQ=logS-ln(2)+logaa,phi[ll]):
## logS = log(4*a2s^3*QL/(2*a2+a2s*a1))
##         log(4*a2*QL/(2*a2s+a1))
 phi[ll] := subs(ln(QL)=logS-2*ln(2)-3*ln(a2s)+logaa,phi[ll]):
 phi[ll] := subs(ln(a2s^3*QL+a2s*QL*a2)=logS-ln(2)+logaa,phi[ll]):
 phi[ll] := subs(a2^(9/2)=a2s^9,
                 a2^(7/2)=a2s^7,
                 a2^(5/2)=a2s^5,
                 a2^(3/2)=a2s^3,
                 a2^(1/2)=a2s,
                 a2=a2s^2,
                 (a2s^2)^(9/2)=a2s^9,
                 (a2s^2)^(11/2)=a2s^11,
                 (a2s^2)^(13/2)=a2s^13,
                 (a2s^2)^(-9/2)=a2s^(-9),
                 (a2s^2)^(-11/2)=a2s^(-11),
              phi[ll]);
 phi[ll] := normal(expand(phi[ll]));
 phi[ll] := subs(ln(QL)=logS-2*ln(2)-3*ln(a2s)+logaa,phi[ll]):
 phi[ll] := normal(expand(phi[ll]));
 if ll=0 then
  num := numer(phi[ll]): num := expand(num):
  den := denom(phi[ll]):
  num := collect(num,[logS,ndN,j1p,j1q,jp]);
  phi[ll] := num/den;
 fi;
od;
```

C.3   MAPLE SCRIPT: VERTEX ADJACENT

```
##
##    Vertex Adjacent LINEAR
##
## define arrays
##
phi   := array(0..2);
flx   := array(0..2);
##
rh   := a2s*Lambda;
##
## kernel
##
hypP :=   Lambda^3*(ndN/rh^3 - 3*(j1p*Lambda)*(j1q*Lambda)/rh^5);
```

```
## powers from shape function product
for ll from 0 to 2 do
 phi[ll] := Lambda^ll*hypP;
 phi[ll] := int(phi[ll],Lambda=0..QL):
 phi[ll] := normal(expand(phi[ll]));
od;
##
hypF :=  Lambda^3*(j1p*Lambda/rh^3);
## powers from shape function product
for ll from 0 to 2 do
 flx[ll] := Lambda^ll*hypF;
 flx[ll] := int(flx[ll],Lambda=0..QL):
 flx[ll] := normal(expand(flx[ll]));
od;
```

REFERENCES

1. R. A. Adey and S. M. Niku. Computer modelling of corrosion using the boundary element method. In *Computer Modelling in Corrosion*, volume ASTM STP 1154, pages 248–269. American Society of Testing and Materials, Philadelphia, 1992.

2. M. H. Aliabadi. Evaluation of mixed mode stress intensity factors using the path independent integral. In *Proceedings 12th International Conference on Boundary Element Methods*, pages 281–292, Southampton, 1990. Computational Mechanics Publications.

3. M. H. Aliabadi. Boundary element formulations in fracture mechanics. *Applied Mechanics Reviews*, 50(2):83–96, 1997.

4. M. H. Aliabadi. *The Boundary Element Method: Vol 2. Application in solids and structures*. John Wiley & Sons, Ltd, England, 2002.

5. T. L. Anderson. *Fracture Mechanics: Fundamentals and Applications*. CRC Press LLC, Boca Raton FL, 1995.

6. H. Andra. Integration of singular integrals for the Galerkin-type boundary element method in 3D elasticity. *Comp. Mech. Appl. Mech. Engng.*, 157:239–249, 1998.

7. W. T. Ang, J. Kusuma, and D. L. Clements. A boundary element method for a second order elliptic partial differential equation with variable coefficients. *Engineering Analysis with Boundary Elements*, 18:311–316, 1996.

8. S. Aoki and K. Kishimoto. Prediction of galvanic corrosion rates by the boundary element method. *Mathematical and Computer Modelling*, 15(3-5):11–22, 1991.

9. S.N. Atluri and T. Zhu. New meshless local Petrov-Galerkin (MPLG) approach in computational mechanics. *Computational Mechanics*, 22:117–127, 1998.

10. S. J. Badger, S. B. Lyon, and S. Turgoose. Modeling the twin probe scanning electrode response. *J. Electrochemical Soc.*, 145(12):4074–4081, 1998.

258 REFERENCES

11. A. A. Bakr. *The Boundary Integral Equation Method in Axisymmetric Stress Analysis Problems.* Springer, Berlin, 1986.

12. J. Balas, J. Sladek, and V. Sladek. *Stress Analysis By the Boundary Element Method.* Elsevier, Amsterdam, 1989.

13. P. K. Banerjee. *The Boundary Element Methods in Engineering.* McGraw Hill, London, 1994.

14. D. M. Barnett. The precise evaluation of derivatives of the anisotropic elastic Green's functions. *Phys. Stat. Sol. (b),* 49:741–748, 1972.

15. R. S. Barsoum. On the use of isoparametric finite elements in linear fracture mechanics. *International Journal for Numerical Methods in Engineering,* 10:25–37, 1976.

16. G. Barton. *Elements of Green's Functions and Propagation.* Oxford University Press, Oxford, 1999.

17. A. A. Becker. *The Boundary Element Method in Engineering.* McGraw-Hill, London, 1992.

18. G. Beer. *Programming the Boundary Element Method: An Introduction for Engineers.* John Wiley & Sons, New Jersey, 2001.

19. T. Belytschko, Y. Y. Lu, and L. Gu. Element-free Galerkin methods. *International Journal for Numerical Methods in Engineering,* 37:229–256, 1994.

20. R. Bialecki and G. Kuhn. Boundary element solution of heat conduction problems in multizone bodies of nonlinear materials. *Int. J. Numer. Meth. Engrg.,* 36:799–809, 1993.

21. J. O'M. Bockris and A. K. N. Reddy. *Modern Electrochemistry.* Plenum Press, New York, 1970.

22. M. Bonnet. *Boundary Integral Equation Methods for Solids and Fluids.* Wiley and Sons, England, 1995.

23. M. Bonnet. Exploiting partial or complete geometrical symmetry in 3D symmetric Galerkin indirect BEM formulations. *International Journal for Numerical Methods in Engineering,* 57(8):1053–1083, 2003.

24. M. Bonnet and H. D. Bui. Regular BIE for three-dimensional cracks in elastodynamics. In T. A. Cruse, editor, *Advanced Boundary Element Methods,* volume 16, pages 41–47, Berlin and New York, 1988. Springer-Verlag.

25. M. Bonnet, G. Maier, and C. Polizzotto. Symmetric Galerkin boundary element method. *ASME Appl. Mech. Rev.,* 51(11):669–704, 1998.

26. S. Bradshaw, T. Canfield, and J. Kokinis abd T. Disz. An interactive virtual environment for finite element analysis. In *Proceedings of High Performance Computing '95,* 1995.

27. C. A. Brebbia, editor. *Topics in boundary element research,* volume 7. Springer Verlag, Berlin and New York, 1990.

28. C. A. Brebbia and J. Dominguez. *Boundary Elements: An Introductory Course.* WIT Press, Boston, Southampton, 1992.

29. C. A. Brebbia, J. C. F. Telles, and L. C. Wrobel. *Boundary Element Techniques.* Springer-Verlag, Berlin and New York, 1984.

30. S. Bryson. Virtual reality in scientific visualization. *Commun. ACM,* 39(5):62–71, 1996.

31. H. D. Bui. An integral equation method for solving the problem of a plane crack of arbitrary shape. *J. Mech. Phys. Solids,* 25:29–39, 1977.

32. J. S. Bullock, G. E. Giles, and L. J. Gray. Simulation of an electrochemical plating process. In C. A. Brebbia, editor, *Topics in boundary element research*, volume 7, chapter 7, pages 121–141. Springer Verlag, Berlin and New York, 1990.

33. C. V. Camp and G. S. Gipson. Overhauser elements in boundary element analysis. *Mathematical and Computer Modelling*, 15(3-5):59–69, 1991.

34. D. Capuani, D. Bigoni, and M. Brun. Integral representations at the boundary for stokes flow and related symmetric Galerkin formulation. *Archives of Mechanics*, 57(5):363–385, 2005.

35. G. E. Cardew, M. R. Goldthorpe, I. C. Howard, and A. P. Kfouri. On the elastic T-term. *Fundamentals of Deformation and Fracture: Eshelby Memorial Symposium*, 1985.

36. A. Carini, M. Diligenti, P. Maranesi, and M. Zanella. Analytical integrations for two dimensional elastic analysis by the symmetric Galerkin boundary element method. *Computational Mechanics*, 23:308–323, 1999.

37. Angelo Carini and Alberto Salvadori. Analytical integrations 3D BEM. *Comp. Mech.*, 28:177–185, 2002.

38. E. M. Carrillo-Heian, C. Unuvar, J. C. Gibeling, G. H. Paulino, and Z. A. Munir. Simultaneous synthesis and densification of niobium silicide/niobium composites. *Scripta Materialia*, 45:405–412, 2001.

39. H. Carslaw and J. Jaeger. *Conduction of heat in Solids*. Oxford University Press, London, 2nd edition, 1959.

40. A. Chandra and S. Mukherjee. *Boundary Element Methods in Manufacturing*. Oxford University Press, New York, 1997.

41. C. Chang and M. Mear. Boundary element method for two-dimensional linear elastic fracture analysis. *Int. J. Fracture*, 74:219–251, 1995.

42. U. Chang, C. S. Kang, and D. J. Chen. The use of fundamental Green's functions for the solution of problems of heat conduction in anisotropic media. *International Journal of Heat and Mass*, 16:1905–1918, 1973.

43. M.K. Chati and S. Mukherjee. Evaluation of gradients on the boundary using fully regularized hypersingular boundary integral equations. *Acta Mechanica*, 135:41–45, 1999.

44. M.K. Chati and S. Mukherjee. The boundary node method for three-dimensional problems in potential theory. *International Journal for Numerical Methods in Engineering*, 47:1523–1547, 2000.

45. M.K. Chati, S. Mukherjee, and Y.X. Mukherjee. The boundary node method for three-dimensional linear elasticity. *International Journal for Numerical Methods in Engineering*, 46:1163–1184, 1999.

46. M.K. Chati, S. Mukherjee, and G.H. Paulino. The meshless hypersingular boundary node method for three-dimensional potential theory and linear elasticity problems. *Engineering Analysis with Boundary Elements*, 25:639–653, 2001.

47. M.K. Chati, G.H. Paulino, and S. Mukherjee. The meshless standard and hypersingular boundary node methods - applications to error estimation and adaptivity in three-dimensional problems. *International Journal for Numerical Methods in Engineering*, 50:2233–2269, 2001.

48. C. S. Chen, R. Krause, R. G. Pettit, L. Banks-Sills, and A. R. Ingraffea. Numerical assessment of T-stress computation using a p-version finite element method. *International Journal of Fracture*, 107(2):177–199, 2001.

260 REFERENCES

49. F. H. K. Chen and R. T. Shield. Conservation laws in elasticity of the J-integral type. *Journal of Applied Mathematics and Physics (ZAMP)*, 28:1–22, 1977.

50. H. T. Chen and C. K. Chen. Hybrid Laplace transform/finite difference method for transient heat conduction problems. *International Journal for Numerical Methods in Engineering*, 26:1433–1447, 1988.

51. H. T. Chen and J.-Y. Lin. Application of the Laplace transform to non-linear transient problems. *Applied Mathematical Modeling*, 15:144–151, 1991.

52. A-H. D. Cheng, Y. Abousleiman, and T. Badmus. A Laplace transform bem for axysymmetric diffusion utilizing pre-tabulated green's function. *Engineering Analysis with Boundary Elements*, 9:39–46, 1992.

53. A.H.-D. Cheng. Darcy's flow with variable permeability: a boundary integral solution. *Water Resources Research*, 20:980–984, 1984.

54. A.H.-D. Cheng. Heterogeneities in flows through porous media by the boundary element method. In C.A. Brebbia, editor, *Topics in Boundary Element Research, Vol. 4: Applications in Geomechanics*, chapter 6, pages 129–144. Springer-Verlag, 1987.

55. Alexander H.-D. Cheng and Daisy T. Cheng. Heritage and early history of the boundary element method. *Engineering Analysis with Boundary Elements*, 29(3):268–302, 2005.

56. J. K. Choi and S. A. Kinnas. Numerical water tunnel in two- and three-dimensions. *J. Ship Research*, 42:86–98, 1998.

57. H. B. Coda and W. S. Venturini. Further improvements on 3D treatment BEM elastodynamic analysis. *Engineering Analysis with Boundary Elements*, 17:231–243, 1996.

58. H. B. Coda and W. S. Venturini. A simple comparison between two 3D time domain elastodynamic boundary element formulations. *Engineering Analysis with Boundary Elements*, 17:33–44, 1996.

59. B. Cotterell and J. R. Rice. Slightly curved or kinked cracks. *International Journal of Fracture*, 16(2):155–169, 1980.

60. S. L. Crouch. Solution of plane elasticity problems by the displacement discontinuity method. *Int. J. Numer. Meth. Engrg.*, 10:301–343, 1976.

61. S. L. Crouch and A. M. Starfield. *Boundary Element Methods in Solid Mechanics*. George Allen and Unwin, London, 1983.

62. K. S. Crump. Numerical inversion of Laplace transforms using a Fourier series approximation. *Journal of the Association for Computing Machinery*, 23:89–96, 1974.

63. T. A. Cruse. Numerical solutions in three-dimensional elastostatics. *Int. J. Solids Struct.*, 5:1259–1274, 1969.

64. T. A. Cruse. *Boundary Element Analysis in Computational Fracture Mechanics*. Kluwer Academic Publishers, Boston, 1988.

65. T. A. Cruse, D. W. Snow, and R. B. Wilson. Numerical solutions in axisymmetric elasticity. *Computers and Structures*, 7(3):445–451, 1977.

66. D. A. S. Curran, M Cross, and B A Lewis. A preliminary analysis of boundary element methods applied to parabolic differential equations. In C. A. Brebbia, editor, *New Developments in boundary element methods*, pages 179–190. Computational Mechanics Publications, Southampton, 1980.

67. Bernard Danloy. Numerical construction of Gaussian quadrature formulas for $\int_0^1 (-\log x)\, x^\alpha f(x)\, dx$ and $\int_0^\infty e_m(x) f(x)\, dx$. *Mathematics of Computation*, 27:861–869, 1973.

68. B. Davies and B. Martin. Numerical inversion of the Laplace transform: A survey and comparison of methods. *Journal of Computational Physics*, 33:1–32, 1979.

69. L. A. De Lacerda and L. C. Wrobel. Hypersingular boundary integral equation for axisymmetric elasticity. *International Journal for Numerical Methods in Engineering*, 52(11):1337–1354, 2001.

70. L. A. De Lacerda and L. C. Wrobel. Dual boundary element method for axisymmetric crack analysis. *International Journal of Fracture*, 113(3):267–284, 2002.

71. F. A. de Paula and J. C. F. Telles. A comparison between point collocation and Galerkin for stiffness matrices obtained by boundary elements. *Engineering Analysis with Boundary Elements*, 6(3):123–128, 1989.

72. A. Deb, Jr. D. P. Henry, and R. B. Wilson. Alternate BEM formulations for 2- and 3-d thermoelasticity. *Int. J. Solids Structures*, 27:1721–1738, 1991.

73. J. Deconinck. Electrochemical cell design. In C. A. Brebbia, editor, *Topics in boundary element research*, volume 7, chapter 8, pages 142–170. Springer Verlag, Berlin and New York, 1990.

74. V. G. DeGeorgi. A review of computational analyses of ship cathodic protection systems. In *BEM XIX*, pages 829–838, Southampton, 1997. Computational Mechanics.

75. M. Denda. Mixed mode I, II and III analysis of multiple cracks in plane anisotropic solids by the BEM: A dislocation and point force approach. *Engineering Analysis with Boundary Elements*, 25(4-5):267–278, 2001.

76. E. Divo and A. J. Kassab. Generalized boundary integral equation for heat conduction in non-homogeneous media: recent developments on the sifting property. *Engineering Analysis with Boundary Elements*, 22:221–234, 1998.

77. E. Divo and A. J. Kassab. A generalized BEM for steady and transient heat conduction in media with spatially varying thermal conductivity. In M. Goldberg, editor, *Boundary Integral Methods: Numerical and Mathematical Aspects*, pages 37–76. Computational Mechanics Publications, 1999.

78. E. Divo and A. J. Kassab. *Boundary Element Method for Heat Conduction: with Applications in Non-Homogeneous Media, Topics in Engineering Series Vol. 44*. WIT Press, Billerica, MA, 2002.

79. Z.-Z. Du and J. W. Hancock. The effect of non-singular stresses on crack-tip constraint. *Journal of the Mechanics and Physics of Solids*, 39(3):555–567, 1991.

80. C.A.M. Duarte and J.T. Oden, An h-p adaptive method using clouds, *Computer Methods in Applied Mechanics and Engineering*, 139:237-262, 1996.

81. H. Dubner and J. Abate. Numerical inversions of Laplace transforms by relating them to the finite Fourier cosine transform. *Journal of the Association for Computing Machinery*, 15:115–223, 1968.

82. D. G. Duffy. On the numerical inversion of laplace transforms: Comparison of three new methods on characteristic problems from applications. *ACM Transactions on Mathematical Software*, 19:333–359, 1993.

83. Ney A. Dumont, Richardo A. P. Chaves, and G. H. Paulino. The hybrid boundary element method applied to problems of potential in functionally graded materials. *International Journal of Computational Engineering Science*, 5(4):863–891, 2004.

84. F. Durbin. Numerical inversion of laplace transforms: Efficient improvement to Dubner and Abate's method. *Journal of the Association for Computing Machinery*, 17:371–376, 1974.

85. J. F. Durodola and R. T. Fenner. Hermitian cubic boundary elements for two dimensional potential problems. *International Journal for Numerical Methods in Engineering*, 30:1051–1062, 1990.

262 REFERENCES

86. Harrouni K. EL, D. Quazar, L. C. Wrobel, and C. A. Brebbia. Dual reciprocity boundary element method for heterogeneous porous media. In C.A. Brebbia and M. S. Ingber, editors, *Boundary element technology VII*, pages 151–159. Computational Mechanics Publication and Elsevier Applied Science, 1992.

87. Harrouni K. EL, D. Quazar, L. C. Wrobel, and A. H.-D. Cheng. Global interpolation function based DRBEM applied to Darcy's flow in heterogenous media. *Engineering Analysis with Boundary Elements*, 16:281–285, 1995.

88. G. Fairweather and A. Karageorghis. The method of fundamental solutions for elliptic boundary value problems. *Advances in Computational Mathematics*, 9:69–95(27), 1998.

89. T. Fett. T-stresses in rectangular plates and circular disks. *Engineering Fracture Mechanics*, 60(5-6):631–652, 1998.

90. A. Frangi. Regularization of boundary element formulations by the derivative transfer method. In V. Sladek and J. Sladek, editors, *Singular Integrals in the Boundary Element Method*, Advances in Boundary Elements, chapter 4, pages 125–164. Computational Mechanics Publishers, 1998.

91. A. Frangi. Fracture propagation in 3D by the symmetric Galerkin boundary element method. *International Journal of Fracture*, 116:313–330, 2002.

92. A. Frangi and M. Bonnet. A Galerkin symmetric and direct BIE method for Kirchoff elastic plates: formulation and implementation. *International Journal for Numerical Method in Engineering*, 41:337–369, 1998.

93. A. Frangi and M. Guiggiani. A direct approach for boundary integral equations with high-order singularities. *International Journal for Numerical Method in Engineering*, 49:871–898, 2000.

94. A. Frangi and G. Novati. Symmetric BE method in two-dimensional elasticity: evaluation of double integrals for curved elements. *Computational Mechanics*, 19:58–68, 1996.

95. A. Frangi, G. Novati, R. Springhetti, and M. Rovizzi. 3D fracture analysis by the symmetric Galerkin BEM. *Comp. Mech.*, 28:220–232, 2002.

96. Q. Fulian, A. C. Fisher, and G. Denuault. Applications of the boundary element method in electrochemistry: scanning electrochemical microscopy. *J. Phys. Chem. B*, 103:4387–4392, 1999.

97. Q. Fulian, A. C. Fisher, and G. Denuault. Applications of the boundary element method in electrochemistry: scanning electrochemical microscopy, Part 2. *J. Phys. Chem. B*, 103:4393–4398, 1999.

98. M. Garzon, D. Adalsteinsson, L. J. Gray, and J. A. Sethian. A coupled integral-level set boundary integral method for moving boundary simulations. *Interfaces and Free Boundaries*, 7:277–302, 2005.

99. L. Gaul, M. Kogl, and M. Wagner. *Boundary Element Methods for Engineers and Scientists*. Springer-Verlag, Berlin Heidelberg New York, 2003.

100. D. P. Gaver. Observing stochastic processes, and approximate transform inversion. *Operational Research*, 14:444–459, 1966.

101. N. Ghosh, H. Rajiyah, S. Ghosh, and S. Mukherjee. A new boundary element method formulation for linear elasticity. *Journal of Applied Mechanics*, 53(1):69–76, 1986.

102. G. E. Giles, L. J. Gray, and J. S. Bullock. Validation of the BEPLATE code. In *1997 Electroforming Course and Symposium*, Orlando, FL, 1998. American Electroplaters and Surface Finishers Society.

REFERENCES **263**

103. G. E. Giles, L. J. Gray, J. S. Bullock, and G. J. Van Berkel. Numerical simulation of anodic reactions. In D. Landolt, M. Matlosz, and Y. Sato, editors, *Fundamental Aspects of Electrochemical Deposition and Dissolution Including Modeling*, volume PV 99-33, New Jersey, 1999. Electrochemical Society.

104. M. A. Goldberg and C. S. Chen. The method of fundamental solutions for potential, helmholtz and diffusion problems. In M. A. Goldberg, editor, *Boundary Integral Methods: Numerical and Mathematical Aspects*, pages 103–176. Pineridge Press, Southampton, Boston, 1999.

105. E. Graciani, V. Mantič, F. Paris, and A. Blazquez. Weak formulation of axisymmetric frictionless contact problems with boundary elements: Application to interface cracks. *Computers and Structures*, 83(10-11 SPEC. ISS.):836–855, 2005.

106. E. Graciani, V. Mantič, F. Paris, and J. Cañas. A critical study of hypersingular and strongly singular boundary integral representations of potential gradient. *Comp. Mech.*, 25:542–559, 2000.

107. L. J. Gray. Boundary element method for regions with thin internal cavities. *Engrg. Analy. Boundary Elem.*, 6:180–184, 1989.

108. L. J. Gray. Symbolic computation of hypersingular boundary integrals. In J. H. Kane, G. Maier, N. Tosaka, and S. N. Atluri, editors, *Advances in boundary element techniques*, pages 157–172. Springer-Verlag, Berlin Heidelberg, 1993.

109. L. J. Gray. Evaluation of singular and hypersingular Galerkin boundary integrals: direct limits and symbolic computation. In V. Sladek and J. Sladek, editors, *Singular Integrals in the Boundary Element Method*, Advances in Boundary Elements, chapter 2, pages 33–84. Computational Mechanics Publishers, 1998.

110. L. J. Gray and S. J. Chang. Hypersingular formulation for 2D elastic wave scattering. *Engrg. Analy. Boundary Elem.*, 10(4):337–344, 1992.

111. L. J. Gray and M. Garzon. On a Hermite boundary integral approximation. *Computers and Structures*, 83(10-11):889–894, 2005.

112. L. J. Gray, J. Glaeser, and T. Kaplan. Direct evaluation of hypersingular Galerkin surface integrals. *SIAM J. Sci. Comput.*, 25(5):1534–1556, 2004.

113. L. J. Gray and B. Griffith. A faster Galerkin boundary integral algorithm. *Comm. Numer. Meth. Engng.*, 14:1109–1117, 1998.

114. L. J. Gray, T. Kaplan, J. D. Richardson, and G. H. Paulino. Green's functions and boundary integral analysis for exponentially graded materials: heat conduction. *ASME Journal of Applied Mechanics*, 70:543–549, 2003.

115. L. J. Gray and E. D. Lutz. On the treatment of corners in the boundary element method. *J. Comp. Appl. Math.*, 32:369–386, 1990.

116. L. J. Gray, L. F. Martha, and A. R. Ingraffea. Hypersingular integrals in boundary element fracture analysis. *International Journal for Numerical Methods in Engineering*, 29(6):1135–1158, 1990.

117. L. J. Gray and G. H. Paulino. Symmetric Galerkin boundary integral formulation for interface and multi-zone problems. *International Journal for Numerical Method in Engineering*, 40(16):3085–3101, 1997.

118. L. J. Gray and G. H. Paulino. Symmetric Galerkin boundary integral fracture analysis for plane orthotropic elasticity. *Computational Mechanics*, 20(1/2):26–33, 1997.

119. L. J. Gray and G. H. Paulino. Crack tip interpolation, revisited. *SIAM Journal on Applied Mathematics*, 58(2):428–455, 1998.

120. L. J. Gray, A.-V. Phan, and T. Kaplan. Boundary integral evaluation of surface derivatives. *SIAM J. Sci. Comput.*, 26:294–312, 2004.

121. L. J. Gray, A.-V. Phan, G. H. Paulino, and T. Kaplan. Improved quarter-point crack tip element. *Engineering Fracture Mechanics*, 70(2):269–283, 2003.

122. L. J. Gray and C. San Soucie. Hermite interpolation algorithm for hypersingular boundary integrals. *Int. J. Numer. Meth. Engrg.*, 36:2357–2367, 1993.

123. M. D. Greenberg. *Foundations of Applied Mathematics*. Prentice-Hall, Englewood Cliffs, New Jersey, 1978.

124. L. Greengard and V. Rokhlin. A new version of the fast multipole method for the Laplace equation in three dimensions. In *Acta numerica, 1997*, pages 229–269. Cambridge University Press, Cambridge, UK, 1997.

125. S. T. Grilli, P. Guyenne, and F. Dias. A fully nonlinear model for three-dimensional overturning waves over arbitrary bottom. *Int. J. Num. Meth. Fluids*, 35(7):829–867, 2001.

126. M. Guiggiani. Error indicators for adaptive mesh refinement in the boundary element method - a new approach. *International Journal for Numerical Methods in Engineering*, 29:1247–1269, 1990.

127. Massimo Guiggiani. The evaluation of Cauchy principal value integrals in the boundary element method - a review. *Mathematical and Computer Modelling*, 15:175–184, 1991.

128. M. R. Gungor, L. J. Gray, and D. Maroudas. Effects of mechanical stress on electromigration-driven transgranular void dynamics in passivated metallic thin films. *Applied Physics Letters*, 73(26):3848–3850, 1998.

129. J. Hadamard. *Lectures on Cauchy's Problem in Linear Partial Differential Equations*. Dover Publications, 1952.

130. W. S. Hall and W. H. Robertson. Advanced boundary element calculations for elastic wave scattering. In M. F. McCarthy and M. A. Hayes, editors, *Elastic Wave Propagation*, pages 459–464. Elsevier Science Publishers B. V., 1989.

131. W.S. Hall. *The Boundary Element Method*. Kluwer Academic Publishers, Dordrecht, The Netherlands, 1994.

132. F. Hartmann, C. Katz, and B. Protopsaltis. Boundary elements and symmetry. *Ingenieur-Archiv*, 55:440–449, 1985.

133. C. Hastings. *Approximations for Digital Computers*. Princeton University Press, Princeton, NJ, 1955.

134. R. D. Henshell and K. G. Shaw. Crack tip finite elements are unnecessary. *International Journal for Numerical Methods in Engineering*, 9:495–507, 1975.

135. D. A. Hills, P. A. Kelly, D. N. Dai, and A. M. Korsunsky. *Solution of Crack Problems: The Distributed Dislocation Technique*. Kluwer Academic Publishers, The Netherlands, 1996.

136. S. M. Hölzer. The symmetric Galerkin BEM for plane elasticity: scope and applications. In C. Hirsch, editor, *Numerical Methods in Engineering '92*. Elsevier, 1992.

137. M. Ikeuchi and K. Onishi. Boundary elements in transient convective diffusive problems. In C. A. Brebbia, T. Futagami, and M. Tanaka, editors, *Boundary Elements V*, pages 275–282. Springer, Berlin, 1983.

138. N. I. Ioakimidis. A natural approach to the introduction of finite-part integrals into crack problems of three-dimensional elasticity. *Engng. Fract. Mech.*, 16:669–673, 1982.

139. N. I. Ioakimidis. Exact expression for a two-dimensional finite part integral appearing during the numerical solution of crack problems in three-dimensional elasticity. *Comm. Appl. Numer. Methods*, 1:183–189, 1985.

140. M. A. Jaswon and G. T. Symm. *Integral equation methods in potential theory and elastostatics*. Academic Press, New York, 1977.

141. K. L. Johnson. *Contact Mechanics*. Cambridge University Press, England, 1985.

142. J. Kane and C. Balakrishna. Symmetric Galerkin boundary formulations employing curved elements. *International Journal for Numerical Method in Engineering*, 36:2157–2187, 1993.

143. J. H. Kane. *Boundary Element Analysis in Engineering Continuum Mechanics*. Prentice Hall, New Jersey, 1994.

144. A. C. Kaya and F. Erdogan. On the solution of integral equations with strongly singular kernels. *Quart. Appl. Math.*, 45:105–122, 1987.

145. Theodor Kermanidis. Numerical solution for axially symmetrical elasticity problems. *International Journal of Solids and Structures*, 11(4):493–500, 1975.

146. A. P. Kfouri. Some evaluations of the elastic T-term using Eshelby's method. *International Journal of Fracture*, 30(4):301–315, 1986.

147. D. R. Khattab, A. H. A. Aziz, and A. S. Hussein. An enhanced www-based scientific data visualization service using VRML. In *VRCAI '04: Proceedings of the 2004 ACM SIGGRAPH international conference on Virtual Reality continuum and its applications in industry*, pages 134–140. ACM Press, 2004.

148. J.-H. Kim and G. H. Paulino. Finite element evaluation of mixed-mode stress intensity factors in functionally graded materials. *International Journal for Numerical Method in Engineering*, 53(8):1903–1935, 2002.

149. J.-H. Kim and G. H. Paulino. An accurate scheme for mixed-mode fracture analysis of functionally graded materials using the interaction integral and micromechanics models. *International Journal for Numerical Methods in Engineering*, 58(10):1457–1497, 2003.

150. J.-H. Kim and G. H. Paulino. T-stress, mixed-mode stress intensity factors, and crack initiation angles in functionally graded materials: A unified approach using the interaction integral method. *Computer Methods in Applied Mechanics and Engineering*, 192(11):1463–1494, 2003.

151. R. Kitey, A. V. Phan, H. V. Tippur, and T. Kaplan. Modeling of crack growth through particulate clusters in brittle matrix by symmetric-galerkin boundary element method. *International Journal of Fracture*, 141(1-2):11–25, 2006. Kitey, R. Phan, A. -V. Tippur, H. V. Kaplan, T.

152. V.S. Kothnur, S. Mukherjee, and Y.X. Mukherjee. Two dimensional linear elasticity by the boundary node method. *International Journal of Solids and Structures*, 36(4):1129–1147, 1999.

153. G. Krishnasamy, L. W. Schmerr, T. J. Rudolphi, and F. J. Rizzo. Hypersingular boundary integral equations : some applications in acoustic and elastic wave scattering. *Journal of Applied Mechanics*, 57:404–414, 1990.

154. L. A. De Lacerda, L. C. Wrobel, and W. J. Mansur. Boundary integral formulation for two-dimensional acoustic radiation in a subsonic uniform flow. *Journal of Acoustic Society of America*, 100:98–107, 1996.

155. J. C. Lachat and J. O. Watson. Effective numerical treatment of boundary integral equations: a formulation for three-dimensional elastostatics. *International Journal for Numerical Method in Engineering*, 10(5):991–1005, 1976.

156. O. E. Lafe and A. H.-D. Cheng. A pertubation boundary element code for steady state groundwater flow in heterogeneous aquifers. *Water Resources Research*, 23:1079–1084, 1987.

157. O. E. Lafe, J. A. Liggett, and P. L.-F. Liu. BIEM solutions to combination of leaky, layered, confined, unconfined, nonisotropic acquifers. *Water Resources Research*, 17:1431–1444, 1981.

158. P. Lancaster and K. Salkauskas. *Curve and Surface Fitting - An Introduction*. Academic Press, London, 1986.

159. S. G. Larsson and A. J. Carlson. Influence of non-singular stress terms and specimen geometry on small-scale yielding at crack tips in elastic-plastic materials. *Journal of the Mechanics and Physics of Solids*, 21(4):263–277, 1973.

160. P. S. Leevers and J. C. D. Radon. Inherent stress biaxiality in various fracture specimen. *International Journal of Fracture*, 19(4):311–325, 1982.

161. L. Lehmann and H. Antes. Dynamic structure-soil-structure interaction applying the symmetric galerkin boundary element method (SGBEM). *Mechanics Research Communications*, 28(3):297–304, 2001.

162. D. Lesnic, L. Elliott, and DB Ingham. Treatment of singularities in time-dependent problems using boundary element method. *Engineering Analysis with Boundary Elements*, 16:65–70, 1995.

163. B. Q. Li and J. W. Evans. Boundary element solution of heat convection-diffusion problems. *Journal of Computational Physics*, 93:255–272, 1991.

164. G. Li and N. R. Aluru. Boundary cloud method: a combined scattered point/boundary integral approach for boundary-only analysis. *Computer Methods in Applied Mechanics and Engineering*, 191(21-22):2337 2370, 2002. 18.

165. S. Li, M. E. Mear, and L. Xiao. Symmetric weak form integral equation method for three-dimensional fracture analysis. *Comp. Meth. Appl. Mech. Engng.*, 151:435–459, 1998.

166. J. A. Liggett and P. L. F. Liu. Unsteady flow in confined aquifers: A comparison of boundary integral methods. *Water Resources Research*, 15:861–866, 1979.

167. J. A. Liggett and P. L. F. Liu. *The boundary integral equation method for porous media flow*. George Allen & Unwin, England, 1983.

168. J. L. Lions and R. Dautray. Approximation des équations intégrales par éléments finis. etude d'erreur. In *Analyse mathématique et calculscientifique pour les sciences et les techniques (Chapter 13)*, pages 953–968. Masson, Paris, 1985.

169. G. R. Liu. *Mesh free methods: Moving beyond the finite element method*. CRC Press LLC, Boca Raton, Florida, 2003.

170. E.D. Lutz. *A novel boundary element method for linear elasticity*. PhD thesis, Cornell University, Ithaca, New York, USA, 1994.

171. Singh K. M. and M. Tanaka. On exponential variable transformation based boundary element formulation for advection-diffusion problems. *Engineering Analysis with Boundary Elements*, 24:225–235, 2000.

172. G. Kuhn M. Haas. A symmetric Galerkin BEM implementation for 3D elastostatic problems with an extension to curved elements. *Comp. Mech.*, 28:250–259, 2002.

173. G. Maier, M. Diligenti, and A. Carini. A variational approach to boundary element elastodynamic analysis and extension to multidomain problems. *Comp. Meth. Appl. Engng.*, 92:193–213, 1991.

174. G. Maier, S. Miccoli, G. Novati, and S. Sirtori. A Galerkin symmetric boundary element methos in plasticity: formulation and implementation. In J. H. Kane, G. Maier,

N. Tosaka, and S. N. Atluri, editors, *Advances in Boundary Element Techniques*, pages 288–328. Springer-Verlag, Berlin Heidelberg, 1993.

175. G. Maier, G. Novati, and S. Sirtori. On symmetrization in boundary element elastic and elastoplastic analysis. In G. Kuhn and H. Mang, editors, *Discretization methods in structural mechanics*, pages 191–200. Springer-Verlag, Berlin and New York, 1990.

176. G. Maier and C. Polizzotto. A Galerkin approach to boundary element elastoplastic analysis. *Computer Methods in Applied Mechanics and Engineering*, 60:175–194, 1987.

177. D. Maillet, S. Andre, J. C. Batsale, A. Degiovanni, and C. Moyne. *Numerical Fracture Mechanics*. Thermal Quadrupoles Solving the Heat Equation through Integral Transforms, John Wiley and Sons, 2000.

178. M. Mayr. On the numerical solution of axisymmetric elasticity problems using an integral equation approach. *Mech. Res. Comm.*, 3:393–398, 1976.

179. B. H. McCormick, T. A. DeFanti, and M. D. Brown. Visualization in Scientific Computing. *Computer Graphics*, 21(6), 1987.

180. G. Menon. Hypersingular error estimates in boundary element methods. Master's thesis, Cornell University, Ithaca, NY, 1996.

181. M.G.Duffy. Quadrature over a pyramid or cube of integrands with a singularity at a vertex. *SIAM J Numer. Analy.*, 19:1260–1262, 1982.

182. J. H. Michell. Elementary distributions of plane stress. *Proceedings of the London Mathematical Society*, 32:35–61, 1900.

183. L. M. Milne-Thomson. Elliptic integrals. In M. Abramowitz and I. A. Stegun, editors, *Handbook of Mathematical Functions*, pages 587–626. Dover Publications, 1965.

184. Y. Miyamoto, W. A. Kaysser, B. H. Rabin, A. Kawasaki, and R. G. Ford. *Functionally Graded Materials: Design, Processing and Applications*. Kluwer Academic Publishers, Dordrecht, 1999.

185. M. N. J. Moore, L. J. Gray, and T. Kaplan. Evaluation of supersingular integrals: second order boundary derivatives. *Int. J. Num. Meth. Engng.*, 69:1930–1947, 2007.

186. G. J. Moridis and D. L. Reddell. The Laplace transform boundary element (LTBE) method for the solution of diffusion-type equations. In C. A. Brebbia and G. S. Gipson, editors, *BEM XIII*, pages 83–97. Computational Mechanics Publications, 1991.

187. K. H. Muci-Küchler and J. C. Miranda-Valenzuela. A new error indicator based on stresses for adaptive meshing with hermite boundary elements. *Engng. Analy. Boundary Elements*, 23:657–670, 1999.

188. K. H. Muci-Küchler and T. J. Rudolphi. Coincident collocation of displacement and tangent derivative boundary integral equations in elasticity. *International Journal for Numerical Method in Engineering*, 36:2837–2849, 1993.

189. K. H. Muci-Küchler and T. J. Rudolphi. Application of tangent derivative boundary integral equations to the formulation of higher order boundary elements. *Int. J. Solid Struct.*, 31:1565–1584, 1994.

190. S. Mukherjee. *Boundary Element Methods in Creep and Fracture*. Elsevier Applied Science Publishers, England, 1982.

191. S. Mukherjee and Y. X. Mukherjee. *Boundary Methods: Elements, Contours, and Nodes*. CRC Press, Boca Raton FL, 2005.

192. S. Mukherjee and Y.X. Mukherjee. The hypersingular boundary contour method for three-dimensional linear elasticity. *ASME Journal of Applied Mechanics*, 65:300–309, 1998.

268 REFERENCES

193. Y.X. Mukherjee and S. Mukherjee. The boundary node method for potential problems. *International Journal for Numerical Methods in Engineering*, 40:797–815, 1997.

194. A. Nagarajan, E. Lutz, and S. Mukherjee. A novel boundary element for linear elasticity with no numerical integration for two-dimensional and line integrals for three-dimensional problems. *ASME Journal of Applied Mechanics*, 61(2):264–269, 1994.

195. A. Nagarajan, S. Mukherjee, and E. Lutz. The boundary contour method for three-dimensional linear elasticity. *Journal of Applied Mechanics*, 63:278–286, 1996.

196. T. Nakamura and D. M. Parks. Determination of T-stress along three-dimensional crack fronts using an interaction integral method. *International Journal of Solids and Structures*, 29(13):1597–1611, 1992.

197. B. Nayroles, G. Touzot, and P. Villon. Generalizing the finite element method : diffuse approximation and diffuse elements. *Computational Mechanics*, 10:307–318, 1992.

198. John S. Newman. *Electrochemical Systems*. Prentice Hall, Englewood Cliffs, New Jersey, 1991.

199. L. Nicolazzi, C. A. Duarte, E. Fancello, and C. Barcellos. Hp-clouds - a meshless method in boundary elements. part I: Formulation. *Boundary Element Communications*, 8:80–82, 1997.

200. L. Nicolazzi, C. A. Duarte, E. Fancello, and C. Barcellos. Hp-clouds - a meshless method in boundary elements. part II: Implementation. *Boundary Element Communications*, 8:83–85, 1997.

201. G. Novati and R. Springhetti. A Galerkin boundary contour method for two-dimensional linear elasticity. *Computational Mechanics*, 23:53–62, 1999.

202. A. J. Nowak. The multiple reciprocity method of solving transient heat conduction problems. In C. A. Brebbia and J. J. Connor, editors, *BEM XI*, pages 81–95. Computational Mechanics Publications, Springer-Verlag, 1989.

203. N. P. O'Dowd, C. F. Shih, and R. H. Dodds Jr. The role of geometry and crack growth on constraint and implications for ductile/brittle fracture. In *Constraint Effects in Fracture Theory and Applications*, volume 2 of *ASTM STP 1244*, pages 134–159. American Society for Testing and Materials, 1995.

204. F. W. J. Olver. Bessel functions of integer order. In M. Abromowitz and I. A. Steegun, editors, *Handbook of mathematical functions*, chapter 9, pages 355–434. National Bureau of Standards, Washington, D.C., 1972.

205. F. París and J. Cañas. *Boundary Element Method*. Oxford University Press Inc., New York, 1997.

206. P. W. Partridge, C. A. Brebbia, and L. C. Wrobel. *The Dual Reciprocity Boundary Element Method*. Computational Mechanics Publications, Southampton and Boston, 1992.

207. R. Pasquetti, A. Caruso, and Wrobel LC. Transient problems using time-dependent fundamental solutions. In L. C. Wrobel and C. A. Brebbia, editors, *Boundary Element Methods in Heat transfer*, pages 33–62. Computational Mechanics Publications, Southampton Boston, 1992.

208. G. H. Paulino. *Novel formulations of the boundary element method for fracture mechanics and error estimation*. PhD thesis, Cornell University, 1995.

209. G. H. Paulino and L. J. Gray. Galerkin residuals for adaptive symmetric-Galerkin boundary element methods. *Journal of Engineering Mechanics (ASCE)*, 125(5):575–585, 1999.

REFERENCES 269

210. G. H. Paulino, L. J. Gray, and V. Zarikian. Hypersingular residuals – A new approach for error estimation in the boundary element method. *International Journal for Numerical Methods in Engineering*, 36(12):2005–2029, 1996.

211. G. H. Paulino, Z.H. Jin, R. H. Dodds. Failure of functionally graded materials. In *Comprehensive Structural Integrity*, Karihaloo B, Knauss WG (eds), vol. 2, Chapter 13. Elsevier: Amsterdam, 2003; 607–644.

212. G. H. Paulino and J.-H. Kim. A new approach to compute T-stress in functionally graded materials using the interaction integral method. *Engineering Fracture Mechanics*, 2003; *Eng Fract Mech*, Volume: 71, Issue: 13–14 (2004), pp. 1907–1950.

213. G. H. Paulino and A. Sutradhar. The simple boundary element method for multiple cracks in functionally graded media governed by potential theory: a three-dimensional Galerkin approach. *International Journal for Numerical Methods in Engineering*, 65:2007–2034, 2006.

214. G. H. Paulino, Alok Sutradhar, and L. J. Gray. Boundary element methods for functionally graded materials. In *Boundary Element Technology*, Brebbia CA, Dippery RE (eds), vol. XV, Section 3. WIT Press: Southampton, 2003; 137–146.

215. J. J. Perez-Gavilan and M. H. Aliabadi. A symmetric Galerkin BEM for harmonic problems and multiconnected bodies. *Meccanica*, 36(4):449–462, 2001.

216. J. J. Perez-Gavilan and M. H. Aliabadi. A symmetric Galerkin boundary element method for dynamic frequency domain viscoelastic problems. *Computers & Structures*, 79(29-30):2621–2633, 2001.

217. J. J. Perez-Gavilan and M. H. Aliabadi. Symmetric Galerkin BEM for shear deformable plates. *International Journal for Numerical Methods in Engineering*, 57(12):1661–1693, 2003.

218. A. V. Phan, L. J. Gray, and T. Kaplan. On some benchmark results for the interaction of a crack with a circular inclusion. *Journal of Applied Mechanics-Transactions of the ASME*, 74(6):1282–1284, 2007. Phan, A. -V. Gray, L. J. Kaplan, T.

219. A.-V. Phan, T. Kaplan, L. J. Gray, D. Adelsteinsson, J. A. Sethian, W. Barvosa-Carter, and M. A. Aziz. Modeling a growth instability in a stressed solid. *Modelling Simul. Mater. Sci. Eng.*, 9:309–325, 2001.

220. A-V. Phan, S. Mukherjee, and Mayer J.R.R. The boundary contour method for two-dimensional linear elasticity with quadratic boundary elements. *Computational Mechanics*, 20:310–319, 1997.

221. A-V. Phan, S. Mukherjee, and Mayer J.R.R. The hypersingular boundary contour method for two-dimensional linear elasticity. *Acta Mechanica*, 130:209–225, 1998.

222. A. V. Phan, J. A. L. Napier, L. J. Gray, and T. Kaplan. Stress intensity factor analysis of friction sliding at discontinuity interfaces and junctions. *Computational Mechanics*, 32(4-6):392–400, 2003.

223. A.-V. Phan, J. A. L. Napier, L. J. Gray, and T. Kaplan. Symmetric-Galerkin BEM simulation of fracture with frictional contact. *International Journal for Numerical Methods in Engineering*, 57(6):835–851, 2003.

224. Joel R. Phillips and Jacob K. White. A precorrected-FFT method for electrostatic analysis of complicated 3-D structures. *IEEE Trans. Comput. Aided Design of Integrated Circuits and Systems*, 16:1059–1071, 1997.

225. A. P. Pierce and J. A. L. Napier. A spectral multipole method for efficient solution of large-scale boundary element models in elastostatics. *International Journal for Numerical Method in Engineering*, 38:4009–4034, 1995.

270 REFERENCES

226. C. Polizzotto. A symmetric Galerkin boundary/domain element method for finite elastic deformations. *Computer Methods in Applied Mechanics and Engineering*, 189(2):481–514, 2000.

227. D. Poljak. Foreword to EABE special issue on electromagnetics. *Engineering Analysis with Boundary Elements*, 27(4), 2003.

228. Pasquetti A. Caruso R. Boundary element approach for transient and non-linear thermal diffusion. *Num Heat Transf, PartB*, 17:83–99, 1990.

229. P. A. Ramanchandran. *Boundary element methods in transport phenomena*. Elsevier, London, 1994.

230. J. N. Reddy and A. Miravete. *Practical Analysis of Composite Laminates*. CRC Press, Boca Raton, 1995.

231. B. Reidinger and O. Steinbach. A symmetric boundary element method for the Stokes problem in multiple connected domains. *Mathematical Methods in the Applied Sciences*, 26(1):77–93, 2003.

232. J. J. Rencis and K. Y. Yong. A self-adaptive h-refinement technique for the boundary element method. *Comp. Meth. Appl. Mech. Eng.*, 73:295–316, 1989.

233. J. R. Rice. Mathematical analysis in the mechanics of fracture. In H. Liebowitz, editor, *Fracture – An Advanced Treatise*, volume II, pages 191–311. Pergamon Press, Oxford, 1968.

234. F. Rizzo and D. J. A. Shippy. Method of solution for certain problems of transient heat conduction. *AIAA Journal*, 8:2004–2009, 1970.

235. F. J. Rizzo. An integral equation approach to boundary value problems of classical elastostatics. *Quart. Appl. Math.*, 25:83–95, 1967.

236. F. J. Rizzo, D. J. Shippy, and M. Rezayat. A boundary integral equation method for radiation and scattering of elastic waves in three dimensions. *Int. J. Numer. Meth. Engrg.*, 21:115–129, 1985.

237. S. Rjasanow and O. Steinbach. *The Fast Solution of Boundary Integral Equations*. Springer, Berlin, 2007.

238. T. J. Rudolphi and K. H. Muci-Küchler. Consistent regularization of both kernels in hypersingular integral equations. In C. A. Brebbia and G. S. Gipson, editors, *Boundary Elements XIII*, pages 875–887, Southampton and Boston, 1991. Computational Mechanics.

239. M. A. Sales and L. J. Gray. Evaluation of the anisotropic Green's function and its derivatives. *Computers and Structures*, 69(2):247–254, 1998.

240. Alberto Salvadori. *Quasi brittle fracture mechanics by cohesive crack models and symmetric Galerkin boundary element method*. PhD thesis, Politecnico Milano, Milan, Italy, 1999.

241. Alberto Salvadori. Analytical integrations of hypersingular kernel in 3D BEM problems. *Comp. Meth. in Appl. Mech. Engng.*, 190:3957–3975, 2001.

242. Alberto Salvadori. Analytical integrations in 2D BEM elasticity. *International Journal for Numerical Method in Engineering*, 53:1695–1719, 2002.

243. A. H. Schatz, V. Thomée, and W. L. Wendland. *Mathematical Theory of Finite and Boundary Element Methods*. Birkhäuser, Basel, Boston, & Berlin, 1990.

244. W. Schroeder, K. Martin, and W. Lorensen. *The Visualization Toolkit: An Object Oriented Approach to 3D Graphics 3rd Edition*. Kitware, Inc., 2003.

245. C. Schwab and W. L. Wendland. On the extraction technique in boundary integral equations. *Math. Comp.*, 68(225):91–122, 1999.

REFERENCES **271**

246. J. A. Sethian. Curvature and the evolution of fronts. *Comm. Math. Phys.*, 101:487–499, 1985.

247. J. A. Sethian. *Level Set Methods and Fast Marching Methods: Evolving Interfaces in Computational Geometry, Fluid Mechanics, Computer Vision, and Materials Science.* Cambridge University Press, Cambridge, 1999.

248. T. L. Sham. The determination of the elastic T-term using higher-order weight functions. *International Journal of Fracture*, 48(2):81–102, 1991.

249. R. Shaw. An integral equation approach to diffusion. *International Journal of Heat and Mass transfer*, 17:693–699, 1974.

250. R. P. Shaw. Green's functions for heterogeneous media potential problems. *Engineering Analysis with Boundary Elements*, 13:219–221, 1994.

251. R. P. Shaw and G. S. Gipson. Interrelated fundamental solutions for various heterogeneous potential, wave and advective-diffusive problems. *Engineering Analysis with Boundary Elements*, 16:29–34, 1995.

252. R. P. Shaw and G. D. Manolis. Two dimensional heat conduction in graded materials using conformal mapping. *Communications in Numerical Methods in Engineering*, 19:215–221, 2003.

253. N. I. Shbeeb, W. K. Binienda, and K. L. Kreider. Analysis of the driving forces for multiple cracks in an infinite nonhomogeneous plate, part II: Numerical solutions. *Journal of Applied Mechanics*, 66(2):501–506, 1999.

254. W. R. Sherman and A. B. Craig. *Understanding Virtual Reality: Interface, Application and Design.* Morgan Kaufmann, 2003.

255. F. Shi, P. Ramesh, and S. Mukherjee. Adaptive mesh refinement of the boundary element method for potential problems by using mesh sensitivities as error indicators. *Computational Mechanics*, 16(6):379–395, 1995.

256. S. Sirtori. General stress analysis method by means of integral equations and boundary elements. *Meccanica*, 14:210–218, 1979.

257. S. Sirtori, G. Maier, G. Novati, and S. Miccoli. A Galerkin symmetric boundary element method in elasticity: formulation and implementation. *International Journal for Numerical Methods in Engineering*, 35(2):255–282, 1992.

258. J. Sladek and V. Sladek. Evaluation of T-stresses and stress intensity factors in stationary thermoelasticity by the conservation integral method. *International Journal of Fracture Mechanics*, 86(3):199–219, 1997.

259. J. Sladek, V. Sladek, and S.N. Atluri. Local boundary integral equation (LBIE) method for solving problems of elasticity with nonhomogeneous material properties. *Computational Mechanics*, 24:456–462, 2000.

260. J. Sladek, V. Sladek, and P. Fedelinski. Contour integrals for mixed-mode crack analysis: effect of nonsingular terms. *Theoretical and Applied Fracture Mechanics*, 27(2):115–127, 1997.

261. D. J. Smith, M. R. Ayatollahi, and M. J. Pavier. The role of T-stress in brittle fracture for linear elastic materials under mixed-mode loading. *Fatigue & Fracture of Engineering Materials & Structures*, 24(2):137–150, 2001.

262. G. D. Smith. *Numerical Solution of Partial Differential Equations: Finite Difference Methods.* Oxford Univ. Press, second edition, 1985.

263. H. Stehfest. Algorithm 368: Numerical inversion of Laplace transform. *Communications of the Association for Computing Machinery*, 13:47–19, 1970.

264. H. Stehfest. Remarks on algorithm 368: Numerical inversion of Laplace transform. *Communications of the Association for Computing Machinery*, 13:624, 1970.

265. E. A. Sudicky. The Laplace transform Galerkin technique: a time-continuous finite element theory and application to mass transport in groundwater. *Water Resources Research*, 25:1833–1846, 1989.

266. S. Suresh and A. Mortensen. *Fundamentals of Functionally Graded Materials*. The Institute of Materials, IOM Communications Ltd., London, 1998.

267. A. Sutradhar. *Galerkin boundary element modeling of three-dimensional graded material systems*. PhD thesis, University of Illinois at Urbana-Champaign, 2005.

268. A. Sutradhar and G. H. Paulino. A simple boundary element method for problems of potential in nonhomogeneous media. *International Journal for Numerical Methods in Engineering*, 60:2203–2230, 2004.

269. A. Sutradhar and G. H. Paulino. A simple boundary element method for transient heat conduction in functionally graded materials. *Computer Methods in applied mechanics and engineering*, 193:4511–4539, 2004.

270. A. Sutradhar and G. H. Paulino. Symmetric galerkin boundary element computation of T-stress and stress intensity factors for mixed-mode cracks by the interaction integral method. *Engineering Analysis with Boundary Elements*, 28(11):1335–1350, 2004.

271. A. Sutradhar, G. H. Paulino, and L. J. Gray. On hypersingular surface integrals in the symmetric Galerkin boundary element method: Application to heat conduction in exponentially graded materials. *International Journal for Numerical Methods in Engineering*, 62:122–157, 2005.

272. A. Talbot. The accurate numerical inversion of Laplace transforms. *Journal of the Institute of Mathematics and Its Applications*, 23:97–120, 1979.

273. M. Tanaka and T.A. Cruse. *Boundary element methods in applied mechanics*. Pergamon Press, Oxford, 1988.

274. M. Tanaka, T. Matsumoto, and Y. Suda. A dual reciprocity boundary element method applied to the steady-state heat conduction problem of functionally gradient materials. *Electronic Journal of Boundary Elements*, BETEQ 2001, No. 1:128–135, 2002.

275. M. Tanaka, V. Sladek, and J. Sladek. Regularization techniques applied to boundary element methods. *Applied Mechanics Reviews*, 47(10):457–499, 1994.

276. J. C. F. Telles. A self-adaptive co-ordinate transformation for efficient numerical evaluation of general boundary element integrals. *International Journal for Numerical Method in Engineering*, 24(5):959–973, 1987.

277. S. P. Timoshenko and J. N. Goodier. *Theory of Elasticity*. McGraw-Hill, New York, 3rd edition, 1976.

278. S. P. Timoshenko and J. N. Goodier. *Theory of Elasticity, 3rd Edition*. McGraw-Hill, New York, 1987.

279. K. Tomlinson, C. Bradley, and A. Pullan. On the choice of a derivative boundary element formulation using Hermite interpolation. *International Journal for Numerical Methods in Engineering*, 39(3):451–468, 1996.

280. Yukio Ueda, Kazuo Ikeda, Tetsuya Yao, and Mitsuru Aoki. Characteristics of brittle fracture under general combined modes including those under bi-axial tensile loads. *Engineering Fracture Mechanics*, 18(6):1131–1158, 1983.

281. G. J. Van Berkel. The electrolytic nature of electrospray. In R. B. Cole, editor, *Electrospray Ionization Mass Spectroscopy*, chapter 2, pages 65–105. John Wiley, New York, 1997.

282. G. J. Van Berkel, G. E. Giles, J. S. Bullock, and L. J. Gray. Computational simulation of redox reactions within a metal electrospray emitter. *Analytical Chemistry*, 71:5288–5296, 1999.

283. Y.-Y. Wang and D. M. Parks. Evaluation of the elastic T-stress in surface-cracked plates using the line-spring method. *International Journal of Fracture*, 56(1):25–40, 1992.

284. J. O. Watson. Hermitian cubic and singular elements for plain strain. In P. K. Banerjee and J. O. Watson, editors, *Developments in Boundary Element Methods - Vol. 4*, chapter 1, pages 1–28. Elsevier Applied Science Publishers, London and New York, 1986.

285. W. T. Weeks. Numerical inversion of Laplace transforms using Laguerre functions. *Journal of the Association for Computing Machinery*, 13:419–429, 1966.

286. P. H. Wen and M. H. Aliabadi. A contour integral for the evaluation of stress intensity factors. *Applied Mathematical Modelling*, 19(8):450–455, 1995.

287. E. T. Whittaker and G. N. Watson. *A Course of Modern Analysis*. Cambridge University Press, Cambridge, MA, 1927.

288. J. G. Williams and P. D. Ewing. Fracture under complex stress - the angled crack problem. *International Journal of Fracture*, 8(4):441–416, 1972.

289. M. L. Williams. On the stress distribution at the base of a stationary crack. *Journal of Applied Mechanics, Transactions ASME*, 24(1):109–114, 1957.

290. R. C. Williams, A. V. Phan, H. V. Tippur, T. Kaplan, and L. J. Gray. SGBEM analysis of crack-particle(s) interactions due to elastic constants mismatch. *Engineering Fracture Mechanics*, 74(3):314–331, 2007.

291. R. B. Wilson and T. A. Cruse. Efficient implementation of anisotropic three dimensional boundary-integral equation stress analysis. *International Journal for Numerical Methods in Engineering*, 12:1383–1397, 1978.

292. L. C. Wrobel. *The boundary element method: Applications in Thermofluids and Acoustics*. Wiley, Chichester, 2002.

293. L. C. Wrobel and C. A. Brebbia. The boundary element for steady state and transient heat conduction. In R. W. Lewis and K Morgan, editors, *Numerical Methods in thermal problems*, pages 58–73. Pineridge Press, Swansea, 1979.

294. L. C. Wrobel and C. A. Brebbia. The dual reciprocity boundary element formulation for non-linear diffusion problems. *Computer Methods in Applied Mechanics and Engineering*, 65:147–164, 1987.

295. T. W. Wu and L. Lee. A direct boundary integral formulation for acoustic radiation in a subsonic uniform flow. *Journal of Sound and Vibration*, 175:51–63, 1994.

296. K. Xu, S. T. Lie, and Z. Cen. Crack propagation analysis with galerkin boundary element method. *International Journal for Numerical and Analytical Methods in Geomechanics*, 28(5):421–435, 2004.

297. J. F. Yau, S. S. Wang, and H. T. Corten. A mixed-mode crack analysis of isotropic solids using conservation laws of elasticity. *Journal of Applied Mechanics, Transactions ASME*, 47(2):335–341, 1980.

298. N. G. Zamani, J. F. Porter, and A. A. Mufti. A survey of computational efforts in the field of corrosion engineering. *International Journal for Numerical Method in Engineering*, 23(7):1295–1311, 1986.

299. X. F. Zhang, Y. H. Liu, and Z. Z. Cen. A solution procedure for lower bound limit and shakedown analysis by SGBEM. *Acta Mechanica Solida Sinica*, 14(2):118–129, 2001.

300. Zhiye Y. Zhao and Shengrui Lan. Boundary stress calculation - a comparison study. *Computers and Structures*, 71(1):77–85, 1999.

301. S. P. Zhu and P. Satravaha. An efficient computational method for nonlinear transient heat conduction problems. *Applied Mathematical Modeling*, 20:513–522, 1996.

302. S. P. Zhu, P. Satravaha, and X. Lu. Solving linear diffusion equations with the dual reciprocity method in laplace space. *Engineering Analysis with Boundary Elements*, 13:1–10, 1994.

303. T. Zhu, J-D. Zhang, and S.N. Atluri. A local boundary integral equation (LBIE) method in computational mechanics, and a meshless discretization approach. *Computational Mechanics*, 21:223–235, 1998.

304. O. C. Zienkiewicz and R. L. Taylor. *The Finite Element Method*, volume 1 & 2. McGraw-Hill, London, 4th Edition, 1991.

INDEX

Adaptive meshing, 164
Approximation
  Collocation, 35
  Galerkin, 37
Approximations and Solution, 2
Automotive Electrocoating, 8
Axis singularity, 136
BEAN, 167
  Contour, 232
  General Instructions, 234
  Main control window, 228
  Plot, 231
  Troubleshooting, 235
Bi-material interface, 147
Boundary Cloud Method, 17
Boundary Conditions
  Dirichlet, 39
  Neumann, 39
  discontinuous, 165
Boundary Contour Method, 16
Boundary element method, 1
Boundary Flux Equation, 28
Boundary limit approach, 27
boundary limit approach, 5
Boundary Node Method, 17
Boundary Potential Equation, 23
Cancellation of divergent terms
  three dimensions, 90
  two dimensions, 56
Cauchy Principal Value, 27, 50, 82

Collocation, 6
  C^1 condition, 40
Comparison of Collocation and Galerkin, 58
Continuity requirement, 40
Corners
  multiple interface, 152
  three dimensions, 97
  two dimensions, 64
Crack-tip elements, 185
  modified quarter-point, 185
  quarter-point, 185
Differentiability condition, 40
Dirac Delta function, 241
Displacement correlation technique, 178
Displacement discontinuity, 176
Divergence, 242
Divergence theorem, 24, 243
Divergent terms, 51, 80
Elasticity
  two dimensions, 173, 33
  anisotropic, 97
  three dimensions, 30
Electrochemistry, 65
Electrocoating Simulation, 10
  Boundary Condition, 10
  Boundary Solution, 10
  Design Iterations, 10
  Large Scale Computation, 11
  Meshing, 10
Element Refinement Criterion, 162

275

INDEX

Engineering Optimization, 9
  Car Frame Design, 9
Equation
  Navier, 30
  Bernoulli, 110
  Diffusion, 217
  Navier-Stokes, 110
Error estimation, 157
  global, 163
  local, 162
Exponential transform technique, 217
Exterior limit boundary integral equation, 25, 70
Fast Methods, 4
Field point, 2
Fracture analysis, 171
Functionally graded material FGM, 197
Galerkin, 6
Galerkin Residuals, 160
Galerkin weight functions, 72
Galerkin
  C^0 condition, 42
Gradient, 242
Gradient Equations, 112
Gradient Evaluation, 139
Green's function, 2, 4, 25, 244
  Axisymmetry, 132
  Transient heat conduction FGM, 218
  Aanisotropic elasticity, 99
  Elastic wave scattering, 63
  Functionally graded material, 199
  Helmholtz equation, 97
  Laplace, 4
Hadamard Finite Part, 28
Hermite Interpolation, 123–124
Higher Order Interpolation, 60, 92, 106
Hypersingular Integration, 39
Integration
  adjacent, 43, 53
  coincident, 38, 43, 48, 73, 75
  edge adjacent, 38, 73, 83
  non-singular, 38, 74
  vertex adjacent, 38, 73, 88
Interacting cracks, 192
Interaction integral, 179
Interface, 145
Interior limit boundary integral equation, 25
Internal boundary, 145
Isoparametric, 33
Iterative solution method, 125
J-integral, 178

Kelvin solution, 31
Kronecker Delta function, 242
Laplace equation, 2
  three dimensional, 24
  two dimensional, 30
Laplace transform boundary element method, 216
Linear triangular element, 34, 71
Local BIE Method, 18
Log Integral Transformation, 136
Meshless and Mesh-Reduction Methods, 16
Method of Fundamental Solutions, 16
Motz Problem, 168
Multiple Interfaces, 151
Multizone, 145
Nonhomogenous media, 199
Nonlinear algorithm, 65
Nonlinear Boundary Conditions, 65
Numerical Inversion of the Laplace Transform, 222
Overhauser elements, 42
Path-independent integral, 182
Quadratic Element, 60, 93, 221
Self Adaptive Strategy, 161
SGBEM formulation
  Axisymmetry, 129
Singular integrals, 5
Singular Integrals
  linear element, 48
Somigliana identity, 174
Source point, 2
Steady state heat conduction, 198
Stress equation, 32
Stress intensity factor, 172
Surface derivatives, 112
Surface flux, 24
Surface Gradient, 109
Surface stress, 118
Symbolic Computation, 51
Symmetric Neighborhood, 27
T-stress, 172
  auxiliary fields, 180
  Determination, 182
Tangential derivatives, 42
Taylor expansion, 211
Thin geometry, 146
Transient heat conduction, 216
Virtual Reality, 12
Visualization, 11
Water wave, 110
Wave Calculation, 110